Crop Growth Simulation Modelling and Climate Change

About the Editors

M. Mohanty: A renowned soil scientist, Joined Indian Institute of Soil Science, Bhopal (ICAR) as Soil Physicist in the year 1999 after completing post graduate studies in soil science from IARI, Pusa New Delhi. Worked extensively in the areas of tillage and crop residues management in rice-wheat region in vertisol, climate change, soil quality and published more than 40 research articles in International Journals of high impact factors. He has published three books and many research bulletins and has more than 40 books chapters to the credit. He has published the work in many journal of international repute like Ecological Modelling, Agricultural Ecosystem and Environment, Soil and Tillage research and many more. Recently he was awarded with FAI award for his excellent works on nutrient management in wheat crop of central India.

Dr. Nishant K. Sinha: Agricultural Physicist. Recently, he is working as scientist in the division of Soil Physics, Indian Institute of Soil Science, Bhopal, Madhya Pradesh. He has published 20 research papers in peer reviewed journals and authored 3 books. He also has good expertise in crop root dynamics, crop simulation modelling and water productivity.

Dr. R.S. Chaudhary: Principal Scientist and Head of Division at Indian Institute of Soil Science, Bhopal. He has worked on various research tasks of soil and water conservation, water-shed development, catchment survey and planning, efficient management of water and nutrients, soil surface manipulations, conservation tillage, land configurations, and quantifi-cation of runoff and soil losses, in various soils and cropping systems of India. As enthusiastic researcher and critical reviewer he has a number of research papers in national and international journals, book chapters, review articles, popular articles, and bulletins to his credit. Presently be is involved in the research projects on conservation agriculture, climate change, assessment of soil quality and soil resilience.

Dr. K. M. Hati: Soil Scientist with Specialization in Soil Physics. Presently he is working as Principal Scientist at IISS, Bhopal in the Division of Soil Physics. He has more than 20 years of experience in research and training in the field of Soil Physics, remote sensing and conservation agriculture. During his period of service he has worked on conservation agriculture, carbon sequestration mechanisms, soil quality, water and nutrient dynamics. He also worked on application of remote sensing principles for improving input use efficiency. He published more than 30 peer reviewed research papers in international and 20 papers in national journals of repute. Besides this, he also authored 3 books, 15 book chapters, 10 review articles and 6 papers in national and international conference proceedings.

Dr. Ashok K. Patra: Director of the Indian Institute of Soil Science, Bhopal. He was a recipient of the prestigious INRA Fellowship (2001-2003) of the French Research Ministry to work on molecular soil ecology in N cycling at the CNRS-Claude Bernard Université Lyon, France. He has contributed more than 200 publications, which has been widely cited throughout the world. He was a faculty of Post Graduate School, IARI, New Delhi and actively involved in teaching and guiding of postgraduate students at IARI for 15 years (1999-2014). He is a recipient of several awards/recognitions and fellowships, namely: British Council TCT Award 1996; DBT Overseas Associateship Award 2008; FAI Dhiru Morarji Memorial Award, 2011; Bharat Jyoti Award, 2012; Rajiv Gandhi Excellence Award, 2012; ISSS Dr. G.S. Sekhon Memorial Lecture Award, 2012 of ISSS, New Delhi; Hooker of IARI (2013), New Delhi. Dr. Patra was Editor, Range Management and Agroforestry 1996-1998; Councillor, Indian Society of Soil Science (2005-2006); President (Delhi Chapter), Indian Society of Soil Science (2012-14); Member, Nature's Reader Panel (2009). He served as expert of several important committees and acted as reviewer for more than 25 international journals. Dr Patra is a Fellow of the National Academy of Agricultural Sciences, Indian Society of Soil Science and Range Management Society of India.

Crop Growth Simulation Modelling and Climate Change

M. Mohanty
Nishant K. Sinha
K.M. Hati
R.S. Chaudhary
Ashok K. Patra

Published by:

Scientific Publishers (India)
5 A, New Pali Road, P.O. Box 91
Jodhpur 342 001 (India)

Branch Office
Scientific Publishers (India)
4806/24, Ansari Road, Daryaganj
New Delhi - 110 002 (India)

E-mail: info@scientificpub.com
Website: www.scientificpub.com

Print : 2021

ISBN: 978-81-7233-941-8

Typeset by Rajesh Ojha
Printed in India

भाकृअनुप
ICAR

डॉ. अलोक कुमार सिक्का
उप महानिदेशक (प्रा सं प्र)

Dr. Alok K. Sikka
Deputy Director General (NRM)

भारतीय कृषि अनुसंधान परिषद
कृषि अनुसंधान भवन-II, पूसा, नई दिल्ली 110 012

INDIAN COUNCIL OF AGRICULTURAL RESEARCH
KRISHI ANUSANDHAN BHAVAN-II, PUSA, NEW DELHI - 110 012

Ph. : 91-11-25848364 (O), 24121571 (R)
Fax : 91-11-25848366
E-mail: aksikka@icar.org.in; aloksikka@yahoo.co.in

Foreword

Crop growth simulation models are widely used as tools to simulate the potential impacts of environmental factors on agricultural ecosystems. In recent times, the most important and active area of application is in research on the potential impacts of climate change on crop productivity and on natural resource management. The Book **"Crop Growth Simulation Modelling and Climate Change"**, a compressively described book on crop simulation models is in that direction. The output from these models has been used as a major data source for Inter-governmental Panel on Climate Change (IPCC) assessments for agriculture. Assuming that a model provides realistic simulations for a given environment and management practices, the effects of crop responses to specific changes in temperature, CO_2 or other abiotic factors are simulated with initial field conditions (e.g., for soil moisture and nitrogen status), crop cultivar characteristics, planting dates, N fertilization and irrigation. Though the models are limited in the number of processes they consider, contrasting with the real-world complexity of cropping systems, use of simulation models in climate change research has been tremendous. Considering the limitations associated with models and their development, simulation studies need to be credible, replicable, and readily comparable as the results ultimately affect the decision of policy makers and stakeholders. The assessments of potential impacts of climate change on crops through simulation models not only consider the economic yield, but also examines the plant characteristics, resource use, environmental parameters, socio-economic analyses, and C-sequestration potentials through cropping systems.

The goal of the present book is to review some of the important models, which are widely used to simulate the impacts of climate change on crops. It is also important to learn the potential limitations in application of models in climate change research and suggest the priority areas for model development, testing, or improvement. Diverse methods, scenarios, and models have been used to characterize the potential impacts of climate change on crop yield and other associated aspects of agricultural production. Crop simulation modelling will see increasing use in climate change research and in applications of research in decision making.

I appreciate the work done by the authors to bring this book to the present format for the students and the researchers. I wish the authors for success of the book.

(Alok K. Sikka)

April 2015

Preface

Profitable crop production to cater to the needs of the growing population with the simultaneous maintenance of quality of the environment in the face of degradation of natural resources under climate change scenario has become a great challenge that agricultural scientists and producers must address in recent days. The anthropogenic enrichment of GHGs in the atmosphere and the cumulative radiative forcing of all GHGs has led to an increase in the global surface temperature of 0.6°C since 19th Century with the current warming rate of 0.17°C/ decade (IPCC, 2001), which is higher than the critical rate of 0.1°C/decade leading to climate change. It has been reported that agriculture contributes to 28% of the global GHG emission. Climate change through its climate variability and climate forcing not only affects crop production but also affects the movement of water and nutrients in soil-plant-atmosphere continuum (SPAC). Crop growth models integrate the effects of soils, weather, management, genetics, and pests on daily growth, and can be used to gain insight into spatial and temporal yield variability. With the help of simulation models, it is possible to simulate various aspects of climate change in crop production. Thus, a crop growth simulation model not only predicts the final state of total biomass or harvestable yield, but also contains quantitative information about major processes involved in the growth and development of a plant and various processes taking place in soil-plant system. These processes govern the utilization of agricultural inputs like water and nutrient and ultimately the crop productivity. It is imperative to understand the various physical processes involved and assess the fluxes of SPAC to manage them efficiently for achieving higher resource use efficiency and sustaining agricultural Productivity at higher level. Crop growth simulation models can help in understanding this complex system and taking critical decisions with respect to improving resource use efficiency under diverse soil and agro climatic situations and management scenarios.

Keeping this in view, the editors have bought out this book on "**Crop Growth Simulation Modelling and Climate Change**". A group of authors have dealt with different aspects of crop modelling viz., Crop growth simulation models in agricultural crop production, Applications of Crop Growth Simulation Models in Climate Change Assessments, Biophysical impacts and priorities for adaptation of agricultural crops in a changing climate, Climate change projections – India's Perspective, Impact of Rising Atmospheric CO_2 concentration on Plant and Soil processes, Modelling the impact of climate change on soil erosion in stabilization and destabilization of soil organic carbon, Simulating Crop Yield, Soil Processes, Greenhouse Gas Emission and Climate Change Impacts with APSIM, InfoCrop Model, CropSyst model and its application in natural resource management, Climate change and crop production system: assessing the consequences for food security, A biophysical model to analyze climate change impacts on rainfed rice productivity in the mid-hills of Northeast India, AquaCrop Modelling: A Water Driven Simulation Model, Conservation Agriculture: A strategy to cope with Climate Change, Effect of climate change on productivity of wheat and possible mitigation strategies using DSSAT model in foot hill of Western Himalayas, Integrating Remote Sensing Data in Crop Process Models, Climate change impact assessment using DSSAT model, Decision Support System for Managing Soil Fertility and Productivity in Agriculture, De-Nitrification De-Composition Model - An Introduction for SOC Simulations, Crop Simulation Modeling for Climate Risk assessment: Adaptation and Mitigation Measures and Rules of Simulations, Rothamsted Carbon (RothC) Model and its Application in Agriculture etc.

The editors are grateful to Dr A.K. Patra, Director, Indian Institute of Soil Science, Bhopal for his guidance and encouragement in bringing out this publication. The editors are thankful to all the authors who have contributed for this book. Information collected from various sources (published and unpublished) has been duly acknowledged.

It is believed that this book will prove useful for the scientific communities and policy makers, for whom it is intended. Authors look forward to receive suggestions for improvement of the book.

March, 2015 **Editors**

IISS, Bhopal

Contributors

A.K. Vishwakarma
ICAR Research Complex for NEH Region, Umiam - 793 103 Meghalaya.

Anup Das
ICAR Research Complex for NEH Region, Umiam, Meghalaya

Ashok K. Patra
Division of Soil Physics, Indian Institute of Soil Science,
Nabibagh, Berasia Road, Bhopal - 462 038, India

B. Chakrabarti
Division of Agricultural Physics,
Indian Agricultural Research Institute, Pusa,
New Delhi - 110 012, India

B.U. Choudhury
ICAR Research Complex for NEH Region, Umiam, Meghalaya

Bharat P. Meena
Indian Institute of Soil Science,
Nabibagh, Berasia Road, Bhopal - 462 038

Brajendra
ICAR Research Complex for NEH Region, Umiam - 793 103 Meghalaya

Debashis Mandal
Central Soil and Water Conservation Research and Training Institute,
218, Kaulagarh Road, Dehradun, India

Dillip Kumar Swain
Agricultural and Food Engineering Department,
Indian Institute of Technology, Kharagpur - 721 302, India

H. Pathak
CESCA, Indian Agricultural Research Institute,
Pusa, New Delhi - 110 017, India

J. Somasundaram
Division of Soil Physics, Indian Institute of Soil Science,
Nabi Bagh, Berasia Road, Bhopal - 462 038, India

K.A. Pathak
ICAR Research Complex for NEH Region, Umiam - 793 103, Meghalaya

K.K. Bandyopadhyay
Division of Agricultural Physics, Indian Agricultural Research Institute, Pusa, New Delhi - 110 012, India

K.M. Bujarbaruah
ICAR Research Complex for NEH Region, Umiam - 793 103 Meghalaya

K.M. Hati
Division of Soil Physics, Indian Institute of Soil Science, Nabibagh, Berasia Road, Bhopal - 462 038, India

K.P. Mohapatra
ICAR Research Complex for NEH Region, Umiam, Meghalaya

M. Mohanty
Division of Soil Physics, Indian Institute of Soil Science, Nabibagh, Berasia Road, Bhopal - 462 038, India

M.L. Dotaniya
Division of Environmental Soil Science, Indian Institute of Soil Science, Nabi Bagh, Berasia Road, Bhopal - 462 038

Monalisha Pramanik
Central Soil and Water Conservation Research and Training Institute, Research Centre, Datia - 475 661, India

N.K. Lenka
Division of Soil Physics, Indian Institute of Soil Science, Nabibagh, Berasia Road, Bhopal - 462 038, India

N.K. Sharma
Central Soil and Water Conservation Research and Training Institute, 218, Kaulagarh Road, Dehradun, India

Nirmal Kumar
National Bureau of Soil Survey & Land Use Planning, Amravati Road, Nagpur - 440 033, India

Nishant K. Sinha
Division of Soil Physics, Indian Institute of Soil Science, Nabibagh, Berasia Road, Bhopal - 462 038, India

P. Bhattacharya
ICAR- Central Rice Research Institute, Cuttack - 753 006, India

P.K. Mishra
Central Soil and Water Conservation Research and Training Institute, 218, Kaulagarh Road, Dehradun, India

Pankaj K Tiwari
Indian Institute of Soil Science, Nabibagh, Berasia Road, Bhopal - 462 038

Pramod Jha
Division of Soil Chemistry & Fertility, Indian Institute of Soil Science, Bhopal - 462 038

R. Elanchezhian
Division of Soil Chemistry and Fertility, Indian Institute of Soil Science,
Bhopal - 462 038, India

R. Saha
Division of Soil Physics, Indian Institute of Soil Science,
Nabibagh, Berasia Road, Bhopal - 462 038, India

R.S. Chaudhary
Division of Soil Physics, Indian Institute of Soil Science,
Nabibagh, Berasia Road, Bhopal - 462 038, India

Rajeev Ranjan
Central Soil and Water Conservation Research and Training Institute,
218, Kaulagarh Road, Dehradun, India

Rohit Patidar
R.A.K. College of Agriculture,
Sehore- 466 001, India

S.V. Ngachan
ICAR Research Complex for NEH Region, Umiam - 793 103, Meghalaya

Sangeeta Lenka
Division of Soil Physics, Indian Institute of Soil Science,
Nabibagh, Berasia Road, Bhopal - 462 038, India

Smita Gupta
Govind Ballabh Pant University of Agriculture and Technology,
Pantnagar - 263 145, India

Sushree Sagarika Satapathy
Agricultural and Food Engineering Department,
Indian Institute of Technology,
Kharagpur - 721 302, India

Taznin A. Choudhury
ICAR Research Complex for NEH Region,
Umiam, Meghalaya, India

Contents

1 Crop Growth Simulation Models in Agricultural Crop Production

M. Mohanty, Nishant K. Sinha and Ashok K. Patra

Division of Soil Physics
Indian Institute of Soil Science, Nabibagh, Berasia Road,
Bhopal-462038, India

Introduction

Agriculture has undergone significant changes since its inception. Scholars have proposed several theories - oasis theory, hilly flanks hypothesis, feasting model, demography theory, evolutionary theory and many more - to understand origin and development of agriculture. All theories reached to consensus that agriculture had originated approximately 10,000 years BC. Nonetheless, determining the exact origin of agriculture is still enigma, because the transition of hunter-gathering society began thousands of years prior to the invention of writing. Transition of agriculture from cultivation of wild grain to present day mechanized and computerized agriculture is the result of various scientific discoveries and innovations. Increase in population, shrinking natural resources, deteriorating soil quality and climate change posed many challenges before the agricultural scientists for sustainable agriculture. During 20th century, scientists anticipated the need of future and tried to develop database that when interpreted may be used for improving the management practices to increase the agricultural productivity. In earlier days, researchers were most interested in the abnormal and unusual occurrence. With the inventions of computational machines, researchers have started to define, establish, and quantify the relationships among determinants of productivity. Further, with the advancement of computer, continuous observations over time became possible. In early 1960s, concept of mathematical modelling evolved, in which biological/natural processes could be explained through mathematical equations referred as "simulation modelling". This was the time when computer applications were started in agriculture leading to the beginning of "modelling era". During the 1970s systems dynamics became formalized. Further improvement has been continued especially through advancement in computer programming techniques and in process of verification, validation, and evaluation (See Box 1). Now a days, simulation of dynamic agricultural systems has become an integral part of agricultural science; which increases the understanding of ecosystem interactions, as influenced by the environment and management practices and has greatly expanded the potential for decision-making (Barrett and Nearing, 1998).

Box 1: History of crop modelling

Infancy Stage (1960s)	(simple) models to estimate light interception and photosynthesis in crop canopies (de Wit 1965); main frames
Juvenile stage (1970s)	comprehensive models; many parameters; complex description of interaction between crop and main growth factors; (improved equipment to measure physiological processes in the field)
Adolescence stage (1980s)	summary models; key variables to capture most of the behaviour of a system; universal models for several crops; parameter crisis
Maturity Stage (1990s- till now)	parsimonious models; keep models simple for user; few key variables to reduce parameter and input needs for user; understanding; limitations of crop models

The word "modelling" originated from the Latin word *modellus*, which means a way to copy the real world. The ability of *Homo sapiens* to abstract the real world is the most important feature that gave them a competitive edge over their precursor. A Model can be graphical, mathematical, physical, or simplified version of a concept, phenomenon, relationship, structure, system, or an expression of the real world (Fig. 1).

The main objectives of a model are: (1) to facilitate understanding by eliminating unnecessary components, (2) to aid in decision making by simulating 'what if' scenarios, (3) to explain, control, and predict events on the basis of past observations. Models range from simple sketches to computer programs with millions of lines of code, but all of them have one thing in common: some elements of the actual 'thing' are abstracted into the model. Models can be divided into four broad classes based on their degree of abstraction (1) *Iconic model*: least abstract, physical, 'look-alike' model, such as a model airplane or train. (2) *Conceptual model*: system is described in natural language. (3) *Analogous model*: more abstract but having some resemblance to what it represents, such as a chart, graph, map, network diagram. (4) *Formal or mathematical model*: most-abstract model with no resemblance but only an approximation to what it represents, such as a mathematical equation or formula, financial statement, language, and set of accounts.

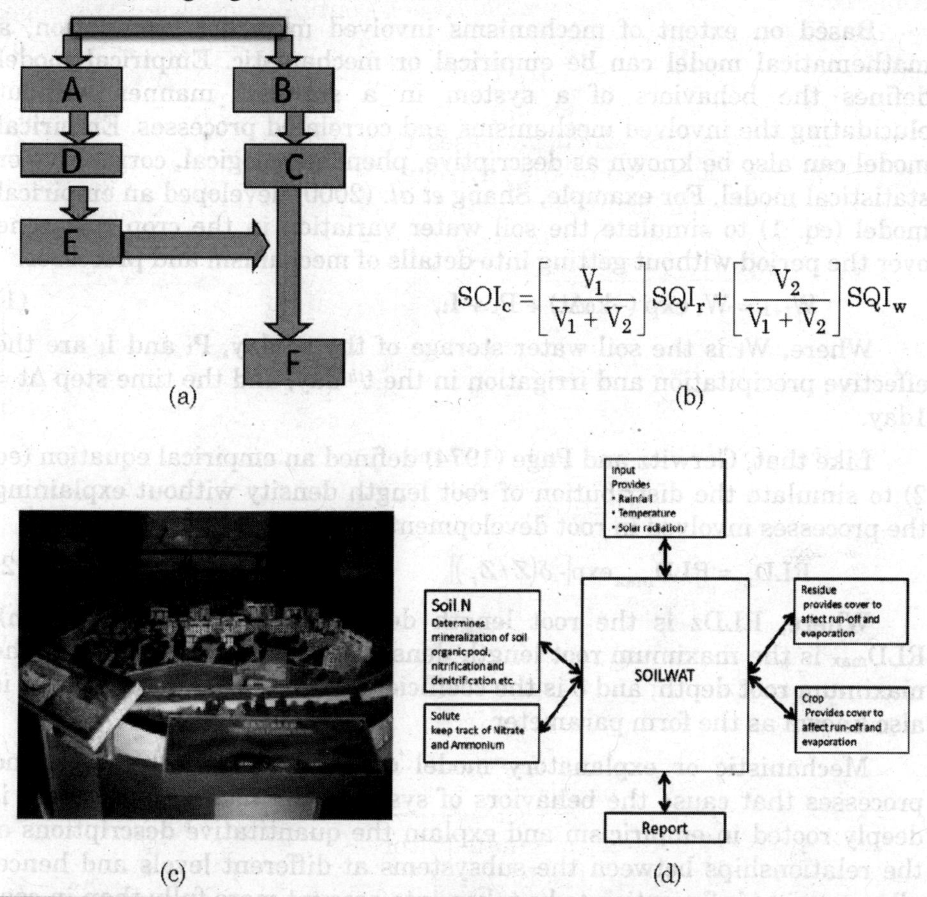

Fig. 1. An example of (a) graphical, (b) mathematical, (c) physical and (d) conceptual model

Types of mathematical model

In agriculture, use of formal/mathematical model is more predominant compared to others. These models can simulate the behavior of a real crop by predicting the growth of its components such as leaves, roots, stems and grains. Thus, a crop growth simulation model not only predicts the final state of the total biomass or harvestable yield, but also contains quantitative information about major processes involved in the growth and development of a plant. A mathematical model can be empirical and mechanistic; static and dynamic; or deterministic and stochastic (Thornley and France, 1984).

Empirical and mechanistic model

Based on extent of mechanisms involved in model formulation, a mathematical model can be empirical or mechanistic. Empirical model defines the behaviors of a system in a simplest manner without elucidating the involved mechanisms and correlated processes. Empirical model can also be known as descriptive, phenomenological, correlative or statistical model. For example, Shang *et al.* (2000) developed an empirical model (eq. 1) to simulate the soil water variation in the crop root zone over the period without getting into details of mechanism and processes.

$$W_{t+1} = W_t \exp(-k_0 \Delta t) + P_t + I_t, \quad (1)$$

Where, W_t is the soil water storage of the tth day, P_t and I_t are the effective precipitation and irrigation in the tth day, and the time step Δt = 1day.

Like that, Gerwitz and Page (1974) defined an empirical equation (eq 2) to simulate the distribution of root length density without explaining the processes involved in root development. The model is,

$$RLD_z = RLD_{max} \exp[-\delta(Z/Z_r)] \quad (2)$$

Where, RLD_z is the root length density (cm^{-3}) at depth Z (cm); RLD_{max} is the maximum root length density at the soil surface; Z_r is the maximum root depth; and δ is the coefficient of root distribution, which is also known as the form parameter.

Mechanistic or explanatory model describes the mechanisms and processes that cause the behaviors of system. Any mechanistic model is deeply rooted in empiricism and explain the quantitative descriptions of the relationships between the subsystems at different levels and hence, allow *a priori* information to be taken into account more fully than in case of an empirical approach (Sirotenko, 2000). The crop growth simulation model like APSIM, DSSAT, CropSyst are some of the recently and widely used mechanistic models in agricultural sciences.

Static and dynamic model

A static model is the model that does not care the element of time. Static model usually algebraic in form and solved for a specific set of conditions, which exist at a set point in time (Mcphee, 2009). It is always an approximation and sufficiently close to equilibrium, e.g. Langmuir and Freundlich Adsorption Isotherm. Contrary to it, a dynamic model contains the time variable explicitly like logistic model (eq. 3).

$$L(t) = \frac{L_o L_f}{L_0 + [L_f - L_0]\exp(\tau_0 t)} \quad (3)$$

Where L (t) is the length in mm of shoot at any time, L_0 the length at the onset of growth (t = 0), Γ 0 the relative growth rate at time 0 (per day).

Deterministic and stochastic model

Deterministic model is one that estimates the exact value of the dependent variable like yield without any associated probability distribution. The Agricultural Trade Policy Simulation Model (ATPSM) - A deterministic model- designed for detailed analysis of agricultural trade policy issues at the global and regional level. The model solution gives estimates of the changes in trade volumes, prices, government revenues and welfare indicators associated with changes in the trade policy environment (http://unctad.org). Whereas, a stochastic model contains some random elements or probability distributions within the model, so that not only can it predict the expected value of a quantity such as rainfall, but also its variance. The greater the uncertainty in the behavior of a system, the more important it may be to construct a stochastic model e.g. rainfall, water movement in highly heterogeneous soil (Datta, 2008).

System and model characteristics

A model is a simplified depiction of a system, and a system is a closed part of reality that contains interrelated elements (de Wit, 1982a). The system we consider here is the agricultural cropping system. In 1982, de Wit and Penning de Vries proposed a classification of this system into four production situations:

Production level I : Potential production. It is an ideal situation in which the crop does not suffer from any shortage of water or nutrients in the entire growth season i.e. Growth occurs in conditions with ample supply of water and nutrients. Here, growth and development depends only on the weather conditions particularly radiation, temperature and crop characteristics. Fig. 2 is a flow diagram showing the processes that are to be considered at this level.

At Production Level 1, light is a driving variable. Assimilated carbohydrates are stored in an easily accessible form, such as starch, which is used for growth. Whereas, temperature as an external variable, modifies growth rates, photosynthesis and other physiological processes. In growth processes, reserves are converted into 'structural biomass' with a specific efficiency. Structural biomass consists of those components that are not mobilized again for maintenance or growth processes elsewhere in the plant. The partitioning of biomass between roots, leaves, stems and storage organs is strongly related to the physiological age of the crop, which itself is a function of temperature (Penning de Vries *et al.,* 1989).

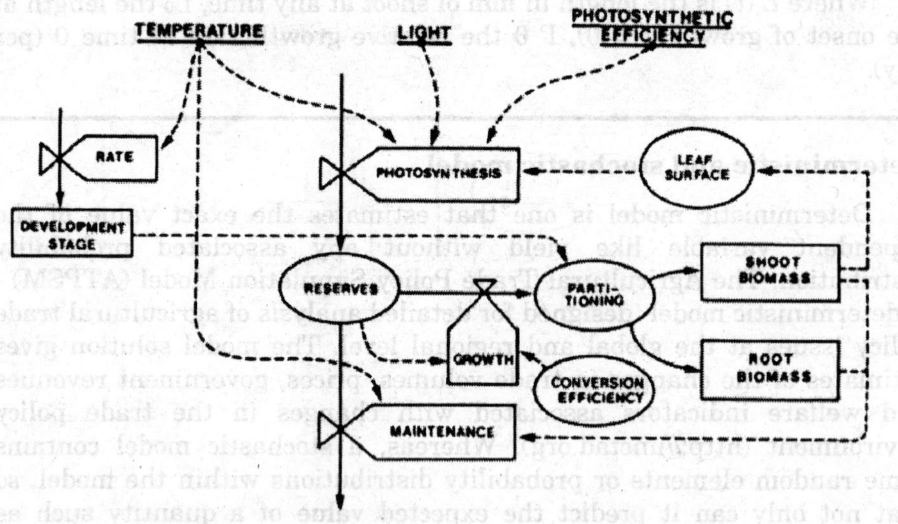

Fig. 2. A relational diagram of a system at Production Level 1 (Penning de Vries *et al.*, 1989).

Production level 2 : Water-limited production. The growth of a crop is limited by shortage of water for at least part of growing period (Fig. 3). This situation frequently occurs in semi-arid regions and also in areas where the rainfall is inadequate and/or poorly distributed part of the growth season (Singh, 2005). This may also occur in other climates under intensive cropping on light soils (Penning de Vries *et al.*, 1989).

At this Level, primeral factors are the degree of extraction of soil water and efficiency of its use by the crop. Water shortage leads to stomatal closure and to a simultaneous reduction of CO_2 assimilation and transpiration. The extent of potential transpiration, and the potential photosynthesis rate, depend on the availability of water. The amount of water stored in the soil is a buffer between rainfall and capillary rise and the processes by which water is lost. This buffering capacity and the simultaneous water loss through transpiration and non-productive processes cause the growth rate to depend indirectly only on rainfall. The relation of plant growth to the principal driving variable of production level 2 is indirect.

Production level 3 : Nitrogen-limited production. In this production system, crops encounter with nitrogen shortage at least part of the growth season, whereas, water shortage or adverse weather conditions occur in the remaining part of the season. This condition is common in the agricultural systems all over the world.

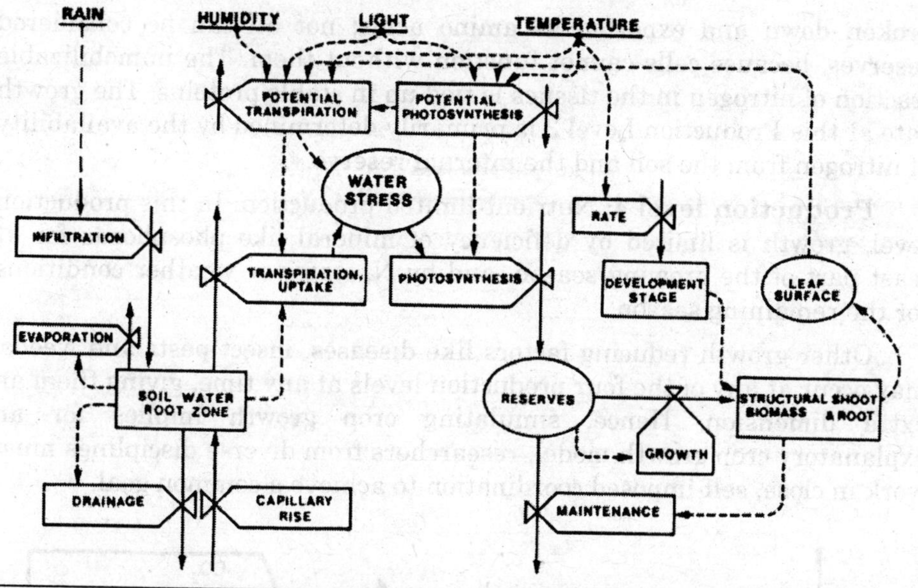

Fig. 3. A relational diagram of a system at Production Level 2 (Penning de Vries *et al.,* 1989).

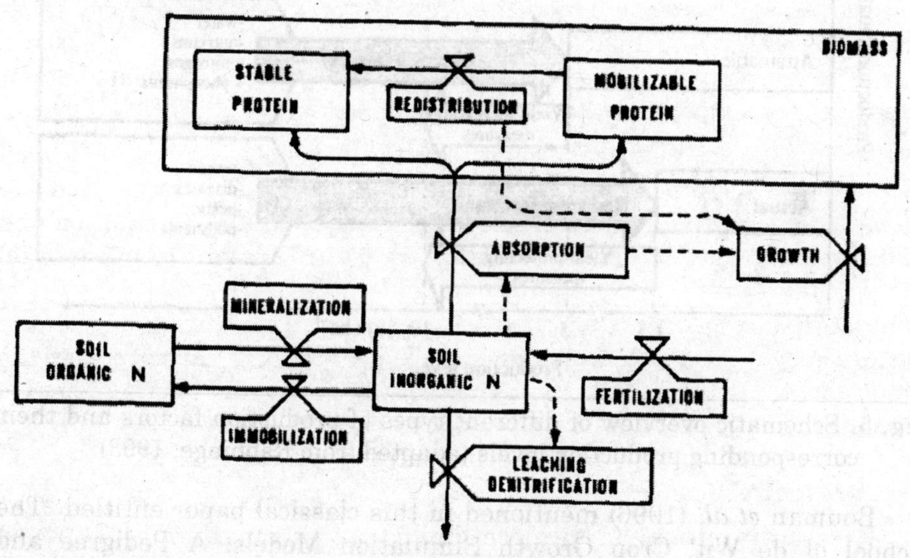

Fig. 4. A relational diagram of a system at Production Level 3 (Penning de Vries *et al.,* 1989).

Nitrogen in plant tissues can be distinguished by two fractions: mobilizable and immobilizable nitrogen (Fig. 4). The relation of nitrogen uptake and growth is different from water uptake and growth. The mobilizable fraction consists of enzymes and membrane proteins that are broken down and exported as amino acids; not all can be considered reserves, because cells cannot function without them. The immobilizable fraction of nitrogen in the tissues is tied up in stable proteins. The growth rate at this Production Level 3 is primarily determined by the availability of nitrogen from the soil and the internal reserve.

Production level 4: Nutrient-limited production. In this production level, growth is limited by deficiency of mineral like phosphorus for at least part of the growing season, and by N, water or weather conditions for the remaining season.

Other growth reducing factors like diseases, insect pests and weeds, may occur at any of the four production levels at any time, giving them an extra dimension. Hence, simulating crop growth implies for an explanatory crop growth model, researchers from diverse disciplines must work in close, self-imposed coordination to achieve a common goal.

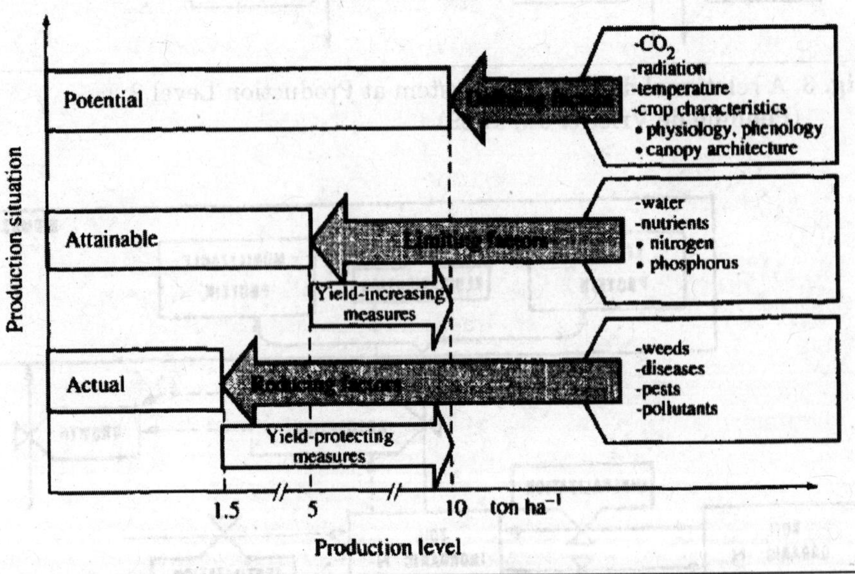

Fig. 5. Schematic overview of different types of production factors and their corresponding production levels (adapted from Rabbinge, 1993).

Bouman *et al.* (1996) mentioned in this classical paper entitled "The 'School of de Wit' Crop Growth Simulation Models: A Pedigree and Historical Overview" that, in practice, actual production situations are difficult to assign to any of these four situations, but this practical simplification of schematizing specific situations allows progress to be made, particularly at the start of a study (de Wit & Penning de Vries, 1982; Rabbinge, 1986; van Duivenbooden & Gosseye, 1990). Recently, a new classification of agricultural production systems was introduced in the C.T. de Wit Graduate School of Production Ecology (Rabbinge, 1993): potential growth is defined by the concentration of atmospheric CO_2, solar radiation, temperature and crop characteristics; attainable growth is determined by the limiting factors of water and nutrients; actual growth is reduced below the attainable by factors such as weeds, pests, diseases and pollutants (Fig. 5). The development of models for each of these situations proceeded at its own rate and in its own direction, depending on the research goals and objectives at the time.

Table 1. Three phases of model development (Rabbinge & de Wit, 1989)

1. Conceptual or Preliminary phase	i) Formulation of objectives. ii) Definition of system limits. iii) Conceptualization of the system (states, rates, auxiliary variables, forcing variables etc.).
2. Comprehensive phase	i) Quantification through literature, process experiment or estimation of the relation between rate and forcing variables, states or auxillary variables. ii) Model construction (definition of computer algorithm) iii) Model verification, i.e. testing the intended behaviour of the model.
3. Summary phase	i) Evaluation of model performance, i.e. testing the model in parts or as a whole, using independent experiments on system level. ii) Sensitivity analysis (numerical or structural) iii) Simplification. iv) Formulation of design rule or forecasting models to be used for practical application.

Penning de Vries (1980) distinguished three phases of model development preliminary, comprehensive and summary phases. These three phases of model development are depicted in Table 1. Preliminary, or conceptual, models reflect current scientific knowledge, and are simple in structure because of incomplete knowledge of the component processes. With increasing insight, preliminary models can evolve into comprehensive models that represent systems in which the essential elements are thoroughly understood and which contain large amounts of information. In plant physiological research the main purpose of both preliminary and comprehensive models is to formalize and integrate knowledge of plant growth processes, to test hypotheses by comparing model results with experiments, to structure research programmes and to extrapolate from the laboratory to the field. In brief, the aim is to increase our understanding of crop performance. After the first comprehensive models had been built and tested, however, they were quickly recognized as powerful tools for exploring situations and possibilities of crop production that were almost impossible to investigate using the conventional methods and techniques of experimentation. Because comprehensive models are typically complex and hardly accessible to potential users, this stimulated the development of summary models. A summary model can be regarded as a model of a (comprehensive) model, in which essential elements are simplified and aspects that are only marginally important are ignored. Summary models have typically been developed in response to application-oriented research questions (e.g. tactical and strategic). Their use, for example, may be in decision support systems for pest and disease management and for plant nutrient management (Bouman *et al.*, 1996).

Forrester diagrams were invented by Jay Forrester- MIT professor- to represent the dynamic system in which a measurable quantity flows between system components. Forrester diagrams are such an abstraction of the basic concepts of system components and material flows to obtain a general tool for qualitative modelling of systems. The following diagrams are used as conventions, when we represents the dynamic systems.

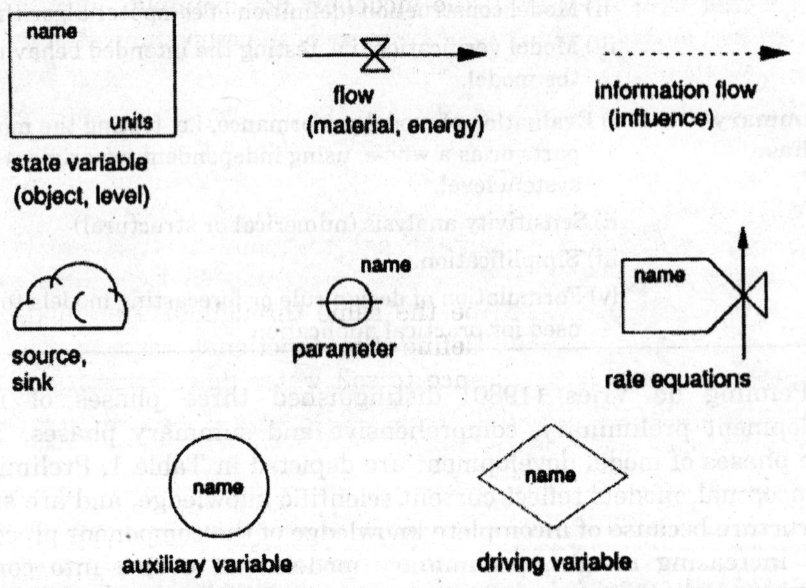

Fig. 6. Forrester diagrams for representation of dynamic system (Adapted from Haefner, 2005).

Terminology

System: A system is a collection of components and their inter-relationships that have been grouped together for the purpose of studying some part of the real world. For example typical crop models define the crop and soil root zone as components that interact in complex ways and are also affected by weather conditions and management practices (Pritsker, 1995; Rabbinge *et al.*, 1994).

Environment and system boundary: The environment of system includes everything except the components of the system. A system boundary is an abstraction of the limits of the system components, separating them from the environment.

Simulation: Simulation includes the processes necessary for operationalizing a model to mimic real system behavior. Developing computer logic and flow diagrams, writing the computer code, and implementing the code on a computer to produce desired outputs are necessary tasks in the simulation process.

Inputs and outputs: Inputs of system are factors that influence the behavior of the system but are not influenced by the system - rainfall, temperature, and light- are examples of inputs to crop systems. These environmental variables affect the soil water and crop growth dynamics in a field but are not themselves affected by the crop system.

Outputs from the system represent the characteristic behavior of the system that is of interest to the modeller. For example, outputs from a crop system model would include biomass of the crop and water content of the soil component.

Parameters and constant: Constants are the quantities which remain the same when experimental conditions are changed or when the model is applied to different genotypes or organisms (France and Thornley 1984). Examples of constants are the molecular weight of glucose, the gravitational constant, and the number of seconds in a day. Parameters, on the other hand, are quantities whose values are less certain but are assumed to be the same throughout a simulation. For example, parameters may define the functional response of photosynthesis to light, the resistance to soil water flow, respiration loss for tissue synthesis, and the number of degree-days from crop emergence to flowering (Jones and Luytem, 1998).

State variable: State variables are quantities that describe the conditions of system components. These state variables change with time in dynamic models as system components interact with each other and with the environment. For example, soil water content and crop biomass are two state variables that change with time in most crop models. A model may have one state variable or it may have many to describe the various characteristics of a system that change with time.

Rate variable: These variables characterize the rate of change of each state variable, e.g. rate of flow of material or energy between state variables.

Auxiliary variables: These variables are used to assist the understanding of the dynamic system. For example, if NO_3^- and NH_4^+ are independent state variable, their ratio $R = NO_3^- / NH_4^+$ is an auxiliary variable which also varies with time.

Modelling agricultural systems

Dynamic crop growth models emerged in the mid-1960s with the pioneering work of De Wit (1965), who introduced Von Bertalanffy's systems theory and Forrester's dynamic model-simulation method into crop science. Crop modelling differs from empirical statistical analysis, just as systems biology differs from bioinformatics. In crop models, constituting elements and processes are put together in mathematical equations. The rules by which the elements or processes interact give rise to systems behavior and emerging properties, which may well be unexpected and even counterintuitive (heuristics). This heuristics, in turn, enhances the understanding of individual processes and improves the next-round modelling and the ability of the model in extrapolating information from one to another environment (Yin and Struik, 2008).

Several types of plant growth models exist, with varying degrees of complexity, depending on their end use and application. Empirical crop and forest growth models, which apply functions fitted to data without considering the physiological processes involved in growth and morphogenesis, continue to be used efficiently for estimating or predicting yield and quality of products. Nevertheless, such models are usually calibrated for a particular species and well-defined site conditions and therefore cannot be valid over a wide range of conditions (Lacointe, 2000). Understanding how plant growth and development interact with environmental factors including light, temperature, soil characteristics and water availability, is extremely important to increase the yield of crop plants, e.g. through breeding programs (Fourcaud *et al.*, 2008). To model these interactions there is need to take into account the physiological processes involved in growth (Fig 7), e.g. water and nutrient uptake, photosynthesis and carbon partitioning (Gifford and Evans, 1981).

A crop can be divided into its constituent parts so that the various processes involved can be modelled separately. These processes may be categorized into above and below ground processes. The above ground processes include (a) Light Interception (LAI, Beer's Law), (b) Carbon Dioxide Fixation (hyperbolic dependence of CO_2-fixation on light interception, empirical), (c) Biomass Partitioning (empirical, priority, functional balance), (d) Tissue Expansion (specific leaf area), (e) Morphology (organ initiation and abortion) and (f) Phenology (Growing Degree Days; photo period) and below ground processes include (g) water movement (saturated and unsaturated), (h) evapotranspiration (mechanistic and empirical) (i) root growth and water uptake (j) soil temperature (k) soil penetration resistance (l) gaseous movement, and (m) nitrogen transformation and movement.

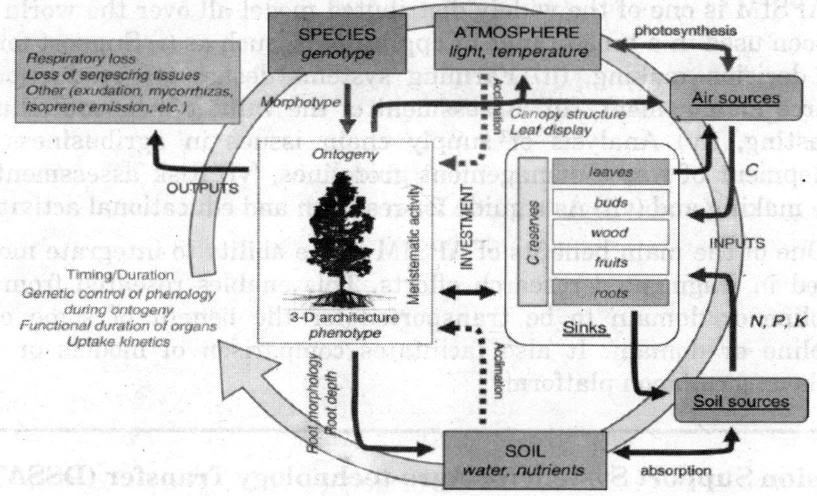

Fig. 7. A schematic representation of interactions between a plant and its environment (Adapted from Fourcaud *et al.*, 2008)

Models in Agriculture

Agricultural production systems simulator (APSIM)

APSIM was developed to simulate biophysical processes in farming systems, particularly as it relates to the economic and ecological outcomes of management practices in the face of climate risk. APSIM is structured around plant, soil and management modules. These modules include a diverse range of crops, pastures and trees, soil processes including water balance, N and P transformations, soil pH, erosion and a full range of management controls. APSIM resulted from a need for tools that can provide accurate predictions of crop production in relation to climate, genotype, soil and management factors, while addressing the long-term resource management issues. The APSIM modelling framework is made up of (i) A set of biophysical modules that simulate biological and physical processes in farming systems, (ii) A set of management modules that allow the user to specify the intended management rules that characterize the scenario being simulated and that control the simulation (iii) Various modules to facilitate data input and output to and from the simulation (iv) A simulation engine that drives the simulation process and facilitates communication between the independent modules. Apart from this the framework also includes (v) various user interfaces for model construction, testing and application, (vi) several interfaces and association database tools for visualization and further analysis of output. (vii) Various model development, testing and documentation tools.

APSIM is one of the widely distributed model all over the world and has been used in a broad range of applications, such as (i) Support for on-farm decision making, (ii) Farming systems design for production or resource management, (iii) Assessment of the value of seasonal climate forecasting, (iv) Analysis of supply chain issues in agribusiness, (v) Development of waste management guidelines, (vi) Risk assessment for policy making and (vii) As a guide for research and educational activities.

One of the main benefits of APSIM is the ability to integrate models derived in fragmented research efforts. This enables research from one discipline or domain to be transported for the benefit of some other discipline or domain. It also facilitates comparison of models or sub-models on a common platform.

Decision Support System for Agro-technology Transfer (DSSAT)

DSSAT is a software package integrating the effects of soil, crop phenotype, weather and management options that allow users to ask "what if" type questions and simulate results by conducting, in minutes on a desktop computer, experiments which would consume a significant part of an agronomist's career. It has been in use for more than 15 years by researchers in over 100 countries. An overview of component and modular structure of DSSAT is depicted in Fig. 8. The DSSAT simulates growth, development and yield of a crop growing on a uniform area of land under prescribed or simulated management as well as the changes in soil, water, carbon, and nitrogen that take place under the cropping system over time. The most important features of DSSAT are:

- It separates modules along disciplinary lines,
- It defines clear and simple interfaces for each module,
- It enables individual components to be plugged in or unplugged with little impact on the main program or other modules, i.e. for comparison of different models or model components,
- It facilitates documentation and maintenance of code,
- It enables modules written in different programming languages to be linked together,
- It allows for easy integration into different types of application packages due to the well-defined and documented interface to the modules,
- It allows for evolution to integrate other components, such as livestock and intercropping, through well-defined module interfaces, and
- It facilitates cooperation among different model development groups where each can focus on specific modules as building

blocks for expanding the scope and utility of the crop simulation model.

- It combines crop, soil and weather databases into standard formats for access by crop models and application programs. The user can then simulate multi-year outcomes of crop management strategies for different crops at any location in the world.
- It also provides for validation of crop model outputs; thus allowing users to compare simulated outcomes with observed results. Crop model validation is accomplished by inputting the user's minimum data, running the model, and comparing outputs. By simulating probable outcomes of crop management strategies, DSSAT offers users information with which to rapidly appraise new crops, products, and practices for adoption (Jones *et al.*, 2003).

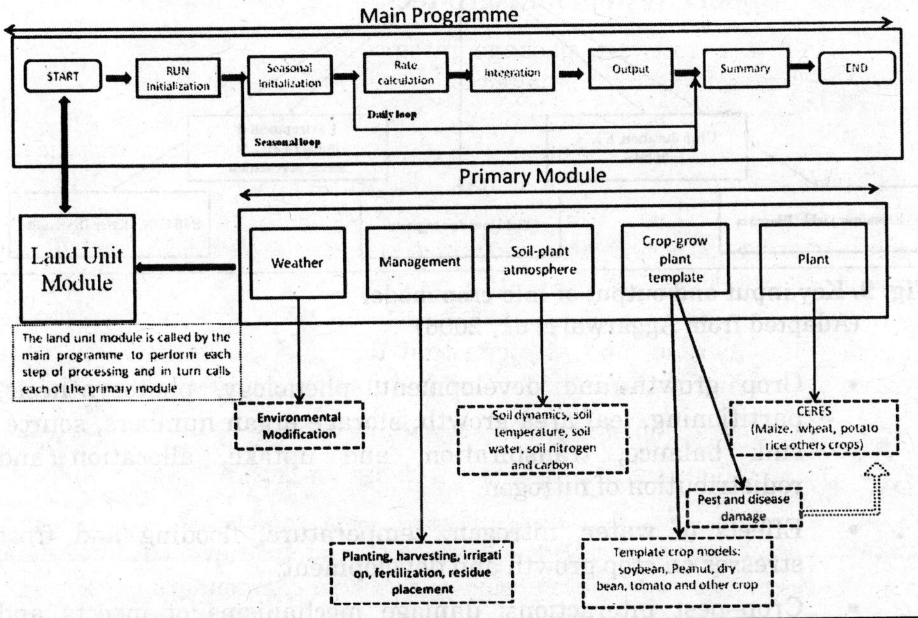

Fig. 8. Overview of the components and modular structure of the DSSAT. (Adapted from Jones *et al.*, 2003).

INFOCROP

Info-Crop, a generic crop model, which simulates the effects of weather, soils, agronomic management (planting, nitrogen, residues and irrigation) and major pests on crop growth, yield, soil carbon, nitrogen and water, and greenhouse gas emissions (Fig. 9). Info-Crop considers various processes such as, crop growth and development, soil water, nitrogen and carbon, and crop–pest interactions. Each process is described by a set of equations, in which the parameters change depending upon the crop/cultivar (Aggarwal *et al.*, 2006). The parameters considered in Info-crop are as follows:-

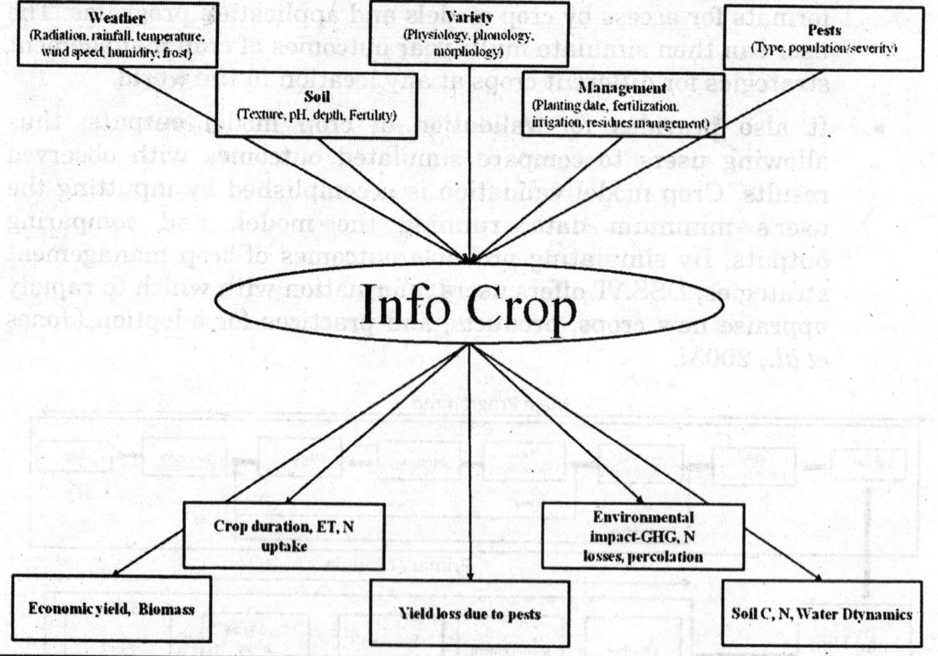

Fig. 9. Key input and output of Info-crop model (Adapted from Aggarwal *et al.*, 2006)

- Crop growth and development: phenology, photosynthesis, partitioning, leaf area growth, storage organ numbers, source: sink balance, transpiration, and uptake, allocation and redistribution of nitrogen.
- Effects of water, nitrogen, temperature, flooding and frost stresses on crop growth and development.
- Crop–pest interactions: damage mechanisms of insects and diseases.
- Soil water balance: root water uptake, inter-layer movement, drainage, evaporation, runoff, ponding.
- Soil nitrogen balance: mineralization, uptake, nitrification, volatilization, interlayer movement, denitrification, leaching.
- Soil organic carbon dynamics: mineralization and immobilization.
- Emissions of greenhouse gases: carbon dioxide (CO_2), methane (CH_4), nitrous oxide (N_2O).

Model calibration and validation

Model calibration involves the modification of some model parameters such that data simulated by the error-free model fit the observed data. In many instances, even if a model is based on observed data, simulated values do not exactly comply with the observed data and minor adjustments have to be made for some parameters (Wegener, 1994). Non-compliance may arise from sampling errors as well as from incomplete knowledge of the system. Alternatively, it may arise when the model is used in a situation that is markedly different from the one under which it was developed. It is a necessary process to identify key parameters and parameter precision required for calibration (Ma *et al.*, 2000). Model calibration is the process of estimating model parameters by comparing model predictions (output) for a given set of assumed conditions with observed data for the same conditions. The model validation stage involves the confirmation that the calibrated model closely represents the real situation. The procedure consists of a comparison of simulated output and observed data that have not been previously used in the calibration stage. Ideally, all mechanistic models should be validated both at the level of overall system output and at the level of internal components and processes. The latter is an important aspect because due to the occurrence of feedback loops in biological systems, good prediction of system's overall output could be attributed to compensating internal errors (Van Keulen, 1976). However, validation of all the components is not possible due to lack of detailed datasets and the option of validating only the determinant ones is adopted. For example, in a soil-water-crop model, it is important to validate the extractable water and leaf area components since biomass accumulated is heavily dependent on these. Model validation involves running model using input parameters measured or determined during the calibration process. According to Refsgaard (1997) model validation is the process of demonstrating that a given site-specific model is capable of making "sufficiently accurate "simulations, although "sufficiently accurate" can change based on project goals. According to the U.S. EPA (2002), the process used to accept, reject, or qualify model results should be established and documented before beginning model evaluation. Although ASCE (1993) emphasized the need to clearly define model evaluation criteria, no commonly accepted guidance has been established, but specific statistics and performance ratings for their use have been developed and used for model evaluation (Moriasi *et al.*, 2007).

An essential part of model validation, probably the first aspect that comes to mind, is comparison of model predictions with observed data. This can help to identify problems with the model and give ideas for improvement. However, one must be careful in drawing general conclusions from the comparison of data and predictions. If the observed situations are not representative of the situations of interest, then comparison with past data may be different from the agreement with future measurements. Also, if the observed data have been used in model development, the degree of fit to that data is in general better than the agreement with future measurements. There is no single best method of comparison between a model predicted and observed data. Different methods highlight different features of the data and of model behavior. Therefore, one should use a number of methods altogether to evaluate model.

a. Graphical comparisons

Graphs are extremely useful for providing a quick visual summary of data and of the comparison between observed and predicted data. Graphical techniques provide a visual comparison of simulated and measured constituent data and a first overview of model performance (ASCE, 1993). According to Legates and McCabe (1999), graphical techniques are essential to appropriate model evaluation. The most widespread graphical presentation of the agreement between measured and calculated values for crop models is a plot as in Fig. 10.

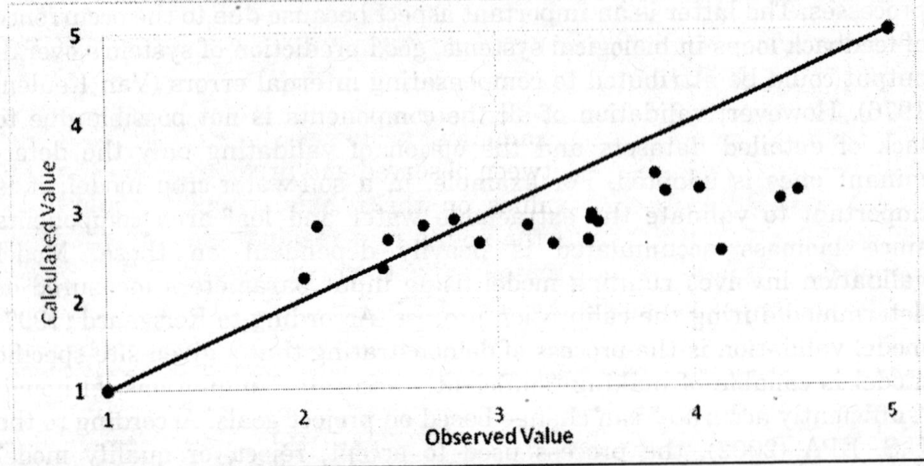

Fig. 10. Observed and predicted value at 1:1 line.

The advantage of this type of graph is that one can see at a glance how accurate model prediction is? However, a word of caution is required. There may be a tendency to underestimate the level of error. The eye tends to evaluate the shortest distance from the point to the 1:1 line, but model error is given by the vertical or equivalently the horizontal distance from the point to the line.

Sometimes, regression line between predicted and observed values may be shown in graph in addition to, or instead of, the 1:1 line (Fig. 11).

The coefficient of interpretation i.e. R^2 indicates the model predictability. A high R^2 i.e. close to 1 shows agreement between observed and predicted values and *vice versa*. However, to include the regression line on a graph of this type seems more misleading than useful. If the points fall close to the regression line one can have the impression that the model is quite good, but in fact that is not a measure of model error.

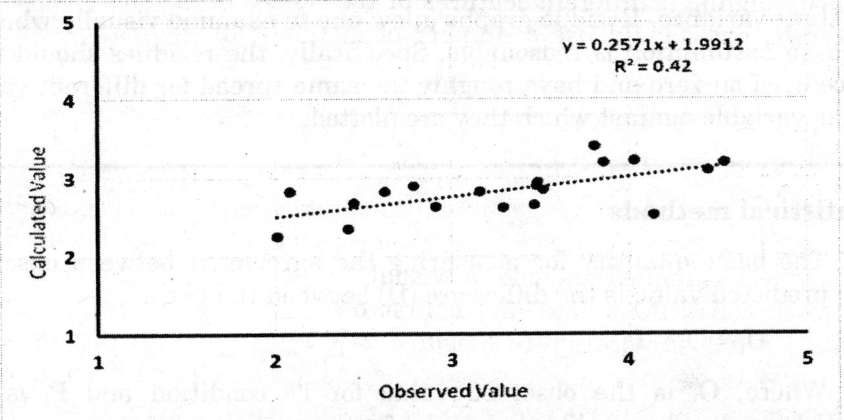

Fig. 11. Observed and predicted value with regression line.

b. Graphing the residues:

The classical method of examining model error in statistics is to plot model error (the difference between observed and predicted values) on the y axis, against measured values on the x axis, popularly known as residual analysis (Fig 12). If the points in a residual plot are randomly dispersed around the horizontal axis, then we can say that model predictability is good enough to predict.

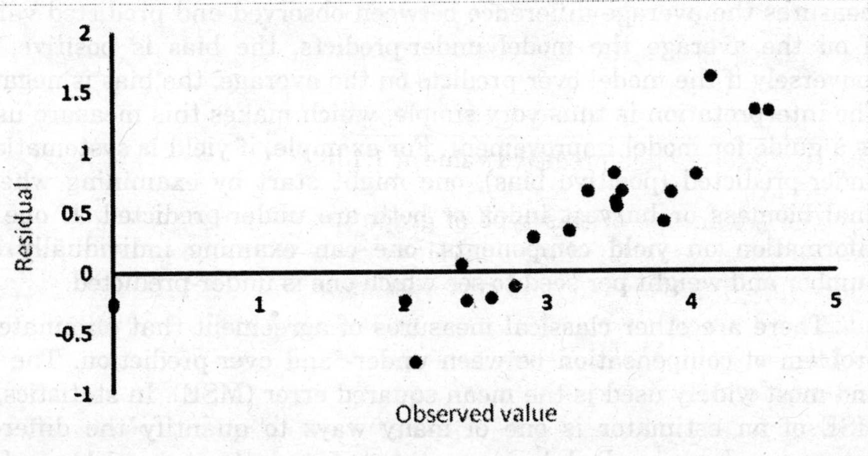

Fig. 12. Residual plot between residual and observed value

It is clearly visualized that points are far from the horizontal line; on the basis of this, we can conclude that predictability of model is poor and model parameter needs to be parameterized again for given situation.

The residue graph will, furthermore, give indications about the variability of model error, which is important for parameter estimation. The simplest assumption is that model errors have zero mean and constant variance. Residue graphs allow one to examine visually whether such an assumption is reasonable. Specifically, the residues should then be centred on zero and have roughly the same spread for different values of the variable against which they are plotted.

Statistical methods

The basic quantity for measuring the agreement between observed and predicted value is the difference (D) between the two i.e.

$$D_i = O_i - P_i \tag{4}$$

Where, O_i is the observed value for ith condition and P_i is the predicted value from calibrated model for the same condition.

A popular way to summarize the D_i values for numerous situations is to calculate their average, also known as "model bias". In statistics, the bias of an estimator is the difference between this estimator's expected value and the true value of the parameter being estimated. An estimator or decision rule with zero bias is called unbiased. Otherwise the estimator is said to be biased.

$$BIAS = \frac{1}{N}\sum_{i=1}^{N} D_i \tag{5}$$

Where, N is the total number of condition/situation/event. The bias measures the average difference between observed and predicted values. If on the average the model under-predicts, the bias is positive, and conversely if the model over-predicts on the average, the bias is negative. The interpretation is thus very simple, which makes this measure useful as a guide for model improvement. For example, if yield is systematically under-predicted (positive bias), one might start by examining whether final biomass or harvest index or both are under-predicted. If one has information on yield components, one can examine individually seed number and weight per seed to see which one is under-predicted.

There are other classical measures of agreement that eliminate the problem of compensation between under- and over-prediction. The first and most widely used is the mean squared error (MSE). In statistics, the MSE of an estimator is one of many ways to quantify the difference between values implied by an estimator and the true values of the quantity being estimated. MSE is a risk function, corresponding to the expected value of the squared error loss or quadratic loss. MSE measures the average of the squares of the "errors." The error is the amount by which the value implied by the estimator differs from the quantity to be estimated. The difference occurs because of randomness or because the estimator doesn't account for information that could produce a more accurate estimate. Mean squared error can be quantified by following equation:

$$MSE = \frac{1}{N}\sum_{i=1}^{N}(O_i - P_i)^2$$

Mean square error is an average of squared differences, large differences are heavily weighted. It is worthwhile to verify if MSE is not essentially due to one or two large differences. If this is the case, it might be more astute to examine those specific cases (problem with the data? exceptional circumstances such as extreme stress?) rather than the overall model. It is always convenient to work with square root of MSE, popularly called root mean square error (RMSE). It is also called the root mean square deviation, (RMSD) and frequently used to measure differences between observed and predicted values by a model. RMSE serves to aggregate residuals into a single measure of predictive power. The calculated RMSE values will have units, thus easier to interpret the result. The RMSE of a model prediction with respect to the observed variable is defined as:

$$RMSE = \sqrt{\frac{\sum_{i=1}^{N}(O_i - P_i)^2}{N}}$$

Lower value of RMSE indicates higher predictability of the model.

Non-dimensional forms of the RMSE are called normalized root mean square error (NRMSE) and it is useful because often one wants to compare RMSE with different units. There are two approaches: normalize the RMSE to the range of the observed data, or normalize to the mean of the observed data.

$$\text{NRMSE} = \frac{\text{RMSE}}{\text{O}_{\text{max}} - \text{O}_{\text{min}}}$$

$$\text{NRMSE} = \frac{\text{RMSE}}{\text{Mean of observed data}}$$

The latter one is also called CV, RMSE for the resemblance with calculating the coefficient of variance.

Correlation – often measured as a correlation coefficient – indicates the strength and direction of a linear relationship between two variables (for example model output and observed values). A number of different coefficients are used for different situations. The best known is the

Pearson product-moment correlation coefficient (also called Pearson correlation coefficient or the sample correlation coefficient), which is obtained by dividing the covariance of the two variables by the product of their standard deviations. If we have a series n observations and n model values, then the Pearson product-moment correlation coefficient can be used to estimate the correlation between model and observations.

$$r = \frac{\sum_{i=1}^{n}(O_i - O_{mean}) \cdot (P_i - P_{mean})}{\sqrt{\sum_{i=1}^{n}(O_i - O_{mean})^2 \cdot \sum_{i=1}^{n}(P_i - P_{mean})^2}}$$

The correlation is +1 in the case of a perfect increasing linear relationship, and –1 in case of a decreasing linear relationship, and the values in between indicate the degree of linear relationship between for example model and observations. A correlation coefficient of 0 means that there is no linear relationship between the variables. The square of the Pearson correlation coefficient (r^2), known as the coefficient of determination, describes how much of the variance between the two variables is described by the linear fit.

Another model evaluation statistics is the "index of agreement", developed by Willmott (1981) as a standardized measure of the degree of model prediction error and varies between 0 and 1. A computed value of 1 indicates a perfect agreement between the measured and predicted values, and 0 indicates no agreement at all. The index of agreement represents the ratio between the mean square error and the "potential error". The potential error as the sum of the squared absolute values of the distances from the predicted values to the mean observed value and distances from the observed values to the mean observed value. The index of agreement can detect additive and proportional differences in the observed and simulated means and variances; however, this index is overly sensitive to extreme values due to the squared differences.

$$\text{Index of agreement} = 1 - \frac{\sum_{i=1}^{N}(O_i - P_i)^2}{\sum_{i=1}^{N}(|P_i - O_{mean}| + |O_i - O_{mean}|)^2}$$

Modelling efficiency (ME) is one of the widely used normalized model statistics. It ranges from 0 to 1, with ME = 1 being the optimal value. ME values should be larger than 0.0 to indicate "minimally acceptable" model performance. It can be explained as:

$$ME = 1 - \frac{\sum_{i=1}^{N}(O_i - P_i)^2}{\sum_{i=1}^{N}(O_i - O_{mean})^2}$$

Model uses and limitations

Crop models have many current and potential uses for answering questions in research, crop management, and policy. Models can assist in synthesis of research understanding about the interactions of genetics, physiology, and the environment, integration across disciplines, and organization of data. It can assist in pre-season and in-season management decisions on cultural practices, fertilization, irrigation, and pesticide use. Crop models can assist policy makers by predicting soil erosion, leaching of agrochemicals, effects of climatic change, and large-area yield forecasts (Boote *et al.*, 1996). But most important is to select the appropriate crop models according to user objectives because models are not universal. Even when a judicious choice is made, it is important that aspects of model limitations be borne in mind such that modelling studies are put in the proper perspective and successful applications are achieved. In this section, we will discuss current uses, potential uses and limitations of crop models in details.

As research tools

Research understanding: Model development ensures the integration of research understanding acquired through discreet disciplinary research and allows the identification of the major factors that drive the system and can highlight areas where knowledge is insufficient. Thus, adopting a modelling approach could contribute towards more targeted and efficient research planning. For example, changing the plant density in a sugarbeet model resulted in model failure (Loomis *et al.,* 1976). This failure stimulated studies that gave additional information concerning biomass partitioning in the sugarbeet (Ropoport and Loomis 1986).

Integration of knowledge across disciplines: Adoption of a modular approach in model coding allows the scientist to pursue his discipline-oriented research in an independent manner and at a later stage to integrate the acquired knowledge into a model. For example, the modular aspect of the APSIM software (Mc Cown *et al.* 1996), allows the integration of knowledge across crops as well as across disciplines for a particular crop. Adoption of a modular framework also allows for the integration of basic research that is carried out in different regions, countries and continents. This ensures a reduction of research costs (e.g., through a reduction in duplication of research) as well as the collaboration between researchers at an international level.

Documentation and data organization in experiemnts: Simulation model development, testing and application demand the use of a large amount of technical and observational data supplied in given units and in a particular order. Data handling forces the modeller to resort to formal data organisation and database systems (Prestwidge *et al.* 1994; IBSNAT 1988, 1990). The systematic organisation of data enhances the efficiency of data manipulation in other research areas (e.g., productivity analysis, change in soil fertility status over time)

Improvement of crop cultivars: As simulation models become more detailed and mechanistic, they can mimic the system more closely. More precise information can be obtained regarding the impact of different genetic traits on economic yields and these can be integrated in genetic improvement programs, e.g., the NTKenaf model (Muchow and Carberry, 1993a). Other workers have used the modelling approach to design crop ideotypes for specific environments (Muchow and Carberry, 1993b).

Analysis of yields: When a model with a sound physiological background is adopted, it is possible to extrapolate to other environments. The use of several simulation models to assess climatically-determined yield in various crops has been mentioned by Boote *et al.* (1996). The CANEGRO model (Inman-Bamber *et al.,* 1993) has been used along the same lines in the South African sugar industry. Through the modelling approach, quantification of yield reductions caused by non-climatic causes (e.g., delayed sowing, soil fertility, pests and diseases) becomes possible. Almost all simulation models have been used for such purposes. Simulation models have also been reported as useful in separating yield gain into components due to changing weather trends, genetic improvements and improved technology (Bell and Fischer, 1994).

As crop system management tools

Input management practices: Management decisions regarding cultural practices and inputs have a major impact on yield. Simulation models, that allow the specification of management options, offer a relatively inexpensive means of evaluating a large number of strategies that would rapidly become too expensive if the traditional experimentation approach were to be adopted. Many publications are available describing the use of simulation models with respect to cultural management (planting and harvest date, irrigation, spacing, selection of variety type) and input application (water and fertiliser). In the Mauritian sugar industry, the evaluation of the impact of alternative harvest dates and irrigation strategies on cane and sucrose yields was achieved using the APSIM-Sugarcane model (Cheeroo-Nayamuth, 1998).

As a Risks assessment tool: Using a combination of simulated yields and gross margins, economic risks and weather-related variability can be assessed. These data can then be used as an investment decision support tool. The APSIM-Sugarcane model successfully demonstrated the variation in economic risk of sugarcane production at Medine, Reduit and

Pamplemousses in Mauritius, and provided good indications of potential irrigation and investment strategy to be adopted at these locations (Cheeroo-Nayamuth, 1998).

Precision farming: Profit maximisation may be achieved by managing farms as sets of sub-units and providing the required inputs at the optimum level to match variation in soil properties across the farm. Such an endeavour is attainable by coupling simulation models with geographic information systems (GIS) to produce maps of predicted yield over the farm. But, one of the prerequisites is a systematic characterisation of units that may prove costly. Such applications of simulation models have been reported by Wegener (1994) and Jones (1993).

As policy analysis tools

Best management practices (BMP): Models having chemical leaching or erosion components can be used to determine the best practices over the long-term. The EPIC model has been used to evaluate erosion risks due to cropping practices and tillage (Williams *et al.* 1984, 1989).

Yield forecasting: Yield forecasting for industries over large areas is important to the producer (harvesting and transport), the processing agent (milling period) as well as the marketing agency. The technique uses weather records together with forecast data to estimate yield across the industry. This technique was illustrated using CERES-Maize over the corn belt of the USA (Hodges *et al.,* 1987).

Introduction of a new crop or crop cultivar: Agricultural research is linked to the prevailing cropping system in a particular region. Hence, data concerning the growth and development of a new crop in that region would be lacking. Developing a simulation model based on scientific data collected elsewhere and a few datasets collected in the new environment helps in the assessment of temporal variability in yield using long-term climatic data. Running the simulations with meteorological data in a balanced network of locations also helps in locating the industry. Such a study was undertaken relative to the proposed introduction of a fibre industry in Northern Australia (Carberry *et al.* 1992).

Climate change and agriculture: Increased levels of CO_2 and other greenhouse gases are contributing to global warming with associated changes in rainfall pattern. Assessing the effects of these changes on crop yield is important at the producer as well as at the government level for planning purposes. Despite objections raised by Passioura (1996), the simulation approach remains the best tool for quantifying these effects. The CROPGRO, CERES and EPIC models, AUSCANE as well as many others have been used to quantify yield change resulting from global climate change. In Mauritius, the APSIM-Sugarcane model was used to assess the vulnerability of the sugar industry to climate change and to quantify the impact of a limited range of mitigation possibilities (e.g., irrigation adoption, variety type and harvest date) on cane and sucrose yields (Cheeroo-Nayamuth and Nayamuth, 2000).

Limitations

The level of complexity needed for a specific model depends on the objectives and questions being asked of the model. It also depends on amount of information (data) and time available for model building and testing. Whisler *et al.* (1986) illustrated the continuum of increasing comprehensiveness in models, starting with totally empirical models and gradually including more mechanism. They proposed that totally empirical models (such as multiple regression) could be used for summarizing past data and may be even interpolating within the range of past data, but that empirical models would be insufficient for extrapolative prediction (outside the range of the data base) or for explanatory interpretation (cause and effect) of experimental results. Further, Whisler *et al.* (1986) proposed that crop simulation models should have a fairly high level of mechanism and comprehensiveness to be able to explain and interpret research results. In reality, most crop simulation models are a mixture of empiricism and mechanism, and even the most mechanistic models make use of empiricisms at some hierarchical level (cellular or biochemical). A Mechanistic crop model is generally considered to be based on physiological and physical processes, and considers cause and effect at the process level. Material (C, N, water) and energy balance are usually included (Boote *et al.*, 1996).

Conclusion

Agricultural systems are complex entities involving a large amount of technical data from various disciplines for adequate elaboration of crop growth models. As investments in model development are high, most researchers are involved in model application rather than in model development. With decreasing costs of computing facilities, models are becoming accessible to a wide range of users with differing degrees of expertise in the modelling field. Knowledge of crop growth and development being incomplete, even the most rigorous model includes some approximations and, therefore, possesses some limitations. Proper understanding of these limitations is important when models are used. Nevertheless, crop models can be used for a wide range of applications. As research tools, model development and application can contribute to identify gaps in our knowledge, thus enabling more efficient and targeted research planning. Models that are based on sound physiological data are capable of supporting extrapolation to alternative cropping cycles and locations, thus permitting the quantification of temporal and spatial variability. Over a relatively short time span and at comparatively low costs, the modeller can investigate a large number of management strategies that would not be possible using traditional methodologies. Despite some limitations, the modelling approach remains the best means of assessing the effects of future global climate change, thus helping in the formulation of national policies for mitigation purposes. Other policy issues, like yield forecasting, industry planning, operations management, consequences of management decisions on environmental issues, are also well supported by modelling. As knowledge is accumulated through further research, models are improved and new models are developed. Also, as more performing models become available and users' experience grows, confidence in the modelling approach will be reinforced and models will be put to more intensive use.

❑❑❑

2 Applications of Crop Growth Simulation Models in Climate Change Assessments

M. Mohanty, Nishant K. Sinha, Rohit Patidar[1] and Ashok K. Patra

Division of Soil Physics, Indian Institute of Soil Science, Bhopal-462038, MP

[1]R.A.K. College of Agriculture, Sehore - 466 001

Climate change

Climate change refers to a statistically significant variation in either the mean state of the climate or in its variability, persisting for an extended period (typically decades or longer). Climate change may be due to natural internal processes or external forcings, or to persistent anthropogenic changes in the composition of the atmosphere or in land use (Ministry of Environment and Forest, Government of India, website). Greenhouse gases (for example, carbon dioxide, methane, nitrous oxide, water vapour, ozone), are the gases that are present in very small quantities in the atmosphere that absorb the heat radiated from the earth's surface to the atmosphere and thus emit some of the heat to the earth's surface. If they did not perform this useful function, most of the heat energy would escape, leaving the earth cold (about –18°C) and unfit to support life. Due to industrial growth in recent times, the atmospheric concentrations of carbon dioxide, methane, and nitrous oxide have grown by about 31%, 151% and 17%, respectively, between 1750 and 2000 (IPCC 2001). An increase in the levels of GHGs could lead to greater warming, which, in turn, could have an impact on the world's climate, leading to the phenomenon known as climate change. Indeed, scientists have observed that over the 20th century, the mean global surface temperature increased by 0.6°C (IPCC 2001). The inter-governmental panel on climate change has projected that the global mean surface temperature is predicted to rise by 1.1–6.4°C by 2100 (IPCC 2007).

Climate change effects on crops

Increased carbon dioxide levels in the atmosphere as a result of climate change will alter global temperatures and rainfall amounts. These factors will influence how well plants grow and affect food production. Higher temperatures cause heat stress in plants. This means they grow less and produce fewer crops. In some cases, the plants do not reproduce at all since excessive heat causes sterility of the pollen (the masculine reproductive part of the flowers). A temperature increase may be beneficial in areas which are very cold at present. For example, in Siberia or Northern Europe it may, in the future, be possible to grow crops for longer periods of the year. But in tropics and subtropics the things will go worse and will be affected much more in terms of crop production than the cold temperate areas. Changes to our climate are happening more quickly now than before in the geological past. Plants will have to adapt to new climate conditions more rapidly than they have ever had to do so before.

Water availability directly affects the growth of plants and how much crop they produce. Excessive rainfall results in floods. Waterlogged soil causes plant roots to rot and heavy rainfall damages tender young plants. Increased rainfall without flooding may be beneficial in very dry areas and allow limited crop growth. So changes in temperature and precipitation patterns as a result of climate change are likely to be bad for large areas of the world but may increase crop production in other regions. However, one of the likely outcomes of climate change is also an increase in the severity of rain storms and droughts and both of these are likely to have large and devastating effects on agriculture.

The increase in atmospheric carbon dioxide (CO_2) levels resulting from fossil fuel combustion has a fertilising effect on most plants since CO_2 is needed for photosynthesis (the biochemical mechanism of plant growth). Photosynthesis converts carbon dioxide and water into the simple sugar glucose and emits oxygen, making it possible for animals to live on earth. Scientific experiments have shown that increasing atmospheric CO_2 levels leads to an increase in plant growth. Therefore, that increases in CO_2 emissions from fossil fuel combustion are good for crop growth and can have a positive influence on photosynthesis; under optimal growing conditions of light, temperature, nutrient and moisture supply, biomass production can increase, especially of plants with C_3 photo-synthetic metabolism, above and even more below ground. However, the negative effects of climate change are usually much larger than the positive ones. Even though, increase in ambient CO_2 does not have significant direct effects on C_4 (C_4 carbon fixation pathway) photosynthesis of maize crop (Leakey *et al.,* 2004, 2006), increase in ambient CO_2 leads to higher water use efficiency in water stress conditions and there by directly influences dry matter production and grain yield (Leakey *et al.* 2004).

The various studies have shown that climate change scenarios decrease simulated yields in many cases, while the direct effects of increasing atmospheric CO_2 mitigate the negative effects primarily in mid and high latitudes. The differences between countries in yield responses to climate change are related to differences in current growing conditions. At low latitudes, crops are grown nearer the limits of temperature tolerance and global warming may subject them to higher stress. In many mid and high latitude areas, increasing temperatures may benefit crops otherwise limited by cold temperatures and short growing seasons in the present climate. Under the estimated effects of climate change and atmospheric CO_2 on crop yields, world cereal production is estimated to decrease between 1 and 7% depending on the GCM climate scenario. The largest negative changes are predicted in developing countries, averaging –9% to –11%. By contrast, in developed countries production is estimated to increase under all but the UKMO scenario (+11% to –3%). Thus existing disparities in crop production between the developed and developing countries are estimated to grow (Fig. 1) (Parry, 2007).

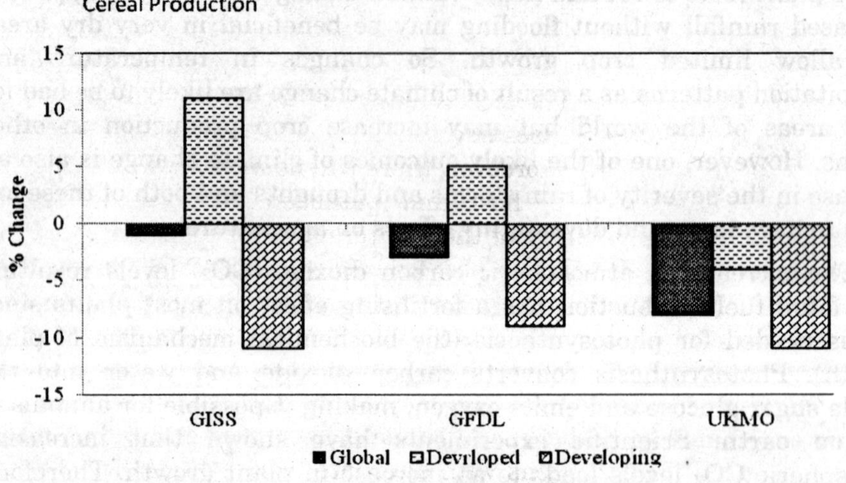

Fig. 1. Cereal crop production in the world under different climate change scenarios

Climate change effect on soil processes

Soils, as a medium for plant growth, would be affected in several other ways. Increased temperatures may lead to more decomposition of soil organic matter;

1. Increased plant growth due to the CO_2 fertilization effect may cause other plant nutrients such as N and P to become in short supply; however, CO_2 increase would stimulate mycorrhizal activity (making soil phosphorus more easily available), and also biological nitrogen fixation (whether or not symbiotic). Through increased root growth there would be extra weathering of the substratum, hence a fresh supply of potassium and micronutrients;
2. The CO_2 fertilization effect would produce more litter of higher C/N ratio, hence more organic matter for incorporation into the soil as humus; litter with high C/N decomposes slowly and this can act as a negative feedback on nutrient availability;
3. The 'CO_2 anti-transpirant' effect would stimulate plant growth in dryland areas, and more soil protection against erosion and lower top soil temperatures, leading to an 'anti-desertification effect'.

Climate change and insect- pest diseases and weed growth

Climate change has the potential to modify host physiology and resistance, as well as to alter stages and rates of development of the pathogen. Elevated concentrations of CO_2 are believed to result in a denser plant canopy. When it is combined with increased humidity, it is likely to promote foliar diseases such as rust, powdery mildew, leaf spot and blights (Råberg, 2008). Moreover, there will be most likely shift in the geographical distribution of host and pathogen. The mechanism of pathogen dispersal, suitability of the environment for dispersal, survival between seasons and changes in host physiology and ecology in the new environment will largely determine how quickly pathogens become established in a new region. Elevated winter temperatures are likely to give higher survival rate for the aphid, which can reproduce early and severe the virus incidence. Changes may occur in the type, amount and relative importance of pathogens affecting a particular crop. It would be more pronounced for pathogens with alternate hosts. Durability of plant resistance may be affected as pathogens will have more time to evolve aggressive races.

Global warming and other climatic changes will affect weed growth, phenology, and geographic distribution in a similar way as for the main crops. Higher CO_2 concentration will stimulate photosynthesis and growth in C_3 weed species, as well as increased water use efficiency in all weed species. Increased rhizome and tuber growth in perennial C_3 weeds is a likely response to elevated CO_2 concentrations. Thus, it would be more difficult to control perennial weeds mechanically as well as chemically. As an increased winter temperature facilitates wintering of insect populations, the effectiveness of biological control of weeds could increase.

Adaptation and mitigation

Types of Adaptation Options in Agriculture with special reference to natural resource management

Agricultural adaptation option with special reference to natural resource management is the technological developments. The main types of adaptations are summarized in Table 1 with examples.

Table 1. Adaptations to climate change

A. Technological developments

Crop development

- Develop new crop varieties, including hybrids, to increase the tolerance and suitability of plants to temperature, moisture and other relevant climatic conditions.

Weather and climate information systems

- Develop early warning systems that provide daily weather predictions and seasonal forecasts.

Resource management innovations

- Develop water management innovations, including irrigation, to address the risk of moisture deficiencies and increasing frequency of droughts.
- Develop farm-level resource management innovations to address the risk associated with changing temperature, moisture and other relevant climatic conditions.

B. Farm production practices

Farm production

- Diversify crop types and varieties, including crop substitution, to address the environmental variations and economic risks associated with climate change.
- Diversify livestock types and varieties to address the environmental variations and economic risks associated with climate change.
- Change the intensification of production to address the environmental variations and economic risks associated with climate change.

Land Use

- Change the location of crop and livestock production to address the environmental variations and economic risks associated with climate change.
- Use alternative fallow and tillage practices to address climate change-related moisture and nutrient deficiencies.

Land topography

- Change land topography to address the moisture deficiencies associated with climate change and reduce the risk of farm land degradation.

Irrigation

- Implement irrigation practices to address the moisture deficiencies associated with climate change and reduce the risk of income loss due to recurring drought.

Timing of operations

- Change timing of farm operations to address the changing duration of growing seasons and associated changes in temperature and moisture.

C. Farm financial management

Crop insurance

- Purchase crop insurance to reduce the risks of climate-related income loss.

Crop shares and future

- Invest in crop shares and future to reduce the risks of climate-related income loss.

Mitigation measures

There is an extensive list of technically possible options for mitigating emission in agriculture and land-use. Measures may be categorised as: reducing emissions via improved farming efficiency, including genetic improvement; displacing fossil fuel emissions via alternative energy sources; and enhancing the removal of atmospheric CO_2 via sequestration into soil and vegetation sinks. Some mitigation options, typically current best management practices, deliver improved farm profitability as well as lower emissions and thus might be adopted without government intervention beyond continued promotion/revision of benchmarking and related advisory information services.

Table 2. Mitigation measures affecting adaptation in agricultural systems

Mitigation measures / Adaptation issues	Soil erosion control	Nutrient loss reduction	Soil and water conservation	Genetic diversity	Microclimate modification	Land use change
Catch crops	+	+				
Reduced tillage	+		+			
Residue management	+		+			
Extensification						+
Fertilizer application		+				
Fertilizer type		+				
Rotation species	+	+		+		
Adding legumes	+	+		+		
Permanent crops	+	+		+		
Agro-forestry	+	+			+	

Source: Adapted from Oleson and Porter (2009).

Oleson and Porter (2009) summarise the effects of a range of mitigation measures on six main factors relevant to adaptation in agriculture (with a particular focus on arable systems). Some of these are shown in Table 2, which is not exhaustive but shows instead some examples of where mitigation measures can have positive (or negative) implications on the adaptive capacity of a system. Mitigation measures generally aim to reduce nutrient losses in the system. This is mainly achieved through increasing the nutrient and water retention in the systems and preventing soil degradation, which also helps to make the system more resilient to droughts and flooding. Similarly, adaptation actions can affect mitigation efforts, particularly those that reduce soil erosion, leaching of nitrogen and phosphorus; conserve soil moisture; increase the diversity of crop rotations; modify the microclimate to reduce temperature extremes and provide shelter, involve land-use change, including the abandonment or extensification of existing agricultural land or the cultivation of new land (Oleson and Porter, 2009).

Simulation models

Crop growth simulation models and biogeochemical and biophysical models have been very helpful in projecting the future crop and soil productivity. These models in connection with different GCM models predict the future agricultural practices that can adapt to different climate change scenarios. Here are a few of the models that can be used for different scenarios analysis to combat impact of climate change on agricultural production of the globe.

Simulation models that are able to assess climate change impact on crop growth, yield and farm economy, still lack complete feedback structures. Only single aspect can be investigated. However, modelling this single aspect increases knowledge on the aspects of expectations from climate change, if interpreted carefully and in the context of the model's abilities. Simulation models are widely used to address "what if" type questions, such as, what if the climate changes, different irrigation or fertilization regimes are used, different sowing dates are used, different cultivars are used, etc. In addressing actual yield predictions required by governments, private corporations, or NGOs, different types of simulation models are used for solving these "what if" type questions. Here, capabilities of different simulation models will be discussed in assessing the impact of climate change on agro ecosystem and what would be the possible mitigation and adaptation.

Agricultural production systems simulator (APSIM)

APSIM was developed to simulate biophysical processes in farming systems, particularly as it relates to the economic and ecological outcomes of management practices in the face of climate risk. APSIM is structured around plant, soil and management modules. These modules include a diverse range of crops, pastures and trees, soil processes including water balance, N and P transformations, soil pH, erosion and a full range of management controls. APSIM resulted from a need for tools that can provide accurate predictions of crop production in relation to climate, genotype, soil and management factors, while addressing the long-term resource management issues.

Decision support system for agro-technology transfer (DSSAT)

DSSAT is a software package integrating the effects of soil, crop phenotype, weather and management options that allows users to ask "what if" type questions and simulate results by conducting, in minutes on a desktop computer, experiments which would consume a significant part of an agronomist's career. It has been in use for more than 15 years by researchers in over 100 countries. The DSSAT simulates growth, development and yield of a crop growing on a uniform area of land under prescribed or simulated management as well as the changes in soil, water, carbon, and nitrogen that take place under the cropping system over time.

Cropping systems simulation model (Crop Syst)

Crop Syst (Cropping Systems Simulation Model) is a multi-year, multi-crop, daily time step crop growth simulation model, developed with emphasis on a friendly user interface, and with a link to GIS software and a weather generator (Stockle, 1996). The link to economic and risk analysis models is under development. The model's objective is to serve as an analytical tool to study the effect of cropping systems management on crop productivity and the environment. For this purpose, Crop Syst simulates the soil water budget, soil-plant nitrogen budget, crop phenology, crop canopy and root growth, biomass production, crop yield, residue production and decomposition, soil erosion by water, and pesticide fate. These are affected by weather, soil characteristics, crop characteristics, and cropping system management options including crop rotation (including fallow years), cultivar selection, irrigation, nitrogen fertilization, pesticide applications, soil and irrigation water salinity, tillage operations (over 80 options), and residue management.

Info Crop

Info Crop, a generic crop model, which simulates the effects of weather, soils, agronomic management (planting, nitrogen, residues and irrigation) and major pests on crop growth, yield, soil carbon, nitrogen and water, and greenhouse gas emissions (Fig. 5). Info Crop considers various processes such as, crop growth and development, soil water, nitrogen and carbon, and crop–pest interactions. Each process is described by a set of equations, in which the parameters change depending upon the crop/cultivar (Aggarwal *et al.* 2006).

These are all the small examples of models that can used for assessing the impact of climate change on crop productivity. There are many other models they can be used for similar purposes. The biogeochemical models like CENTURY, DNDC and RothC are used for carbon sequestration studies also. Are all the models suitable for climate change assessment? It is possible that the models those taking into account the effects of temperature, and CO_2 concentration on photosynthesis and radiation use efficiency are suitable. Most of the crop simulation models accommodate these effects though not for the field crops. The model needs to be robust and which has wider applicability to all geographical situations can be a good tool for climate change assessments.

Simulation models in climate change research

Several previous studies have used crop models to study the impact or sensitivity of climate change on agricultural production. The effects of climate change and elevated CO_2 concentration on wheat production using APSIM model in Australia has been investigated by several studies. These include the assessment on individual effects of rising temperature and CO_2 concentration ([CO_2]) and rainfall changes and the compound impacts of temperature, CO_2 and rainfall changes, c) or multiple probabilistic climate change scenarios indicated that impact of climate change differed significantly between soil types and locations, and in the higher rainfall region of south-western Australia wheat yield will increase for all likely future climate scenarios. They reported that higher [CO_2] increased yield especially at drier sites while higher temperatures had a positive effect in the cooler and wetter southern part of the region. The main difference between soil types was that heavier clay soils are most vulnerable to reduced rainfall while sandy soils were more vulnerable to higher temperature. Higher temperature could both increase and decrease protein concentrations. Wang *et al.* (2009) used the APSIM model to simulate the growth and water balance processes for a 117 year period of baseline, 2050 and 2070 climatic conditions. The results showed that wheat yield reduction caused by 1°C increase in temperature and 10% decrease in rainfall could be compensated by a 266 ppm increase in [CO_2] assuming no interactions between the individual effects. Temperature increase had little impact on long-term average water balance, while [CO_2] increase reduced evapotranspiration and increased deep drainage. Length of the growing season of wheat decreased 22 days in 2050 and 35 days in 2070 conditions as a consequence of 2.3°C and 3.8°C increase in temperature, respectively.

Using DSSAT, Jones and Thornton (2003) simulated the impact of climate change on maize production in Africa and Latin America and showed that there is 10 % decrease in aggregate maize production by 2055. Alexandrov & Hoogenboom (2000) simulated the impact of climate variability for the major crops, including maize and winter wheat, and assessed possible adaptation measures for Bulgarian agriculture under an expected climate change. Four transient GCM (ECHAM4, HadCM2, CSIRO-Mk2b and GFDL-R15) had been used in this study for climate change scenario. These GCM models had been used in DSSAT crop model and projected the shorter vegetative and reproductive growth season for maize and winter wheat during the 21st century. Projected yield reduction for winter wheat varied between 0 and 7% during the 2020s and 2050s, and between 4 and 20% in the 2080s. DSSAT also showed that by shifting the sowing date, yield loss can be reduced under the different

GCM climate change scenarios in this study. DSSAT crop model is widely used in climate change research over the globe.

The CENTURY ecosystem model is used to investigate how land use and climate affect SOM and plant growth. Bhattacharyya *et al.* has evaluated CENTURY model using two long-term fertilizer trials representing Indo-Gangetic plain (2007), humid and semi-arid sites from India (2010). He also modeled soil organic carbon stocks and changes in the Indo-Gangetic plains and predicated that, there will be a 21% decrease in SOC stocks in the IGP from 1967 to 2030. Many authors have tested CENTURY model and estimated the carbon turn over in climate changing scenario.

Info-Crop provides integrated assessment of the effect of weather, variety, pests, soil and management practices on crop growth and yield, as well as on soil nitrogen and organic carbon dynamics in aerobic as well as anaerobic conditions, and greenhouse gas emissions. The model considers the key processes related to crop growth, effects of water deficit, flooding, nitrogen management, temperature and frost stresses, crop-pest interactions, soil water and nitrogen balance and (soil) organic carbon dynamics. Its general structure relating to basic crop growth and yield is largely based on several earlier models, especially SUCROS series, and is written in Fortran Simulation Environment (FSE) programming language. Krishnan *et al.* (2007) used Info-crop to show impact of elevated CO_2 and temperature on rice yield in eastern India for GDFL, GISS and UKMO scenarios and predicted changes of –9.02, –11.30 and –21.35%, respectively. She also suggested that limitations on rice yield imposed by high CO_2 and temperature can be mitigated, at least in part, by altering the sowing time and the selection of genotypes that possess higher fertility of spikelets at high temperatures. Like that, Hebbar *et al.* (2008) for cotton; Srivastava *et al.* (2010) for sorghum; Boomiraj *et al.* (2010) for Indian mustard used info crop for assessment of respective crop under different climate change scenario.

The climate change scenarios study

The Intergovernmental Panel on Climate Change (IPCC, 2001) reported that the average global surface temperature will increase between 1.4 and 3°C above 1990 levels by 2100 for low emission scenarios and between 2.5 and 5.8°C for higher emission scenarios of greenhouse gases and aerosols in the atmosphere. Over the land regions of the Indian subcontinent, the projected (area-averaged) annual mean surface temperature rise by the end of 21st century has been estimated to range between 3.5 and 5.5°C depending upon the future trajectory of anthropogenic radiative forcing (Lal *et al.*, 2001). The projected temperature increase has a large seasonal and spatial dependency over India. During the monsoon season, the temperature rise over south India is projected to be less than 1.5°C by 2050s while the increase in surface temperature is more pronounced over north, central and east India (>2°C). Probable changes in precipitation, cloudiness and solar radiation under the climate changes scenarios were not taken into consideration in this analysis in view of the significant uncertainties associated with non-linear, abrupt and threshold rainfall events projected by GCMs over the Indian subcontinent.

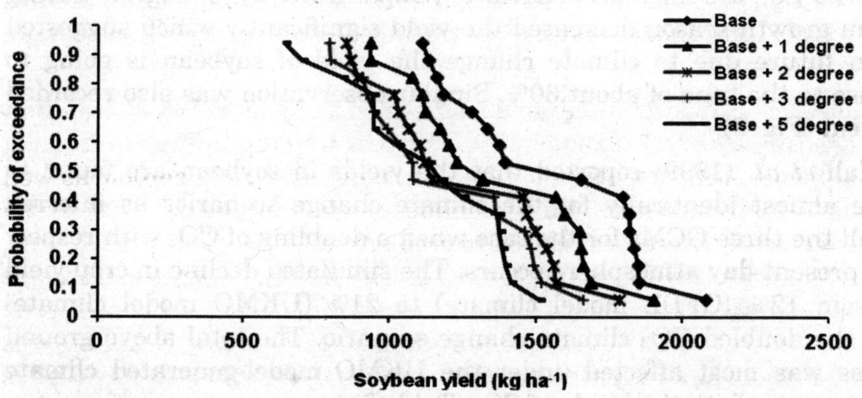

Fig. 2. Effect of increase in surface temperature on soybean grain yield as simulated by the APSIM model.

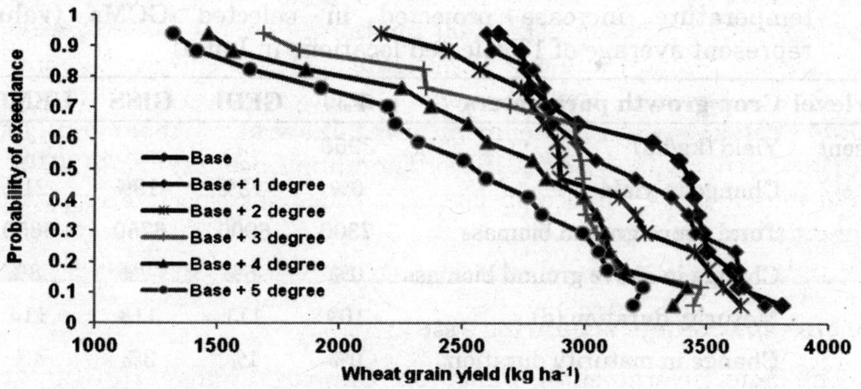

Fig. 3. Probability distribution yield under climate change (elevated temperature in Bhopal)

Impact of increase in surface air temperature on soybean and wheat yield

The increase in surface air temperature has tremendous effect on soybean as observed in long-term simulation by the APSIM model. The results revealed that there was significant decrease in yield of soybean from the normal (base line followed) (Fig. 2). The base line yield for this study was the long-term simulated yield of 16 years under recommended management practices followed for soybean. The probability distribution of soybean yield, described the probability of obtaining a specific amount of grain yield under the increased temperature scenario. There is 50% probability of obtaining <1 t ha^{-1} of soybean grain yield under climate change scenario as predicted by the model, whereas the probability of getting same amount of grain yield under normal condition is observed to be <10%. So, the increased surface temperature by 3 degree during soybean growth season decreased the yield significantly which suggested that in future due to climate change the yield of soybean is going to decrease to the tune of about 30%. Similar observation was also recorded for wheat (Fig. 3).

Mall *et al.* (1999) reported that the yields in soybean are found to decline almost identically for the climate change scenarios as inferred from all the three GCMs for the case when a doubling of CO_2 with respect to the present-day atmosphere occurs. The simulated decline in crop yield was from 12% (GFDL model climate) to 21% (UKMO model climate) under the doubled CO_2 climate change scenario. The total above ground biomass was most affected under the UKMO model-generated climate scenario under both the levels of CO_2 (Table 3).

Table 3. Some crop growth parameters of the soybean variety 'Bragg' as simulated by crop simulation model for two CO_2 levels and temperature increase projected in selected GCMs (values represent average of 13 selected locations in India)

CO_2 level	Crop growth parameters	T_{obs}	GFDL	GISS	UKMO
Present	Yield (kg/ha)	3950			
	Change in yield	0%	-13%	-19%	-21%
	Total above ground biomass	7300	6900	6750	6650
	Change in above ground biomass	0%	-5%	-7%	-8%
	Maturity duration (d)	109	111	113	114
	Change in maturity duration	0%	1%	3%	4%
Double	Yield (kg/ha)	5400	4750	4450	4250
	Change in yield	0%	-12%	-18%	-21%
	Total above ground biomass	10000	9450	9200	9100
	Change in above ground biomass	0%	-5%	-7%	-8%
	Maturity duration (d)	109	111	113	114
	Change in maturity duration	0%	1%	3%	4%

Table 4. Mean predicted change in soybean yield under the fixed temperature and CO_2 scenarios (Bhopal)

CO_2 concentration (ppm)	Temperature (° C)			
	Base	+2	+4	+5
390	0%	– 19%	– 26%	– 27%
450	+10%	– 19%	– 23%	– 23%
550	+23%	– 18%	– 19%	– 20%
650	+32%	–15%	– 17%	– 18%

Combined effect of CO_2 and temperature on soybean yield

Increase in CO_2 concentration increased whereas increase in temperature decreased soybean yield. The decrease was less when temperature in CO_2 in combination was used for simulation (Table 4).

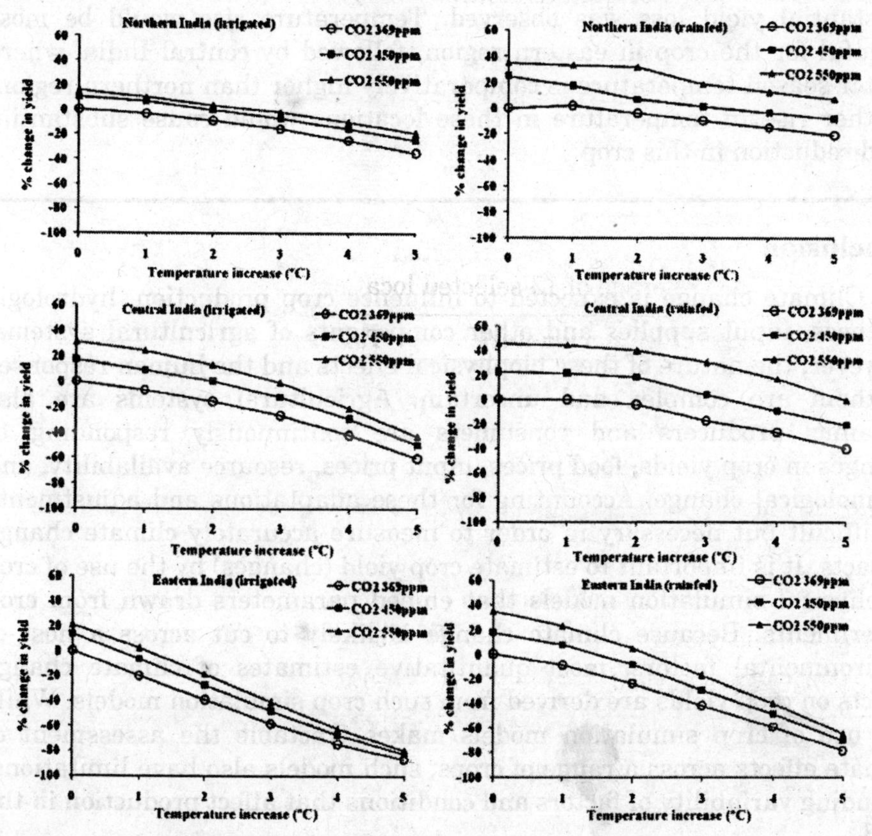

Fig. 4. Effect of CO_2 and temperature on simulated yield of irrigated and rainfed mustard in different locations in India.

Increasing temperature reduced, while increase in CO_2 concentration increased crop yield. Increase in CO_2 from 369 to 550 ppm with no change in temperature has resulted in 15.8–31% increase in yield of irrigated mustard across different regions. Positive yield response of mustard to elevated carbon dioxide was due to, increased photosynthetic activity resulting in increased specific leaf area, leaf weight, biomass production and grain number. But the positive effect of increase in CO_2 concentration was nullified by temperature rise. Under irrigated condition, the grain yield dropped steeply with rise in temperature in eastern India. In this region, yield reduction was maximum (86.6%) with 5ºC rise in temperature. Rise in temperature coupled with rise in CO_2 to 450 and 550 ppm decreased yield reduction to 82.4 and 79.4% respectively. Yield reduction of mustard was moderate in northern part of the country (Fig. 4). In north India, temperature rise by 5ºC, with no rise in CO_2 reduced mustard yield by 34.7%. Rise in CO_2 along with temperature caused less yield reduction of mustard in this region. Mustard crop grown in central part of the country was also vulnerable to temperature rise, where substantial yield loss was observed. Temperature rise would be most harmful for the crop in eastern region, followed by central India, where winter season temperature is comparatively higher than northern region. Further rise in temperature in these locations would cause substantial yield reduction in this crop.

Conclusion

Climate change is expected to influence crop production, hydrologic balances, input supplies and other components of agricultural systems. However, the nature of these biophysical effects and the human responses to them are complex and uncertain. Agricultural systems are also dynamic; producers and consumers are continuously responding to changes in crop yields, food prices, input prices, resource availability, and technological change. Accounting for these adaptations and adjustments is difficult but necessary in order to measure accurately climate change impacts. It is important to estimate crop yield (changes) by the use of crop biophysical simulation models that embed parameters drawn from crop experiments. Because climate change is likely to cut across a host of environmental factors, most quantitative estimates of climate change effects on crop yields are derived from such crop simulation models. While the use of crop simulation models makes tractable the assessment of climate effects across a range of crops, such models also have limitations, including variability of factors and conditions that affect production in the field.

Future research needs

- Development of regional climate change models by downscaling global climate projections for climate change research for a country.
- Development of a Geographical Information System (GIS) -based framework for assessing the risk of climate change for primary production systems in agriculture.
- Development of crop production systems with lower emissions of GHGs, that are sustainable, with respect to impacts on other environmental factors or attributes, and capacity to adapt to climate change.
- Research into methods to reduce nitrous oxide emissions from applied fertiliser.
- Research into methods to manage emissions from manure in intensive livestock industries.
- Research into use of bio-char and recycled organics as a soil amendment to sequester carbon, improve soil organic carbon, improve water holding capacity and nutrient cycling.
- Quantification of the impacts of management practices on soil carbon and parameterisation of models of soil carbon dynamics for agricultural and forest systems.
- Research into the interactive effects of increased atmospheric carbon dioxide in a water and nutrient limited environment on growth of major crop, pasture and forest species. Good understanding of the impacts of climate change will inform adaptation strategies.
- Research into the impacts of climate change on product quality, in all agricultural and forest systems, to inform breeding programs and development of adaptation strategies.
- Do plants acclimatize to changing temperatures? What plants are most vulnerable and where?

3 Biophysical Impacts and Priorities for Adaptation of Agricultural Crops in Changing Climate

R. Elanchezhian
Division of Soil Chemistry and Fertility,
Indian Institute of Soil Science, Bhopal, 462038, India

The amount of CO_2, the major green house gas (GHG), in the air has remained stable at 280 µmol mol^{-1} for most of the previous thousand years and was relatively stable between 200 to 300 µmol mol^{-1} for prehistoric times (150,000 years before present). However, the atmospheric CO_2 concentration is projected to double before the end of this century, which has been rising since the industrial revolution due to various anthropogenic activities. It has risen exponentially from about 280 µmol.mol^{-1} in 1850 to current concentrations of larger than 385 µmol mol^{-1} (IPCC 2007a). Although CO_2 concentration has previously reached 400 µmol mol^{-1} about 3 million year ago, perhaps a few years from now, the global CO_2 concentration (averaged over an entire year) may cross the similar mark of 400 µmol mol^{-1} (Monastersky, 2013). In April 2013 alone, the CO_2 levels have been found to be hovering over and above 400 µmol mol^{-1} mark on quite a few occasions at Mauna Loa, Hawaii (Fig. 1). This increase in atmospheric CO_2 has been accompanied by a coincident increase in the even more potent green house forcing gases like methane, ozone, and nitrous oxide such that combined ambient GHG concentrations are now expected to exceed concentrations of 550 µmol mol^{-1} by 2050 (Brouder and Volenec, 2008).

The increase in major GHGs like CO_2 and associated rise in mean temperature levels have led to global warming and change in climate as evident from change in physical climate and thereby affecting the biological systems like agriculture (Aggarwal *et al.,* 2010; IPCC 2007a; Gimenez, 2006). Climate change is already affecting agricultural systems in several regions of the world. Nevertheless the global change in climate would have a direct bearing on production of agricultural food crops like cereals, further the demand for food crops will be acute from the ever-limited land as the world population is projected to reach 9.1 billion by 2050 (UN 2009). Under such scenario agricultural production systems will require effective adaptation strategies to overcome these expected pressures in the immediate future. Meeting the food demand of a global population expected to reach over 10 billion by the end of the century will require major changes in agricultural production systems (UNFPA, 2011). Under such scenario, improving cropland management is key to increasing crop productivity without further degrading soil and water resources. At the same time, sustainable agriculture has the potential to deliver co-benefits in the form of reduced GHG emissions and increased carbon sequestration, therefore contributing to climate change mitigation.

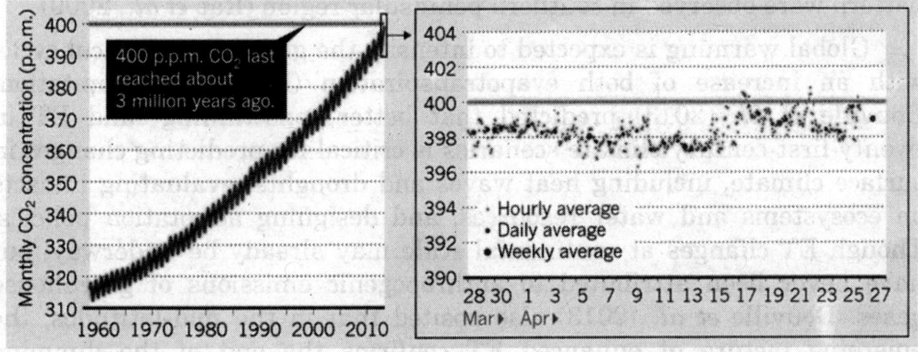

Fig 1. Atmospheric levels of CO_2 in Mauna Loa, Hawaii as adapted from Monastersky (2013).

Global trend of climate change

The changes in climate parameters are being felt globally in the form of changes in temperature and rainfall pattern. Besides CO_2, the global atmospheric concentration of methane and nitrous oxides, other important GHGs, have also increased considerably resulting in the warming of the climate system by 0.74°C between 1906 and 2005 (IPCC, 2007b). Of the last 12 years (1995–2006), 11 years have been recorded as the warmest in the instrumental record of global surface temperature (since 1850). The global average sea level rose at an average rate of 1.8 mm per year over 1961 to 2003. This rate was faster over 1993 to 2003, about 3.1 mm per year (IPCC, 2007a). There is also a global trend of an increased frequency of droughts as well as heavy precipitation events over many regions. Cold days, cold nights and frost events have become less frequent, while hot days, hot nights and heat waves have become more frequent. It is also likely that future tropical cyclones will become more intense with larger peak wind speeds and heavier precipitation. The IPCC (2007b) projected that temperature increase by the end of this century is expected to be in the range 1.8 to 4.0°C. IPCC projected 0.5 to 1.2°C rise in temperature by 2020, 0.88 to 3.16°C by 2050 and 1.56 to

5.44°C by 2080 for the Indian region (South Asia), depending on the future development scenario (IPCC 2007b).

Analysis of long-tem monthly rainfall data from 1871 to 2004 spread across India for the rainfed areas indicated significant negative trends in annual rainfall especially in the eastern parts of Madhya Pradesh, Chhattisgarh and parts of Bihar, Uttar Pradesh, parts of north-west and north-east India and a small patch in Tamil Nadu. However significant increase in rainfall trends was noticed in Jammu and Kashmir and in some parts of southern peninsular region of India. Moreover, the probabilities of occurrence of deficit rainfall conditions have increased between 1991 and 2003 and also the shifts in monthly peak rainfall pattern were observed in southern peninsular region (Rao *et al.*, 2009).

Global warming is expected to intensify the global hydrological cycle, with an increase of both evapotranspiration (ET) and precipitation. Douville *et al.* (2013) predicted that better constraining land ET in twenty-first-century climate scenarios is critical for predicting changes in surface climate, including heat waves and droughts, evaluating impacts on ecosystems and water resources, and designing adaptation policies. Though ET changes at continental scale may already be underway, but have never been attributed to anthropogenic emissions of greenhouse gases. Douville *et al.* (2013) also posited that in the mid-latitudes, the emerging picture of enhanced ET confirms the end of the dimming decades and highlights the possible threat posed by increasing drought frequency to managing water resources and achieving food security in a changing climate.

Impact of abiotic stresses

It is predicted that rising trends in various abiotic stresses such as drought, salinity, submergence, and nutrient deficiencies are the major impediments to the agricultural crop yield throughout the world and especially in the developing world. One of the major challenges facing agriculture is the global water shortage caused by the increasing world population and worldwide climate change. This global water scarcity threatens sustainable crop farming, as agricultural activities account for about 75% of global water consumption, and, in particular, irrigation represents over 90% of water used in many developing countries. Stresses such as drought, flooding and salinity affect the productivity of most agricultural crops to different degrees, depending on the onset time, duration, and intensity of the stress. Water stressed hydrological basins already affect approximately 1.5-2.0 billion people leading to problems of soil salinity and sodicity. Climate change will also bring inundation in low lying areas associated with increased runoff from tropical storms. On the other hand sea level rise will increase soil salinity and waterlogging in coastal regions of the world. Exploiting yield potential and maintaining yield stability of crops in such limited environments are urgent tasks that must be undertaken in order to guarantee food supply for the ever increasing world population.

Climate change is also affecting the agro-silvicultural systems which include forest tree species. Climate-driven forest tree die-off from drought and heat stress has occurred around the world and is expected to increase with climate change and probably has distinct consequences from those of other forest disturbances. Anderegg *et al.* (2013) examined the consequences of drought- and climate-driven widespread forest loss and found strong implications for alterations in biodiversity, ecosystem structure, ecosystem functions, ecosystem services and land–atmosphere interactions. They postulated that more studies are needed that examine changes in a wide array of ecosystem functions across a variety of ecosystems which affect nutrient cycling, decomposition, hydrological cycles and habitat on other tropic levels.

Impact on Indian agriculture

Agriculture is a critical sector of the Indian economy forming the backbone of development. Though its contribution to the overall Gross Domestic Product (GDP) of the country has fallen from about 30 percent in 1990-91 to 14 percent in 2012-13, still more than half of the population (52%) is dependent upon this farm sector for sustenance (State of Indian Agriculture 2012-2013). With the declining share of agriculture to GDP, the continuing high pressure of population on agriculture and the increasing fragmentation of land holdings leading to decreasing availability of cultivated land area per household, the agriculture sector is under tremendous pressure to sustain the livelihood of the rural households under changing climate.

Though an increase in ambient CO_2 is usually considered beneficial as it results in increased photosynthesis of crops, yet the combined increase in temperature and variability of rainfall would considerably affect food production. Studies indicated a probability of 10-40% loss in crop production in India with increase in temperature by 2080–2100 (Aggarwal *et al.*, 2010, Elanchezhian *et al.*, 2012). Similarly studies in wheat crop indicate the possibility of loss of 4-5 million tonnes in production with every rise of 1ºC temperature throughout the growth period even after considering carbon fertilization benefits (Aggarwal 2008; Haris *et al.*, 2013). These assumptions are based on a business as usual scenario, with no new technology development and with either no or limited adaptation by all stakeholders. It was also revealed that climate change may reduce the productivity of maize in some regions of India during both monsoon and winter seasons. With a rise in temperature, the reduction in yields is projected to be larger in warmer locations than the other locations (Aggarwal *et al.*, 2010; Srivastava *et al.*, 2010). However, in areas like sub humid climatic environment of eastern Indo-Gangetic plains, the winter maize crop is projected to benefit with better yields (Haris *et al.*, 2013).

Trends of drought and temperature in India

It was observed that the probability of droughts of 20-30% coincides with the semi-arid regions of the country and higher probabilities are seen in extreme north-western parts based on the analysis of past 100 years data. The probability of severe droughts is low over the country, except in western parts of Rajasthan where the probability is around 20 to 25% (Rao *et al.*, 2009).

Trend analysis on annual mean maximum, minimum and average temperature indicated that the maximum temperature has been increasing in 75% of the stations in south zone followed by central (67%), east (60%) and west zone (57%). No significant trend was observed in 60% of stations in north zone followed by west zone (43%), central zone (37%), south zone (17%) and east zone (14%). Minimum temperature showed increasing trend in 88% of stations in central and east zones followed by north zone (80%), south zone (75%), and west zone (57%). Decreasing trend in minimum temperature was seen in 29% of stations in west zone followed by 20% and 8% stations in north and south zones, respectively. No trend was noticed in 20% of stations in north zone followed by 17% in south zone, 14% in west and 12% each in east and central zone (Rao *et al.*, 2009).

Priorities for adaptation

Adaptation of agriculture to climate change should improve the outcome of the enterprises and IPCC defined adaptation as the adjustment of agronomic practices, agricultural processes, and capital investments in response to climate change threats (Easterling *et al.*, 2007). Examples of adaptations to climate change include adjustment of sowing or planting and harvest dates/times, expansion of croplands to more permissive areas, changing genotypes or species to those with more appropriate thermal time or heat stress tolerance, developing new germplasm with improved traits, altering fertilization rates and irrigation practices, and using climate forecasting to reduce production risks (Howden *et al.*, 2007). Adaptation of the most important food crops in their primary production regions should be supported for food security.

Effective strategies to adapt agricultural production to global climate change require improved predictions of crop responses to global change scenarios. This issue has been achieved by growing crops artificially under enhanced CO_2 conditions like Open top chambers or Free Air CO_2 enrichment facilities for several important agricultural crops like rice, wheat, maize and soybean. These technologies provide in-field fumigation of CO_2 to simulate future atmospheric conditions while growing crops under natural field conditions using current agronomic practices.

Rice provides a staple food for more than half of the world's current population and is predominantly produced in south, southeast, and east Asia (IRRI, 2002). Rice production is sensitive to increasing temperature (Peng *et al.*, 2004), which may limit the positive yield response to rising CO_2 concentration (Ainsworth, 2008). Current temperatures are already approaching critical levels during susceptible stages of rice development in many Asian countries, and drought stress will be exacerbated as temperatures rise (Wassmann *et al.*, 2009). Hence rice varieties tolerant to reproductive stage drought or aerobic rice varieties will perform better under such constrained ecosystem.

Wheat is the second most important source of calories for human consumption, and global wheat yields declined by 1% from 1997 to 2007 (Long and Ort, 2010). By 2050, anticipated climate changes will likely cause reclassification of the Indo-Gangetic Plains of India from a favorable, high-potential wheat production environment to a heat-stressed, short-season production environment. The Indo-Gangetic plains produce approximately 15% of the world's wheat crop, and identifying and developing wheat varieties with improved high-temperature stress tolerance will be critical to maintaining food security for the approximately 200 million people who depend upon local food production in this region (Ortiz *et al.*, 2008). However, the same climate change scenarios that will make land in India less suitable for wheat may cause an expansion of suitable wheat-growing areas in North America and Eurasia (Ortiz *et al.*, 2008), and developing varieties to take maximum advantage of new growing areas could also help sustain global wheat production.

Maize and soybean are the other two most widely produced crops in the world (FAO, 2009) in addition to rice and wheat. U.S.A. produces approximately 40% of the world's maize and soybean and is a leading exporter of both crops and, therefore, an important determinant of the global market for both crops. Additionally, there has been significant investigation of impacts of climate change on U.S.A. production of maize and soybean; therefore, investing in adaptation efforts for maximizing the production of these crops in the United States may be more likely to achieve favorable outcomes. The assumption that warming of temperate regions will increase crop production is in contrast to evidence of adverse effects of recent temperature changes on north American U.S. maize and soybean production (Lobell and Asner, 2003; Kucharik and Serbin, 2008).

Global maize yields are forecast to decline in response to increasing temperature, particularly as the upper range of growing season temperatures become hotter (Schlenker *et al.*, 2009). As a C_4 crop, maize is unlikely to directly benefit from rising atmospheric CO_2 concentration (Long *et al.*, 2006). Moreover, maize is the most widely grown C_4 grain crop in both food secure and food-insecure regions of the world (Leakey, 2009), so adapting maize to higher temperatures would likely be a priority investment. Butler and Huybers (2013) predicted that losses to average US maize yields from a 2°C warming would be reduced from 14% to only 6% and that loss in net production is wholly averted. However, this result does not account for possible changes in temperature variability or water resources, nor does it account for all possible forms of adaptation, but it does show that adaptation is of first-order importance for predicting future changes in yield. Hence further research should be undertaken regarding the ability to adapt to a changing climate, including analysis of other crops and regions, the application of more sophisticated models of crop development, and field trials employing artificially increased temperature.

Clearly, different countries and regions will have their own specific priorities for adaptation. In order to increase global food production and security in a warmer future, broad investment in many of these priorities is an urgent matter.

In short run that is during our lifetime, CO_2 is added to the atmosphere by respiration of plants, microbes, burning of fossil fuels and by land clearing. Photosynthesis is one of the two important mechanisms for removal of CO_2 from the atmosphere and the other being the dissolution of CO_2 in the oceans and seas where solid and dissolved carbonates equilibrate with CO_2. Approximately 13% of carbon in the atmosphere is used in photosynthesis every year. Moreover, the combined effects of elevated CO_2 and warmer temperature did not lead to significant increases in photosynthesis compared with an increase in CO_2 alone. In fact, the combined increases in CO_2 and temperature led to reduced photosynthesis relative to elevated CO_2 (Ruiz-Vera *et al* 2013). This suggests that in future climate change conditions, the interactive effects of elevated CO_2 and warmer temperatures will likely not benefit physiology, growth, and development as predicted from theory, due to overriding environmental factors.

The effects of warmer temperature on photosynthesis will be one of the most important determinants of the impact of global warming on crop yield. The reactions catalyzed by Rubisco are directly affected by temperature. Even though Rubisco itself is quite thermal stable and the rate of carboxylation continues to increase beyond 50°C, decreased discrimination by Rubisco for its alternative substrate oxygen and increased solubility of oxygen relative to CO_2 with rising temperature conspire to inhibit net photosynthesis in C_3 plants due to increased photorespiration (Ainsworth and Ort, 2010). It may seem that a form of Rubisco better able to discriminate against oxygen would be advantageous in the face of rising temperatures. In fact, because there is an inverse relationship of specificity with catalytic rate, to maximize the daily integral of canopy photosynthesis it would be advantageous to trade off specificity for catalytic rate regardless of temperature (Zhu *et al.*, 2010). Though, attempts at manipulating Rubisco have so far yielded poor results, improving the temperature response of photosynthesis by manipulating RuBP regeneration appears much more promising (Ainsworth and Ort 2010).

At current CO_2 levels, light-saturated photosynthesis operates at the transition between Rubisco-limited and RuBP-limited photosynthesis; thus, as CO_2 increases, photosynthesis is increasingly limited by the capacity for RuBP regeneration. Additionally, RuBP regeneration limitation increases at higher temperature (Sage and Kubien, 2007) and RuBP-limited photosynthesis benefits from lower photorespiration at elevated CO_2, implying that increases in the RuBP regeneration rate should raise the temperature optimum of photosynthesis at elevated CO_2. Thus, while attempts at manipulating Rubisco have so far yielded poor results, improving the temperature response of photosynthesis by manipulating RuBP regeneration appears much more promising. Also the genetic basis for heat stress tolerance in crop plants is poorly understood. Identification of adaptive quantitative trait loci (QTLs) for high-temperature tolerance is one approach being taken to try to close these knowledge gaps.

Conclusions

The Global Carbon Project suggests that humans contributed around 10.4 billion tonnes of carbon into the atmosphere in 2011. About half of that is taken up each year by carbon sinks such as the ocean and terrestrial vegetation on land; the rest remains in the atmosphere and raises the global concentration of CO_2. However the behavior of such carbon sinks in near future is unclear, which may force further increase in warming. Moreover, there is now overwhelming evidence that present "business as usual" crop development will be insufficient to adapt crops over the wide range of growing regions that will be required to meet expanding global agricultural demand. Moving crops pole-ward seems an inevitable element of the multifaceted adaptation to increasing global temperatures that must be implemented, but it would be misleading to believe that this alone can maintain yields. For example, migration of the North American Corn Belt into Canada vacates the high-quality prairie soils for the less productive soils farther north. And in many important agricultural areas of the world, pole-ward migration is not possible, such as the Wheat Belt of Australia, where an ocean lies to the south (Long and Ort, 2010). Even adapting crops in the highest priority regions will require broad investment, the integration of new technologies with conventional selection-based breeding, and the coordinated involvement of public and private sectors of the agricultural enterprise. Current and future increases in temperature indicate perhaps the most significant and most urgent challenge for the adaptation of crops to global change.

It is shown that adaptation could decrease the average fractional losses in the USA by roughly a factor of two and could negate losses with respect to total production, at least for a modest 2°C warming (Butler and Huybers 2013). The prospect that adaptation could have such a significant influence on future yields provides impetus for further study. Trials growing crop varieties in different conditions of temperature and water availability, analysis of the sensitivity and adaptability of other major food crops and other growing regions, and the application of more complete biophysical models of crop interactions with environmental variations would all be prudent undertakings for adequately predicting the ecological response of crops to a changing climate.

On the contrary, much less attention has been paid by policymakers to monitoring, modelling and managing the impacts of climate change on the dynamics of Earth surface systems, including glaciers, rivers, mountains and coasts (Knight and Harrrison, 2013). This is a critical omission, because these systems provide water and soil resources, sustain ecosystem services and strongly influence biogeochemical climate feedbacks. Hence the policy gap regarding the management of Earth surface systems' impacts under climate change also needs to be closed to facilitate the sustainability of cross-national Earth surface resource use. Moreover, as the GHG emissions scenarios emerged so far illustrate, even well-intentioned and effective international efforts to limit climate change must face the hard physical reality of certain temperature targets that can no longer be achieved if too much carbon has already been emitted to the atmosphere (Stoker 2013). Both delay and insufficient mitigation efforts close the door on limiting global warming permanently. This warrants an ultimate commitment from every nation of the world to reduce the emission to save our planet earth.

❑❑❑

4 Climate Change Projections – India's Perspective

K. M. Hati, M. Mohanty and, R.S. Chaudhary
Division of Soil Physics, Indian Institute of Soil Science, Nabibagh, Berasia Road, Bhopal - 462 038, India

Introduction

Climate change refers to a change in the state of the climate that can be identified (e.g., by using statistical tests) by changes in the mean and/or the variability of its properties, and that persists for an extended period, typically decades or longer. It is a serious global environmental concern and is primarily caused by the building up of Green House Gases (GHG) in the atmosphere. Climate change may be due to natural internal processes or external forcings such as modulations of the solar cycles, volcanic eruptions and persistent anthropogenic changes in the composition of the atmosphere or in land use. United Nations Framework Convention on Climate Change (UNFCC) defined Climate change as 'a change of climate which is attributed directly or indirectly to human activity that alters the composition of the global atmosphere and which is in addition to natural climatic variability observed over comparable time periods. The UNFCCC thus makes a distinction between climate change attributable to human activities altering the atmospheric composition, and climate variability attributable to natural causes. The average increase in the most important climatic parameter i.e. temperature of earth's near surface air and oceans is referred to as Global Warning. Human activity since the industrial revolution has increased the amount of greenhouse gases in the atmosphere, leading to increased radiative forcing from CO_2, methane, tropospheric ozone, CFCs and nitrous oxide. The global increases in carbon dioxide concentration are primarily due to fossil fuel burning and land use changes, while those of methane and nitrous oxide are primarily due to agricultural practices. The concentrations of CO_2 and methane have increased by 36% and 148% respectively since 1750. These levels are much higher than at any time during the last

650,000 years, the period for which reliable data have been extracted from ice cores (Pearson and Palmer, 2000). Over the past 20 years fossil fuel burning has produced about three-quarters of the increase in CO_2 from human activity. Most of the rest is due to land-use changes, particularly deforestation. In India, out of total GHG emission from the anthropogenic activities, agriculture sector contributes 28%. The emissions are primarily due to methane emission from rice fields (23%), enteric fermentation in ruminant animals (59%) and nitrous oxides from application of manures and fertilizers (Das and Hati, 2013).

It is now well-known that the surface of the earth is warming as a result of mankind's activities. The obvious question then arises is Can we predict changes in climate that will happen in the future?" The chaotic behaviour of the atmosphere, which limits our ability to predict weather to only a week or so, also limits our ability to predict detailed variations in climate. It is impossible to predict, for example, the actual precipitation of a region in the summer of 2050. However, projections of climate change can be made - that is, we can examine scenarios of possible climate change that can come about because of our continuing emissions of greenhouse gases and other radiatively important substances. Such projections are very helpful for politicians and policy-makers when assessing what to do about the problem of anthropogenic climate change, as they give an estimate of the likelihood of change given certain pollution scenarios. However, projections of climate change in future are a very difficult proposition which will depend on many interlinked natural processes and developmental scenarios, thus it will be not like weather forecasts. It is extremely difficult to make a deterministic and definitive prediction of how climate will evolve over the next century and beyond as it is with short-term weather forecasts. Projections of climate change are uncertain, first because they are dependent primarily on scenarios of future anthropogenic and natural forcings that are uncertain, second because of incomplete understanding and imprecise models of the climate system and finally because of the existence of internal climate variability. The term climate projection tacitly implies these uncertainties and dependencies. Nevertheless, with the advent of modern and more powerful computing capabilities, elaborate data collection facilities over larger parts of the world and development of more powerful and process based global circulation models the ability to predict or project a future climatic scenario is improving fast with less variability and more precision in prediction. As greenhouse gas concentrations continue to rise, we expect to see future changes to the climate system that are greater than those already observed and attributed to human activities. Because we can't know the future for certain, scientists use computer-based climate models to project plausible scenarios for the coming century and beyond. It is possible to understand future climate change using models and to use models to characterize outcomes and uncertainties under specific assumptions about future forcing scenarios (Collins *et al.*, 2013). Future climate change predictions are carried out using different climate models or global circulation models. Climate models are used to understand how the climate will evolve over time and typically represent the atmosphere, ocean, land surface, cryosphere, and biogeochemical processes, and solve the equations governing their evolution on a geographical grid covering the globe (IPCC, 2013). Some processes are represented explicitly within climate models, like large-scale circulations, while others are represented by simplified parameterisations. The use of these parameterisations is sometimes due to processes taking place on scales smaller than the typical grid size of a climate model or sometimes to the current limited understanding of these processes. Different climate modelling institutions use different plausible representations of the climate system, which is why climate projections for a single greenhouse gas emissions scenario differ between modelling institutes. This gives rise to "climate model structural uncertainty".

Climate change and India

India is a large country which extends from 8° to 33°N. The variety in terrain, from the high mountains of the Himalayas in the north to tropical coastlands in the south, makes for a wide range of climatic conditions. In the northern mountain regions, winters are cool at lower levels, and increasingly cold at higher altitudes. In the summer, intermediate levels around 2000 m above sea level are pleasantly cool, but it can get quite hot at lower levels. The Himalayas act as a barrier to the cold winds flowing down from Central Asia. Northern inland areas have a continental climate with a high seasonal and diurnal temperature range. Further south, it is hot throughout the year, with annual mean temperatures of 27°C at Mumbai on the mid-west coast, and 29°C at Chennai on the south-east coast. The hottest months here are April and May, prior to the onset of the monsoon. Inland at Hyderabad, the mean temperature reaches 33°C in May, with an average daily maximum of 39°C. In the coastal regions, the heat and humidity can be very oppressive throughout the year, especially in the hot season, and the monsoon season from June to September. The Indian climate is dominated by the great Asiatic monsoon wind system which is totally unlike any other country's prevailing wind system. The driest period over most of India is from December to February when light north-easterly winds bring clear skies and virtually dry weather. From March to May the dry conditions continue, but the intense summer heat causes a reversal of winds over the region, so that from June to October India is influenced by the moist rain-bearing monsoon from the south-west and on some mountain ranges facing the sea, rainfall can be very heavy. Usually, the monsoon reaches the south during late May or early June, reaching the north about six weeks later. More than 80 per cent of the average annual rainfall of the country is received between the months of June and September due to the south-west monsoon. Agriculture production of the country particularly in the rainfed region is mostly dependent on the monsoonal rainfall. Slight change in the amount, intensity and spatial and temporal distribution of the monsoonal rainfall in future will have severe implication on the agricultural production and livelihood security of the country. Climate change can therefore have a profound impact on different sectors like agriculture, human health, water resources, forestry, etc. through its influence on the monsoon-dominant climate over India. Hence there is a widely felt need for understanding the nature of climate change over this part of the globe.

Change in temperature in different seasons over the last century

Due to various anthropogenic activities like, over exploitation of the natural resources viz., land overuse, deforestation, tapping of vulnerable water sources, destruction of the ecosystems, environmental pollutions, and excessive release of GHGs into the atmosphere the climatic trend over the country has been changing in the last few decades. Dash *et al.* (2007) on examination of the trends in the annual temperature anomalies averaged over India during the last century reported that the annual mean temperature over India has increased by about 0.7°C and the maximum temperature has increased by about 0.8°C. On the other hand the annual minimum temperature over India does not show any appreciable enhancement. After analysis of temperature changes in different seasons over the last century, they also showed that the winter-time increase in the mean temperature over India is about 1.0°C whereas during post-monsoon months there has been a rise of about 1.1°C in the last century. During pre-monsoon and monsoon months the mean temperature has increased less by about 0.3 and 0.4°C respectively. Thus, the temperature rise is the highest in post-monsoon months followed by winter. Also, during winter and post-monsoon months the temperature enhancement is more than double of that in pre-monsoon and monsoon months. Such large difference in the temperature rise between the two halves of the year may lead to changes in spatial position of heating sources over the Indian region and adjoining area and hence in the atmospheric circulation patterns.

Change in temperature in different regions in India over the last century

The impact of climate change may differ from one region to the other, especially for a geographically complex terrain in India. Of the seven geographical zones namely, northwest, western Himalaya, north central, northeast, interior peninsula, east coast and west coast, west coast shows maximum increase in the maximum temperature by about 1.2°C. Next to the west coast, maximum temperature has increased by 1°C in the northeast, 0.9°C in the western Himalaya, 0.8°C in the north central, 0.6°C in the northwest, 0.6°C in the east coast region and by the least amount of 0.5°C in the interior peninsula. Unlike the maximum temperature, the minimum temperature in the four northern regions of India shows some peculiar changes. There is sharp decrease in the minimum temperature by 1.9°C in the western Himalaya, 1.4°C in the north east, 1.1°C in the north central and 0.7°C in the northwest India during 1955 to 1972 and then equally sharp rise over the past 3-4 decades (Dash and Hunt, 2007). The minimum temperatures in southern regions such as interior peninsula, east coast and west coast show almost equal enhancement during the last century. The interior peninsula shows maximum increase in the minimum temperature by about 0.5°C. Next to the interior peninsula, minimum temperature increased by 0.4°C in the east coast and by the least amount of 0.2°C in the west coast of India. The changes in minimum temperatures in the northern and southern regions of India are almost in opposite phases although the magnitudes of fluctuation in the northern India are large during last 4-5 decades (Dash and Hunt, 2007).

Possible consequences of temperature rise

The regional and seasonal discrepancies in the rise in surface air temperature has large bearing on agriculture, water availability and power distribution besides on human comforts, health and effective working hours. In addition, the difference in the change in inter-seasonal temperatures will have profound impact on the cropping pattern and hence a challenge to the farmers and agricultural scientists. Occurrences of some diseases, whether of human beings, animals or plants are very sensitive to small perturbations in the delicate seasonal temperature balance. Since temperature is the most important parameter controlling the overall weather and climate of a place, once it undergoes non-uniform temporal and spatial changes, the consequences are the likely intensification of the general impacts of climate change.

Projected climate change scenarios over Indian subcontinent

As we have discussed that the climate change is already impacting the natural ecosystems and is also expected to have substantial adverse effects in India, mainly on agriculture on which 58 per cent of the population still depends for livelihood, water storage in the Himalayan glaciers which are the source of major rivers and groundwater recharge, sea-level rise, and threats to a long coastline and habitations. Climate change will also cause increased frequency of extreme events such as floods, and droughts. These in turn will impact India's food security problems and water security.

The Intergovernmental Panel on Climate Change (IPCC), a specialised body jointly established by the United Nations Environmental Programme (UNEP) and the World Meteorological Organisation, is mandated to prepare scientific assessments on various aspects of Climate Change. The IPCC is currently engaged in the preparation of Fifth Assessment Report on Climate Change through three working groups, viz. Working Group I on Climate Change will present the physical science basis; Working Group II on Climate Change deals with impacts, adaptation and vulnerability; and Working Group III deals with assessment for mitigation of Climate Change. In 2000, the IPCC developed the global and regional emission pathways in its special report on emissions scenarios (SRES) (Nakicenovic, *et al.*, 2000). It developed four families of emission pathways, namely A1, B1, A2 and B2 based on different socio-economic development assumptions. Global climate models under the Coupled Model Inter-comparison Project 3 (CMIP3) or better known as IPCC climate models projected future climate change based on these emission pathways. The Fourth Assessment Report of IPCC used the SRES-based emission scenarios and climate projections from CMIP3 for characterizing future climate change and its impacts on society and ecosystems (IPCC, 2007). The IPCC first published the SRES scenarios in 2000 while the underlying economic and policy assumptions for these scenarios were fixed as early as by 1997 (Moss *et al.*, 2010). SRES scenarios are nearly 17 years old. Now, the scientific community has developed a set of new-emission scenarios termed as representative concentration pathways (RCPs). In contrast to the SRES scenarios, RCPs represent pathways of radiative forcing, not detailed socio-economic narratives or scenarios. Central to the process is the concept that any single radiative forcing pathway can result from a diverse range of socio-economic and technological development scenarios. There are four RCP scenarios: RCP2.6, RCP4.5, RCP6.0 and RCP8.5. These scenarios are formulated such that they represent the full range of stabilization, mitigation and baseline emission scenarios available in the literature. The naming convention reflects socioeconomic pathways that reach a specific radiative forcing (RF) by 2100. For example, RCP8.5 leads to a radiative forcing of 8.5 Wm^{-2} by 2100. The four RCP scenarios used in CMIP5 lead to a total radiative forcing (RF) at 2100 that spans a wider range than that estimated for the three Special Report on Emission Scenarios (SRES) scenarios (B1, A1B, A2) used in the Fourth Assessment Report (AR4), RCP2.6 being almost 2 W m^{-2} lower than SRES B1 by 2100. The magnitude of future aerosol forcing decreases more rapidly in RCP

scenarios, reaching lower values than in SRES scenarios through the 21st century. Carbon dioxide (CO_2) represents about 80 to 90% of the total anthropogenic forcing in all RCP scenarios through the 21st century. The ensemble mean total effective RFs at 2100 for CMIP5 concentration-driven projections are 2.2, 3.8, 4.8 and 7.6 W m^{-2} for RCP2.6, RCP4.5, RCP6.0 and RCP8.5 respectively, relative to about 1850, and are close to corresponding Integrated Assessment Model (IAM)-based estimates (2.4, 4.0, 5.2 and 8.0 W m^{-2}). The individual RCPs are briefly described in Table 1.

Table 1. Description of representative concentration pathway (RCP) scenarios

RCP	Description	Concentration (p.p.m.)	Developed by
RCP2.6	Its radiative forcing level first reaches a value around 3.1 W/m^2 mid-century, returning to 2.6 W/m^2 by 2100. Under this scenario greenhouse gas (GHG) emissions and emissions of air pollutants are reduced substantially over time.	1,370 CO_2-equiv. in 2100	IMAGE modelling team of The Netherlands Environmental Assessment Agency23.
RCP4.5	It is a stabilization scenario where total radiative forcing is stabilized before 2100 by employing a range of technologies and strategies for reducing GHG emissions.	850 CO_2-equiv. (at stabilization after 2100)	MiniCAM modelling team at the Pacific Northwest National Laboratory's Joint Global Change Research Institute.
RCP6.0	It is a stabilization scenario where total radiative forcing is stabilized after 2100 without overshoot by employing a range of technologies and strategies for reducing GHG emissions.	650 CO_2-equiv. (at stabilization after 2100)	AIM modelling team at the National Institute for Environmental Studies, Japan25, 26.
RCP8.5	It is characterized by increasing GHG emissions over time representative of scenarios in the literature leading to high GHG concentration levels.	Peak at 490 CO_2-equiv. before 2100 and then declines	MESSAGE modelling team and the IIASA Integrated Assessment Framework at the International Institute for Applied Systems Analysis (IIASA), Austria27.

New climate projections are being developed by different modelling groups based on these new RCP scenarios. Climate projections for the Fifth Assessment Report of the IPCC are made using the newly developed representative concentration pathways (RCPs) under the Coupled Model Inter-comparison Project 5 (CMIP5). The CMIP5 includes a broader variety of experiments and application of more comprehensive models compared to the earlier CMIP3. CMIP5 models are generally of higher resolution compared to their CMIP3 counterparts. The CMIP5 models also benefit from several years of development in their parameterisations of small scale processes, which, together with resolution increases, are expected to result in a general improvement in the accuracy of their simulations of historical climate, and in the credibility of their projections of future changes. The CMIP5 programme also includes a number of comprehensive Earth System Models (ESMs) which explicitly simulate the earth's carbon cycle and key aspects of atmospheric chemistry, and also contain more sophisticated representations of aerosols compared to CMIP3 models. The projections for the climate change variables for the 21st century by different scientists groups differ considerably based on the scenarios and model assemblage used. Rupakumar *et al.* (2006) project a temperature rise of 2.9°C and 4.1°C for India under the B2 and A2 scenarios of SRES respectively, in 2080s relative to 1970s while Krishankumar *et al.* (2011) projected a warming of 3.5–4.3°C over the same period for the A1B scenario. A1B scenario of SRES is generally considered close to the business as usual emission scenario. The CMIP5-based model ensemble projection of Chaturvedi *et al.* (2013) reported a warming of 2.8°C and 4.3°C under the RCP6.0 and RCP8.5 scenarios respectively, for 2080s compared to the 1970s baseline. In terms of atmospheric CO_2 concentration build-up, the A1B scenario is closer to RCP8.5 until 2050, but then diverges and is closer to RCP6.0 between 2050 and 2100. Climate change projections made by Rupakumar *et al.* (2006) and Krishnakumar *et al.* (2011) for India are based on a single climate model HadRM3 for specific scenarios, whereas the projections made by Chaturvedi *et al.* (2012) is based on 18 GCMs based multiple model ensemble for multiple scenarios.

The IPCC in their forth assessment report using the CMIP3 model output (IPCC, 2007) projected an increase of global surface temperature of the order of 0.74±0.18°C during the 20th century while the global surface temperature is likely to rise a further 1.1 to 6.4°C (2.0 to 11.5^{0}F) by the end of the 21st century. For South Asia region including India a rise in temperature of 0.5 to 1.2°C by 2010, 0.88 to 3.16°C by 2050 and 1.56 to 5.44°C by 2080 has been projected by IPCC under different scenarios. The temperature increase is likely to be much higher in winter (*rabi*) season than in rainy (*kharif*) season. The most widely used set of projections available, the CMIP3 ensemble output for temperature and precipitation for the A1B emission scenario for India and the surrounding

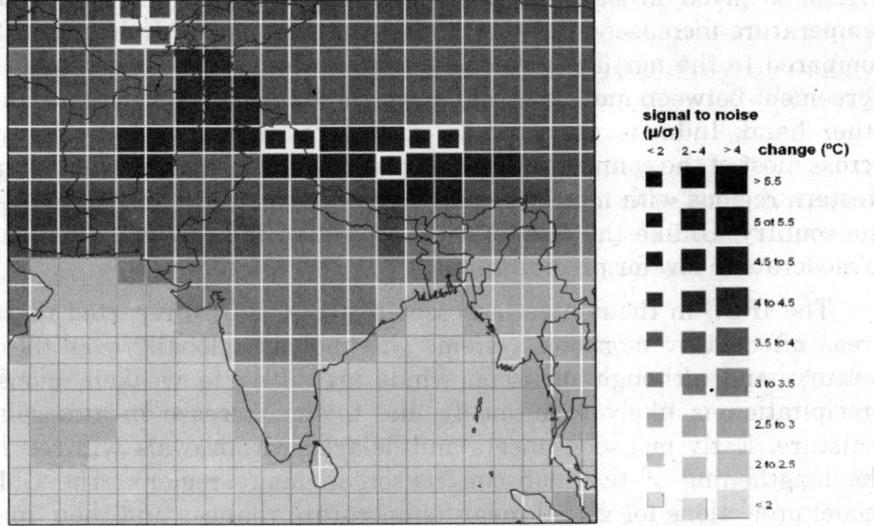

Fig. 1. Percentage change in average annual temperature by 2100 from 1960-1990 baseline climate, averaged over 21 CMIP3 models. The size of each pixel represents the level of agreement between models on the magnitude of the change (Huhne and Slingo, 2012)

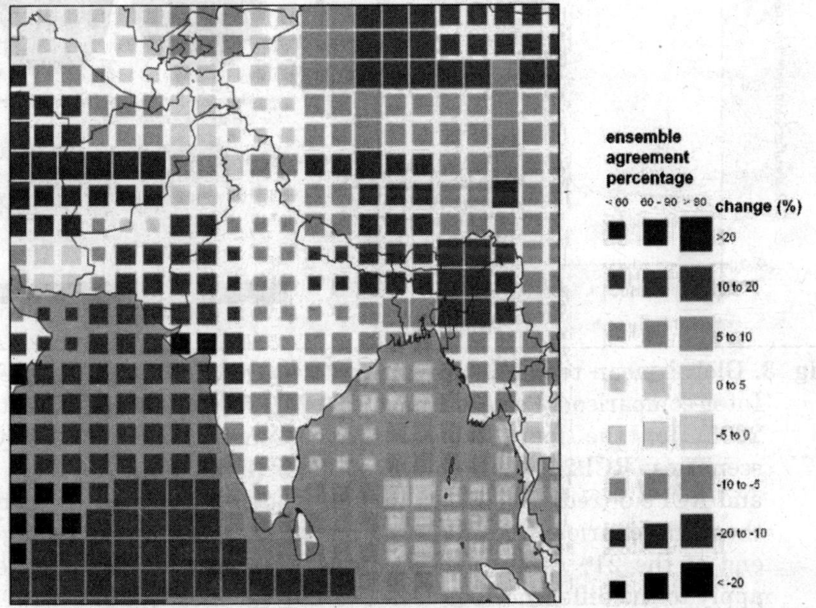

Fig. 2. Percentage change in average annual precipitation by 2100 from 1960-1990 baseline climate, averaged over 21 CMIP3 models. The size of each pixel represents the level of agreement between models on the sign of the change (Huhne and Slingo, 2012).

region, is given in Fig. 1 and 2 (Huhne and Slingo, 2012). Projected temperature increases are lower in the southern part of India, up to 3°C, compared to the north where changes of up to 4.5°C are projected. The agreement between models is high over almost the entire country. On the other hand, India is projected to experience increases in precipitation across most of the country. Increases of up to 20% or higher could occur in western regions with more widespread increases of 5-10% over the rest of the country. Unlike the temperature projection the ensemble agreement is moderate to low for precipitation.

The IPCC in their latest assessment (IPCC, 2013) reported that the areas affected by monsoon systems will increase globally over the 21st century, and although monsoon winds are likely to weaken, monsoon precipitation is likely to intensify due to the increase in atmospheric moisture. Early monsoon onsets and delayed withdrawals will result in the lengthening of the monsoon season in many regions. The CMIP5 model projections for global mean temperature changes and their spatial distribution for the four RCPs scenarios by the end of 21st century are presented in Fig. 3.

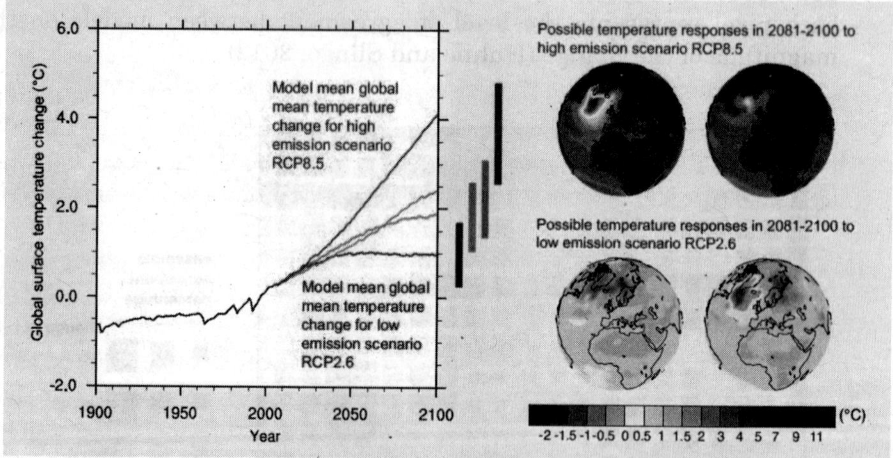

Fig. 3. Global mean temperature change averaged across all Coupled Model Inter-comparison Project Phase 5 (CMIP5) models (relative to 1986–2005) for the four Representative Concentration Pathway (RCP) scenarios: RCP2.6 (dark blue), RCP4.5 (light blue), RCP6.0 (orange) and RCP8.5 (red); 32, 42, 25 and 39 models were used respectively for these 4 scenarios. Likely ranges for global temperature change by the end of the 21st century are indicated by vertical bars. These ranges apply to the difference between two 20-year means, 2081–2100 relative to 1986–2005, which accounts for the bars being centred at a smaller value than the end point of the annual trajectories. For the highest (RCP8.5) and lowest (RCP2.6) scenario, illustrative maps of surface temperature change at the end of the 21st century (2081–2100 relative to 1986–2005) are shown for two CMIP5 models.

Occurrence of climate extremes

Climate extremes encompass both extreme weather, with durations of minutes to days (the synoptic timescale), and extreme climate events with durations of months, in the case of periods of wet/stormy weather, or years, in the case of drought. The frequency of occurrence of climate extremes is expected to change during the next century, with increases in the frequency of heat waves and heavy precipitation events, and decreases in the frequency of frost days, as a consequence of anthropogenically-forced climate change (Easterling *et al.*, 2000). Several studies showed that generally, the frequency of occurrence of more intense rainfall events in many parts of South Asia has increased, causing severe floods, landslides, and debris and mud flows, while the number of rainy days and total annual amount of precipitation has decreased (Mirza, 2002; Lal, 2003). Increasing frequency and intensity of droughts in many parts of South Asia are attributed largely to a rise in temperature, particularly during the summer and normally drier months, and during El Niño/Southern Oscillation (ENSO) events (Lal 2003). A future increase in occurrence of extreme weather events including heat wave and intense precipitation events is projected in South Asia (Lal, 2003) along with an increase in the interannual variability of daily precipitation in the Asian summer monsoon (May, 2004; Giorgi and Bi, 2005). For tropical cyclones, there is a projected increase of 10–20% in the intensity of storms with an increase in sea-surface temperature of 2–4°C relative to the current temperatures in South Asia. An increase in heights of storm-surges could result from the occurrence of stronger winds, increases in sea-surface temperatures and low pressures associated with tropical cyclones resulting in an enhanced risk of coastal disasters along the coastal regions of the country. Tendencies of increase in intense rainfall with the potential for heavy rainfall events spread over few days are likely to impact water recharge rates and soil moisture conditions. A warmer climate, with its increased climate variability, will increase the risk of both floods and droughts (Wetherald and Manabe, 2002). The areas suitable for rainfed agriculture are expected to significantly decrease affecting adversely land productivity potential of the continent (Sivakumar and Stefanski, 2011).

Impact of anthropogenic forcing on the Indian monsoon

Numerous studies with coupled ocean–atmosphere models have explored the potential impact of anthropogenic forcing on the Indian monsoon. When forced with anticipated increases in GHG concentrations, the majority of these studies show an intensification of the rainfall associated with the Indian summer monsoon (Kripalani *et al.*, 2007;

Stowasser *et al.*, 2009; Cherchi *et al.*, 2010). Despite the intensification of precipitation, several of these modelling studies show a weakening of the summer monsoon circulation (Kripalani *et al.*, 2007; Stowasser *et al.*, 2009; Cherchi *et al.*, 2010). The net effect is nevertheless an increase of precipitation due to enhanced moisture transport into the Asian monsoon region (Ueda *et al.*, 2006). In recent years, studies with GCMs have also explored the direct effect of aerosol forcing on the Indian monsoon (Lau *et al.*, 2006; Meehl *et al.*, 2008; Randles and Ramaswamy, 2008). Considering absorbing aerosols (black carbon) only, Meehl *et al.* (2008) found an increase in pre-monsoonal precipitation, but a decrease in summer monsoon precipitation over parts of South Asia. In contrast, Lau *et al.* (2006) found an increase in May–June–July precipitation in that region. If an increase in scattering aerosols only is considered, the monsoon circulation weakens and precipitation is inhibited (Randles and Ramaswamy, 2008). More recently, Bollasina *et al.* (2011) showed that anthropogenic aerosols played a fundamental role in driving the recent observed weakening of the summer monsoon. Given that the effect of increased atmospheric regional loading of aerosols is opposed by the concomitant increases in GHG concentrations, it is unlikely that an abrupt transition to the dry summer monsoon regime will be triggered in the 21st century.

The observed recent weakening tendency in seasonal rainfall and the regional re-distribution has been partially attributed to factors such as changes in black carbon and/or sulphate aerosols (Chung and Ramanathan, 2006; Lau *et al.*, 2008; Bollasina *et al.*, 2011), land use (Niyogi *et al.*, 2010) and sea surface temperatures (Annamalai *et al.*, 2013). An increase in extreme rainfall events occurred at the expense of weaker rainfall events (Goswami *et al.*, 2006) over the central Indian region, and in many other areas (Krishnamurthy *et al.*, 2009). With a declining number of monsoon depressions (Krishnamurthy and Ajayamohan, 2010), the upward trend in extreme rainfall events may be due to enhanced moisture content (Goswami *et al.*, 2006) or warmer SSTs in the tropical Indian Ocean (Rajeevan *et al.*, 2008). CMIP3 projections show suppressed rainfall over the equatorial Indian Ocean (Cai *et al.*, 2011e; Turner and Annamalai, 2012), and an increase in seasonal mean rainfall over India (Ueda *et al.*, 2006; Annamalai *et al.*, 2007; Turner *et al.*, 2007a; Kumar *et al.*, 2011b; Sabade *et al.*, 2011). The CMIP5 models project a clear increase in temperature over India especially in winter, with enhanced warming during night than day (Kumar *et al.*, 2011) and over northern India (Kulkarni, 2012). In summer, extremely hot days and nights are projected to increase. Table 2 summarizes the projected temperature and precipitation changes for south Asian region in the RCP4.5 scenario based on CMIP5. There is high confidence in projected rise in temperature. However, the confidence in the summer monsoon precipitation increases in the future over South Asia is medium.

Table 2. Projections of changes in the temperature and precipitation over the south Asian (SAS) region by the CMIP5 global models for the RCP4.5 scenario (average of a set of 42 global models). (*The area mean temperature and precipitation responses are first averaged for each model over the 1986–2005 period from the historical simulations and the 2016–2035, 2046–2065 and 2081–2100 periods of the RCP4.5 experiments. Based on the difference between these two periods, the table shows the 25th, 50th and 75th percentiles, and the lowest and highest response among the 42 models, for temperature in degrees Celsius and precipitation as a percent change. The temperature responses are averaged over the boreal winter and summer seasons; December, January and February (DJF) and June, July and August (JJA) respectively. The precipitation responses are averaged over half year periods, boreal winter; October, November, December, January, February and March (ONDJFM) and summer; April, May, June, July, August and September (AMJJAS).*

		Temperature (⁰C)						Precipitation (%)				
Month	**Year**	**Min**	**25%**	**50%**	**75%**	**Max**	**Month**	**Min**	**25%**	**50%**	**75%**	**Max**
DJF	2035	0.1	0.7	1.0	1.1	1.4	ONDJFM	-18	-6	-1	4	8
	2065	0.6	1.6	1.8	2.3	2.6		-17	-3	4	7	13
	2100	1.4	2.0	2.3	3.0	3.7		-14	0	8	14	28
JJA	2035	0.3	0.6	0.7	0.9	1.3	AMJJAS	-3	2	3	6	9
	2065	0.9	1.1	1.3	1.7	2.6		-3	5	7	11	33
	2100	0.7	1.4	1.7	2.2	3.3		-7	8	10	13	37
Annual	2035	0.2	0.7	0.8	1.0	1.3	Annual	-2	1	3	4	7
	2065	0.8	1.4	1.6	1.9	2.5		-2	3	7	9	26
	2100	1.3	1.7	2.1	2.7	3.5		-3	6	10	12	27

Adapted from Seneviratne *et al.* (2012).

Chaturvedi *et al.* (2012) from a CMIP5 model ensemble-based annual precipitation change (%) projected a precipitation increase by 1.2–2.4% by 2030s under different RCP scenarios and by 3.5–11.3% by 2080s, relative to the pre-industrial base. Precipitation increase is projected almost all over India except for a few regions in thier short-term projections (2030s). The RCP2.6 experiences the least increase in precipitation, whereas RCP8.5 experiences the highest precipitation increase, and the precipitation changes are larger for each subsequent period (i.e. short, mid and long term) (Fig. 4). The model agreement in precipitation projections increases from RCP2.6 to RCP8.5, and also the model agreement gets stronger from short- to long-term projections; thereby indicating that long-term projections are generally more reliable than their short-term counterparts.

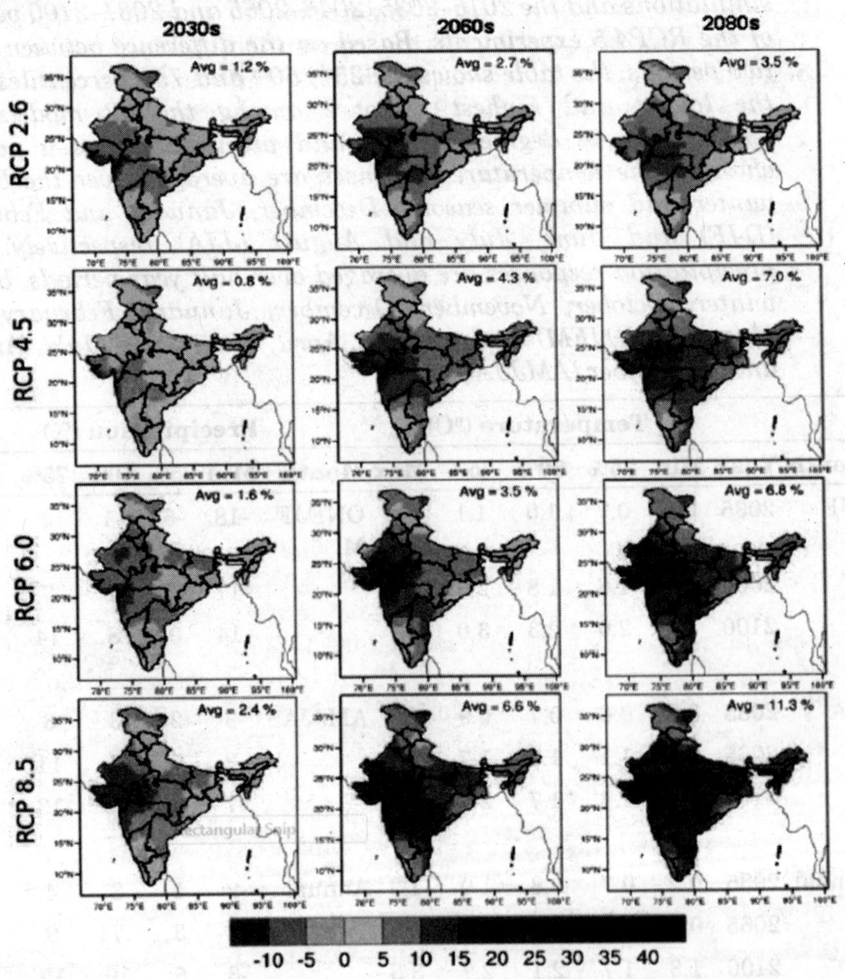

Fig. 4. CMIP5 model ensemble mean precipitation change (%) projected for 2030s (2021–2050), 2060s (2046–2075) and 2080s (2070–2099) relative to the pre-industrial period (1880s, i.e. over 1861–1900) (Chaturvedi *et al.* 2012).

Climate change and India's actions

India's emissions are estimated to be of the order of 1331.6 million tonnes of the carbon dioxide equivalent Green House Gas (GHG) emissions in 2007 with an annual growth of 4.2% from the levels in 1994.

The CO_2 emissions from India are only about 4% of total global CO_2 emissions and much less if the historical concentrations are taken into account. Still the country has been conscious of the global challenge of Climate Change (Anonymous, 2013). In pursuance of the obligations cast on parties to the United Nations Framework Convention on Climate Change (UNFCCC), India has undertaken to communicate information about the implementation of the convention, taking into account the common but differentiated responsibilities and respective capabilities and their specific regional and national development priorities, objectives and circumstances. India has submitted the Second National Communication (NATCOM) to the UNFCCC in 2012. As per the Second national Communication submitted by India to the UNFCCC, it is projected that the annual mean surface air temperature rise by the end of the century ranges from 3.5°C to 4.3°C whereas the sea level along the Indian coast has been rising at the rate of about 1.3 mm/year on an average. These climate change projections are likely to impact human health, agriculture, water resources, natural ecosystems, and biodiversity. India's strategy for addressing Climate Change is reflected in many of its social and economic development programmes (Anonymous, 2013). Wary of the threats imposed by climate change and pressures on natural resources, sustainability and environment, policy makers of the country are taking increased contingence for addressing this problem. India has also been part of 94 multilateral environmental agreements and has voluntarily agreed to reduce its emission intensity of its GDP by 20-25 per cent over 2005 levels by 2020. However, emissions from the agriculture sector would not form part of the assessment of its emissions intensity. Indian economy is already moving along a lower carbon and sustainable path in terms of declining carbon intensity of its GDP which is expected to fall further through lower carbon strategies. It is estimated that India's per capita emission in 2031 will still be lower than the global per capita emission in 2005 (in 2031, India's per capita GHG emissions will be under 4 tonnes of carbon dioxide equivalent (CO_2 eq.) which is lower than the global per capita emissions of 4.22 tonnes of CO_2 eq. in 2005).

Conclusions

Climate change is now well recognized as the biggest challenge facing the planet. Different projections of climate change are associated with a range of limitations and uncertainties – driven mainly by the model and scenario uncertainties. It is evident from the broad range of temperature projections for India, ranging from 1°C to 8°C over the period 1880–2099 under different RCP scenarios and even broader range for the precipitation projections. India is one of the most vulnerable regions in the world to climate change in view of the huge population, the large number of poor facing food insecurity, inappropriate soil and management practices on marginal lands in the semi-arid regions leading to increasing rates of land degradation and the projected impacts of climate change on the agricultural, forestry and fisheries sectors. Improved understanding of the climate change impacts, vulnerability and the adaptation practices to cope with climate change is urgently needed to reduce the risk on human welfare.

❑❑❑

5 Impact of Rising Atmospheric CO_2 Concentration on Plant and Soil Process

M.L. Dotaniya

Division of Environmental Soil Science, Indian Institute of Soil Science, Nabi Bagh, Berasia Road, Bhopal-462 038

The climate on earth is a dynamic one promoting the evolution of various living forms and changing the structure and chemical composition of the atmosphere. Over the past few decades, acceleration in the human-induced changes in the climate of the earth has become the focus of scientific and social scrutiny. The gaseous composition of the atmosphere has undergone a significant change mainly through increased industrial emissions, fossil fuel combustion, widespread deforestation and burning of biomass as well as changes in land use and land management practices. These anthropogenic activities have resulted in an increased emission of radiatively active gases, e.g., carbon dioxide (CO_2), methane (CH_4) and nitrous oxide (N_2O), popularly known as the 'greenhouse gases'. The atmospheric concentrations of carbon dioxide, methane and nitrous oxide were 280±6 ppm (parts per million), 700±60 ppb and 270±10 ppb (parts per billion) between the period 1000 and 1750 AD (IPCC, 2001). Today, these values have become 390 ppm, 1750 ppb and 316 ppb, respectively. These greenhouse gases trap the outgoing infrared radiation from the earth's surface. The process, generally referred to as the greenhouse effect, adds to the net energy of the lower atmosphere and, therefore, results in atmospheric warming. According to the Intergovernmental Panel on Climate Change (IPCC, 2007), the global surface temperature is predicted to increase between 1.4 and 5.8°C by the year 2100. Diurnal temperature range will also decrease, with night time temperature is increasing at twice the rate of day time maximum temperature.

The main greenhouse gases

The most important Greenhouse Gases (GHGs) directly emitted by humans include CO_2, CH_4, nitrous oxide (N_2O), and several others. The sources and recent trends of these gases are detailed below.

Carbon dioxide

It is the primary greenhouse gas that is contributing to recent climate change. CO_2 is absorbed and emitted naturally as part of the carbon cycle, through animal and plant respiration, volcanic eruptions, and ocean-atmosphere exchange. Human activities, such as the burning of fossil fuels and changes in land use, release large amounts of carbon to the atmosphere, causing CO_2 concentrations in the atmosphere to rise. Atmospheric CO_2 concentrations have increased by almost 40% since pre-industrial times, from approximately 280 parts per million by volume (ppmv) in the 18th century to 390 ppmv in 2010. The current CO_2 level is higher than it has been in at least 800,000 years (NRC, 2010). Some volcanic eruptions released large quantities of CO_2 in the distant past. However, the U.S. Geological Survey (USGS) reports that human activities now emit more than 135 times as much CO_2 as volcanoes each year. Human activities currently release over 30 billion tons of CO_2 into the atmosphere every year. This build-up in the atmosphere is like a tub filling with water, where more water flows from the faucet than the drain can take away.

Methane

Methane (CH_4) is produced through both natural and human activities. For example, natural wetlands, agricultural activities, and fossil fuel extraction and transport all emit CH_4. Methane is more abundant in Earth's atmosphere now than at any time in at least the past 650,000 years. Due to human activities, CH_4 concentrations increased sharply during most of the 20th century and are now more than two-and-a-half time's the pre-industrial levels. However, in recent decades, the rate of increase has slowed considerably (Jansen *et al.*, 2007).

Nitrous oxide

Nitrous oxide is produced through natural and human activities, mainly through agricultural activities and natural biological processes. Fuel burning and some other processes also create N_2O. Concentrations of N_2O have risen approximately 18% since the start of the industrial revolution, with a relatively rapid increase towards the end of the 20th century. In contrast, the atmospheric concentration of N_2O varied only slightly for a period of 11,500 years before the onset of the industrial period, as shown in ice core samples (Solomon *et al.*, 2007).

Other greenhouse gases

- Water vapor is the most abundant greenhouse gas and also the most important in terms of its contribution to the natural greenhouse effect, despite having a short atmospheric lifetime. Some human activities can influence local water vapor levels. However, on a global scale, the concentration of water vapor is controlled by temperature, which influences overall rates of evaporation and precipitation. Therefore, the global concentration of water vapor is not substantially affected by direct human emissions.

- Tropospheric ozone (O_3), which also has a short atmospheric lifetime, is a potent greenhouse gas. Chemical reactions create ozone from emissions of nitrogen oxides and volatile organic compounds from automobiles, power plants, and other industrial and commercial sources in the presence of sunlight. In addition to trapping heat, ozone is a pollutant that can cause respiratory health problems and damage crops and ecosystems.
- Chlorofluorocarbons (CFCs), hydrochlorofluorocarbons (HCFCs), hydrofluorocarbons (HFCs), perfluorocarbons (PFCs), and sulfur hexafluoride (SF_6), together called F-gases, are often used in coolants, foaming agents, fire extinguishers, solvents, pesticides, and aerosol propellants. Unlike water vapor and ozone, these F-gases have a long atmospheric lifetime, and some of these emissions will affect the climate for many decades or centuries (http://www.epa.gov/climatechange/science/causes.html)

Role of elevated CO_2 in global climate change

The incoming radiation of sun rays on earth is short wave and is converted to long wave during the reflection from the earth's atmosphere. This long wave radiation is trapped with atmospheric CO_2. Since the Industrial revolution began around 1750, human activities have contributed substantially to climate change by adding CO_2 and other heat-trapping gases to the atmosphere. These greenhouse gas emissions have increased the greenhouse effect and caused the Earth's surface temperature to rise.

Atmospheric CO_2 level stimulates GHG emission?

Rising levels of atmospheric carbon dioxide (CO_2) are likely to cause some soils to release large quantities of two potent greenhouse gases (GHGs), nitrous oxide and methane, according to a recent analysis. The results suggest that the contribution of soils and terrestrial ecosystems to slow climate change has been over estimated. The analysis suggested that, as CO_2 levels rise, emissions of nitrous oxide from upland soils would increase by 18.8%. Methane emissions from wetlands and rice paddies would also rise by 13.2% and 43.4%, respectively. By including soil water content and root biomass, the researchers could also explore the mecha-nisms behind the increased emissions of nitrous oxide and methane. Higher soil water content, possibly caused by more efficient use of water by plants, and increased root biomass in soils both encourage anoxic – low oxygen – conditions in soils. Anoxic conditions encourage microbes to break down nitrate; one of the major sources of nitrous oxide from soils. Rice paddies and wetlands, anaerobic conditions favour the production of methane by methanogenic Archaea. These microbes rely on plants to supply the carbon they need, so increased soil carbon storage as CO_2 levels rise will boost soil methane emissions (van Groenigen *et al.*, 2011).

Global-Warming potential

Global-warming potential (GWP) is a relative measure of how much heat a greenhouse gas trap in the atmosphere. It compares the amount of heat trapped by a certain mass of the gas in question to the amount of heat trapped by a similar mass of carbon dioxide. A GWP is calculated over a specific time interval, commonly 20, 100 or 500 years (Table 1). The GWP depends on the following factors: the absorption of infrared radiation by a given species; the spectral location of its absorbing wavelengths; the atmospheric lifetime of the species.

Table 1 Global warming potential of major GHGs

Species	Chemical formula	Lifetime (years)	Global Warming Potential (Time Horizon)		
			20 years	100 years	500 years
CO_2	CO_2	variable §	1	1	1
Methane	CH_4	12±3	56	21	6.5
Nitrous oxide	N_2O	120	280	310	170
HFC-23	CHF_3	264	9100	11700	9800
HFC-32	CH_2F_2	5.6	2100	650	200
Sulphur hexafluoride	SF_6	3200	16300	23900	34900

The Intergovernmental Panel on Climate Change (IPCC) provides the generally accepted values for GWP, which changed slightly between 1996 and 2001. An exact definition of how GWP is calculated is to be found in the IPCC's 2001 Third Assessment Report. The GWP is defined as the ratio of the time-integrated radiative forcing from the instantaneous release of 1 kg of a trace substance relative to that of 1 kg of a reference gas:

$$GWP(x) = \frac{\int_0^{TH} a_x \cdot [x(t)]\, dt}{\int_0^{TH} a_r \cdot [r(t)]\, dt}$$

where TH is the time horizon over which the calculation is considered; a_x is the radiative efficiency due to a unit increase in the atmospheric abundance of the substance (i.e., $Wm^{-2}\ kg^{-1}$) and [x(t)] is the time-dependent decay in abundance of the substance following an instantaneous release of it at time t=0. The denominator contains the corresponding quantities of the reference gas (i.e. CO_2). The radiative efficiencies a_x and a_r are not necessarily constant over time. While the absorption of infrared radiation by many greenhouse gases varies linearly with their abundance, a few important ones display non-linear behaviour for current and likely future abundances (e.g., CO_2, CH_4, and N_2O). For those gases, the relative radiative forcing will depend upon abundance and hence upon the future scenario adopted. Carbon dioxide has a GWP of exactly 1 (since it is the baseline unit to which all other greenhouse gases are compared).

Causes of elevated CO_2 concentration

Anthropogenic activities have caused the concentration of atmospheric CO_2 to increase from about 280 parts per million (ppm) at the beginning of the industrial revolution to over 390 ppm at the present time. Estimates suggest that the atmospheric CO_2 concentration range will lie between 450 ppm and 600 ppm by the year 2050 (Kattenburg *et al.*, 1995). As microorganisms in soils regulate the dynamics of organic matter decomposition and plant nutrient availability, they play a key role in the responses of ecosystems to global climate changes. Elevated CO_2 would affect soil microorganisms indirectly through increased root growth and rhizodeposition rates (Rogers *et al.*, 1994; Rouhier *et al.*, 1994; Paterson *et al.*, 1997; Sadowsky and Schortemeyer, 1997), because the CO_2 concentration in soil is much greater than the atmospheric CO_2 (van Veen *et al.*, 1991). Impacts of elevated CO_2 on soil ecosystems, focusing primarily on plants and a variety of microbial processes. The processes considered include changes in microbial biomass of C and N, soil enzyme activity, microbial community composition, organic matter decomposition, and functional groups of bacteria mediating trace gas emission in terrestrial and wetland ecosystems. Except from CO_2, other gases i.e. CH_4, N_2O and other gases also playing a significant role in global climate phenomena (Fig. 1). Studies revealed that the increased temperature accelerates rate of microbial decomposition, thereby increasing CO_2 emitted by soil respiration and producing a positive feedback to global warming (Allison *et al.*, 2010). Response of biological systems with reference to temperature are expressed as Q_{10} function.

$Q_{10} = k_2/k_1\ ^{10}(T_2\text{-}T_1)$

Q_{10} is the relative change in decomposition rate with a 10°C rise in temperature. k_2, k_1 are the rate constants for a process of interest at two observed temperatures, T_2 and T_1. Temperature increase soil respiration exponentially.

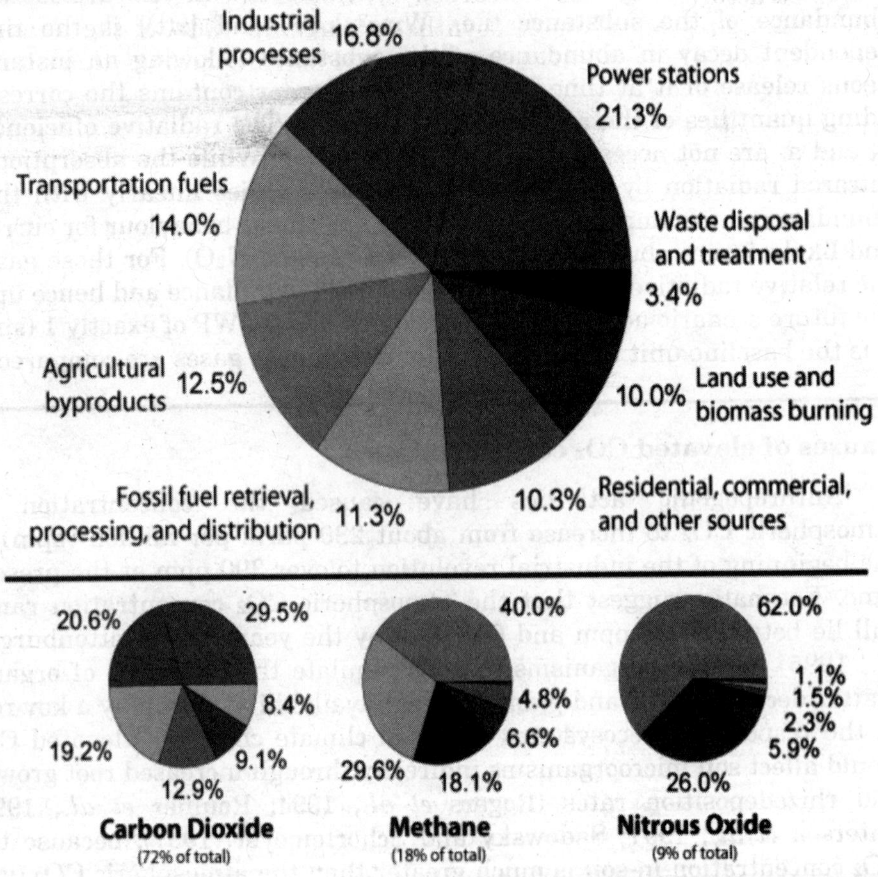

Fig. 1 Annual greenhouse gas emissions by different sectors

Impact of rising CO_2 on plant

Atmospheric concentrations of carbon dioxide have been steadily rising, from approximately 315 ppm in 1959 to a current atmospheric average of approximately 385 ppm (Keeling *et al.*, 2009). Current projections are for concentrations to continue to rise to as much as 500–1000 ppm by the year 2100 (IPCC, 2007a). While a great deal of media and public attention has been focused on the effects that such higher concentrations of CO_2 are likely to have on global climate, rising CO_2 concentrations are also likely to have profound direct effects on the growth, physiology, and chemistry of plants, independent of effects on climate (Ziska, 2008). These effects result from the central importance of CO_2 to plant metabolism. As photosynthetic organisms, plants take up atmospheric CO_2, chemically reducing the carbon. This represents not only an acquisition of stored chemical energy for the plant, but also provides the carbon skeletons for the organic molecules that make up a plant's structure. Overall, the carbon, hydrogen and oxygen assimilated into organic molecules by photosynthesis make up ~96% of the total dry mass of a typical plant (Marschner, 1995). Photosynthesis is therefore at the heart of the nutritional metabolism of plants, and increasing the availability of CO_2 for photosynthesis can have profound effects on plant growth and many aspects of plant physiology.

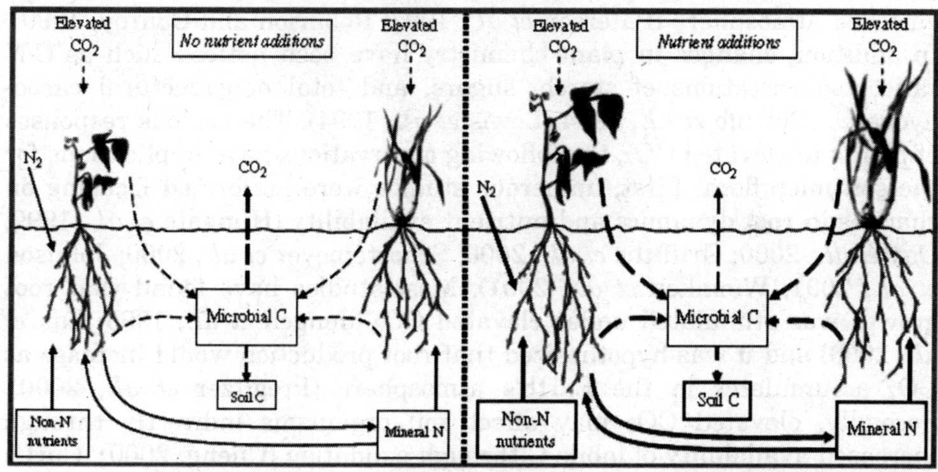

Fig. 2 Conceptual diagram depicting the flows of C and N between plants and soil under elevated atmospheric CO_2 and low vs. high nutrient availability (Adapted de Graaff *et al.*, 2006).

Across a range of Free-Air Carbondioxide Enrichment (FACE) experiments, with a variety of plant species, growth of plants at elevated CO_2 concentrations of 475–600 ppm increases leaf photosynthetic rates by an average of 40%. As the CO_2 concentrations increase, plants can maintain high photosynthetic rates with relatively low stomatal conductance. Across a variety of FACE experiments, growth under elevated CO_2 decreases stomatal conductance of water by an average of 22% (Ainsworth and Rogers, 2007). This would be expected to decrease overall plant water use, although the magnitude of the overall effect of CO_2 will depend on how it affects other determinants of plant water use, such as plant size, morphology, and leaf temperature. Overall, FACE experiments show decreases in whole plant water use of 5–20% under elevated CO_2. This in turn can have consequences for the hydrological cycle of entire ecosystems, with soil moisture levels and runoff both increasing under elevated CO_2 (Leakey *et al.*, 2009). The availability of additional photosynthate enables most plants to grow faster under elevated CO_2, with dry matter production in FACE experiments being increased on an average by 17% for the above ground, and more than 30% for the below ground, portions of plants (Ainsworth and Long, 2005; de Graaff *et al.*, 2006). This increased growth is also reflected in the harvestable yield of crops, with wheat, rice and soybean all showing increases in yield of 12–14% under elevated CO_2 in FACE experiments (Ainsworth, 2008; Long *et al.*, 2006). Photosynthetic rate will increase under elevated CO_2 leading to more plant growth (Fig. 2).

The direct effects of increasing CO_2 on plants include quantitative and qualitative changes in above- and below ground production (Norby, 1994). For example, C_3 plant photosynthetic rates increase under a CO_2-enriched atmosphere (Paterson *et al.*, 1997; Robinson and Conroy, 1999). In addition, changes in plant chemistry have been noticed such as C/N ratio, concentrations of starch, sugars, and total nonstructural carbohydrates (Cotrufo *et al.*, 1994, Lewis *et al.*, 1994). The various responses of plants to elevated CO_2, the following observations have implications for the soil microflora. First, numerous studies were performed focusing on changes in root dynamics and nutrient availability (Hungate *et al.*, 1999; Day *et al.*, 2000; Griffiths *et al.*, 2000; Schortemeyer *et al.*, 2000; Johnson *et al.*, 2001; Wiemken *et al.*, 2001). Most studies have found that root growth was stimulated under elevated CO_2 (Jongen *et al.*, 1995; Zak *et al.*, 2000) and it was hypothesized that root production would increase as CO_2 accumulates in the Earth's atmosphere (Pregitzer *et al.*, 2000). Secondly, elevated CO_2 may affect soil organisms indirectly through increased availability of labile C through exudation (Cheng, 2000). Curtis *et al.* (1994) and Diaz *et al.* (1993) showed that the elevated CO_2 led to an increase in carbon flux from plants into the soil and quantitative and qualitative changes in rhizodeposition linked to CO_2 enrichment (van Veen *et al.*, 1991). Lekkerkerk *et al.* (1990) showed that the root-derived, easily biodegradable compounds of wheat roots increased under a CO_2-enriched condition. This increased input of C in the soil may in turn stimulate mineralization (Baggs *et al.*, 2003) or N_2 fixation (Diaz *et al.*, 1993). Finally, elevated CO_2 could alter the biochemical composition of plant tissue above and below ground (Cotrufo and Ineson, 1995), which then affect soil microorganisms (Ball, 1997). For example, root decomposition was retarded in plants grown under elevated CO_2 (Gorissen *et al.*, 1995; Arnone and Hirschel, 1997). Similarly, it has been found that litter decomposition of plants grown on NPK amended plots were retarded by elevated CO_2. It was thought that changes in chemical quality of plant matter caused retardation of its decomposition (Cotrufo and Ineson, 1995). Different soil and plant parameters were affected by elevated CO_2 concentration in atmosphere, but the effect is not same in all parameters, it varied in a wide range (Table 2).

Table 2. Analysis of variance, showing differences in percentage response to elevated CO_2 within the categorized response variables (Modified from de Graaff *et al.*, 2006).

Responsible variable	**Categorical variable**	***P*-Value**
APB (Aboveground standing plant biomass)	Facility	0.032
APB	Vegetation	0.0012
APB	Soil N availability	0.048
BPB (Belowground standing plant biomass)	Facility	0.170
BPB	Vegetation	0.180
BPB	Soil N availability	0.190
C	Vegetation	0.710
C	Soil N availability	0.050
C:N	Vegetation	0.081
MicC (Microbial C content)	Soil N availability	0.250
rCO_2 (Microbial respiration)	Soil N availability	0.150
N	Vegetation	0.038
N	Soil N availability	0.160
MicN (Microbial N content)	Soil N availability	0.280
N_2	Soil N availability	0.190

*Difference in response are considered significant when $P<0.05$.

In general, trees growing in elevated CO_2 photosynthesize more rapidly and have at least modest increased growth compared to trees growing at ambient CO_2 (DeLucia *et al.*, 1999; Norby *et al.*, 1999; Gielen *et al.*, 2005; Hyvonen *et al.*, 2007). Because of this CO_2 fertilization effect and other factors, model scenarios from the Intergovernmental Panel on Climate Change suggest that terrestrial ecosystems may sequester 20% of anthropogenic CO_2 emissions this century (IPCC 2001, 2007). The fraction of net primary production (NPP) allocated below ground in temperate forests is large but varies widely, ranging from 20% to 65% (Waring *et al.*, 1998). Of the carbon that is allocated below ground in forests, one-half to three-quarters is typically respired back to the atmosphere (Palmroth *et al.*, 2005). Previous results from elevated CO_2 experiments have shown that belowground net primary production and fine-root biomass both tend to increase with CO_2 enrichment (Finzi *et al.*, 2007). Greater fine root mass and mycorrhizal infection promote enhanced P uptake in mycorrhizal plants grown under elevated CO_2 concentrations (Hodge, 1996). Changes in atmospheric CO_2 concentrations indirectly affect mycorrhizae through changes in C allocation from their host plants (Allen *et al.*, 2005). In response to an increase in atmospheric CO_2 there is a greater below the ground C biomass allocation through root exudation and turnover, which is likely to lead to change in size and the activity of soil microflora (Rilling *et al.*, 2001). Rates of soil respiration also typically increase with increased atmospheric CO_2, coupling faster outputs with the faster inputs observed through increased photosynthesis (Luo *et al.*, 1996, Hoosbeek *et al.*, 2007, Wan *et al.*, 2007). Previous research has found that increasing levels of CO_2 in the atmosphere will lead to increased plant growth. This in turn means that more CO_2 is drawn from the atmosphere by plants and, ultimately, some of that carbon would be locked away in the soil. Researchers had predicted that such increases in carbon sequestration in soils would partly counteract human emissions of CO_2, slowing the rate of climate change over the next century. Soil warming can increase nutrient uptake from 100 to 300% by enlarging the root surface area and increasing rates of nutrient diffusion and water influx (Ching and Barber, 1979; Mackay and Barber, 1985) (Table 3). Water soluble nutrients including nitrate, sulfate, Ca, Mg primarily move towards the roots through transpiration-driven mass flow (Barber 1995). Since warmer temperatures increase rates of transpiration, plants tend to acquire water soluble nutrients more readily as temperature increases. Temperature increases in the rhizosphere can also stimulate nutrient acquisition by increasing nutrient uptake via faster ion diffusion rates and increased root metabolism (Bassirirad, 2000). Emerging evidence suggests that warmer temperatures have the potential to significantly affect the nutrient status by altering plant phenology (Nord and Lynch, 2009).

Table 3 Potential interactions of global change variables on soil processes with mineral stress (adapted from Samuvel and Jonathan, 2010)

Process	Global change variables	Interaction with mineral stress
Erosion	Heavy precipitation and drought	Losses of soil nutrients and SOC
Mass flow	Drought, temperature, RH, CO_2	NO_3, SO_4, Ca, Mg, and Si
Root growth & architecture	Drought, temperature, RH, CO_2	All nutrients, especially P and K
Mycorrhizas	CO_2	P, Zn (VAM) and N
Soil microbes (N cycling)	Drought, soil temperature	N
Biological N fixation	Drought, soil temperature	N
Soil redox status	Flooding	Mn, Fe, Al and B
Soil leaching	Heavy precipitation	NO_3, SO_4, Ca, Mg
Plant phenology	Soil temperature	P, N, K
Soil organic carbon status	Soil moisture, temperature, CO_2	All nutrients
Salinization	Precipitation, soil temperature	Na, K, Ca, Mg

Elevated CO_2 also leads to changes in the chemical composition of plant tissues. Due to increased photosynthetic activity, leaf nonstructural carbohydrates (sugars and starches) per unit leaf area increase on average by 30–40% under FACE elevated CO_2 (Ainsworth, 2008). Leaf nitrogen concentrations in plant tissues typically decrease in the FACE under elevated CO_2, with nitrogen per unit leaf mass decreasing on average by 13% (Ainsworth and Long 2005). This decrease in tissue nitrogen is likely due to several factors: dilution of nitrogen from increased carbohydrate concentrations; decreased uptake of minerals from the soil, as stomatal conductance decreases and plants take up less water (Taub and Wang, 2008); and decreases in the rate of assimilation of nitrate into organic compounds (Bloom *et al.*, 2010). In FACE experiments, protein concentrations in grains of wheat, rice and barley, and in potato tubers, are decreased by 5–14% under elevated CO_2 (Taub *et al.*, 2008). Crop concentrations of nutritionally important minerals including calcium, magnesium and phosphorus may also be decreased under elevated CO_2 (Loladze, 2002). Across studies using all types of CO_2 fumigation technologies, there is a lower enhancement of biomass production by elevated CO_2 under low-nutrient conditions (Poorter and Navas, 2003). Crops grown with low amounts of N fertilization also show a greater decrease in protein concentrations under elevated CO_2 than crops grown with higher N fertilization (Taub *et al.*, 2008). While there is little FACE data available on the effects of elevated CO_2 on plant nitrogen and protein concentrations, data from chamber experiments shows C_4 plants to be much less responsive than C_3 plants in this regard (Cotrufo *et al.*, 1998). Many studies in controlled environments have shown that, compared to other plant species, legumes show greater enhancement of photosynthesis and growth by elevated CO_2 (Rogers *et al.*, 2009). Under elevated CO_2 conditions, legumes may be able to shunt excess carbon to root nodules where it can serve as a carbon and energy source for the bacterial symbionts.

Impact of rising CO_2 on soil process

Climate change affecting and modifying the soil forming process as well as nutrient dynamics in soil. It may be due to changing micro organism diversity or mineralization rate in a particular soil. The climate affecting the rate of chemical reactions, physical process and biological transformation of soil. The most rapid processes of chemical or mineralogical change under changing climatic conditions would be loss of salts and nutrient cations by leaching, and salinization. The rising CO_2 concentration directly and indirectly affects the soil process like organic matter dynamics, nutrient dynamics epecially nitrogen and biological parameters *i.e.* soil biodiversity and enzyme activities, modifications in chemical reactions i.e. pH, salinization and alkalinization, physical process i.e. Soil water dynamics, run off, variability in rainfall pattern and intensity, soil erosion, soil structure etc. All processes directly and sometimes indirectly play a crucial role in plant growth and biomass production on beautiful earth. Some of the soil processes affected by elevated CO_2 concentration can be broadly classified into structural (physical) and hydrological processes, and biogeochemical processes. The soil is a complex system, in which many changes operating simultaneously and influence soil health or soil quality. Physical soil constraints also significantly reduce crop yield potential in developing countries. They include poor soil texture, rockiness, compaction (Lal, 1987), and slope steepness (Lal, 1998). These physical soil characteristics determine the water holding capacity of the soil, the degree of root contact with the soil matrix and cation exchange capacity, all of which influence plant nutrient acquisition (Marchner, 1995).

Weathering: The concentration of CO_2 in soil is 10-50 times higher than atmosphere. Through changes in productivity, respiration, and other variables, elevated CO_2 can also influence weathering of carbonates, silicates, and other soil minerals by altering the concentration of CO_2 in soil pore spaces (Sposito, 1989; Andrews and Schlesinger, 2001; Pendall *et al.*, 2001; Karberg *et al.*, 2005). Soil CO_2 is produced primarily by respiration from belowground plant tissue and from heterotrophic organisms in the soil. Some of this CO_2 dissolves in water and forms carbonic acid (H_2CO_3), which reacts with soils via cation exchange and mineral dissolution. For instance, calcium carbonate ($CaCO_3$) reacts with protons in acidic soil solutions to produce Ca-P and bicarbonate ions (HCO_3). An increase in the concentration of CO_2 in soil pore spaces could therefore increase rates of weathering. Changes in the clay mineral surfaces or the bulk composition of the clay fraction of soils are brought about by a small number of transformation processes, listed below (Brinkman, 1982). Each of these processes can be accelerated or inhibited by changes in external conditions due to global change.

- Hydrolysis by water containing carbon dioxide, which removes silica and basic cations;
- Cheluviation, which dissolves and removes especially aluminium and iron by chelating organic acids;
- Ferrolysis, a cyclic process of clay transformation and dissolution mediated by alternating iron reduction and oxidation, which decreases the cation exchange capacity by aluminium interlayering in swelling clay minerals;
- Dissolution of clay minerals by strong mineral acids, producing acid aluminium salts and amorphous silica;
- Reverse weathering, i.e., clay formation and transformation under neutral to strongly alkaline conditions, which may create, e.g., montmorillonite, palygorskite or analcime.

pH: Rising atmospheric CO_2 directly did not affect the pH of most world soils. The indirect effect of global climate change affected the hydrological cycle. High rainfall speed up the leaching process of base cation from soil, alter the soil pH by acidification process. This phenomenon is limited by the buffering capacity of soil. In calcareous soils, soil reaction may range between about 8.5 and 7 depending on the partial pressure of CO_2 in the soil. It maintained against the leaching of basic cations by the different soil processes as long as a few per cent of finely distributed lime remained. Soil buffering capacity is less prominent in calcareous soil, so that cation exchange capacity plays a vital role. It should be noted that the simple modelling of accelerated $CaCO_3$ leaching under a doubled atmospheric CO_2 concentration generally does not hold true. In most soils, the ongoing decomposition of organic matter maintains CO_2 concentrations in the soil air far above atmospheric concentration even now, and $CaCO_3$ solubility is determined by the partial pressure of CO_2 in soil air and its activity in soil water, rather than in the atmosphere. Leaching of lime is thus positively related to rate of organic matter decomposition, negatively to gas diffusion rate, and positively to amount of water percolating through the soil.

Fertility: Rising atmospheric CO_2 is able to change the physiological activities of plants and soil micro-organisms which control the flow of carbon (C) and nitrogen (N) in terrestrial ecosystems. Barrett *et al.* (1998) demonstrated that a doubling of atmospheric CO_2 content under continuous phosphorus deficiency increased wheat root activity by 30-40%, thus increasing the inorganic phosphorus supply for plant utilization, even in areas where growth is currently limited by phosphorus deficiency (Narolia *et al.*, 2013). The root exudates enhanced mineralization process and converted immobile P into plant available (Dotaniya *et al.*, 2013a). Staddon *et al.* (1999) reported that *Plantago lanceolata* and *Trifolium repens* effectively increased their phosphorus use efficiency under elevated CO_2 conditions by reducing shoot phosphorus content as a component of CO_2-induced photosynthetic acclimation. In low soil phosphorus, biomass N did not increase in grasses, but biomass N increased by 28% when soil was high in P under elevated CO_2 condition. The change in climate factors also alters the beneficial plant microbe interactions. The survival and activity of microorganisms in the rhizosphere mainly depend on the flow of carbon in the root exudates. The composition of these root exudates is affected by various environmental factors i.e. soil temperature, CO_2 concentration in the atmosphere as well as soil nutrient availability, soil moisture and pH. N fixation activities may be increased in soil surface layers by elevated CO_2 concentrations due to increased algal growth (Chang *et al.*, 2000). Incubation studies showed that net mineralization of soil S was affected by organic matter addition as well as temperature (Tabatabai and Al-Khafaji, 1980). Rise in soil temperature is likely to increase soil nutrient mineralization, which may lead to C and N losses and thereby disrupt C and N cycles. Runlon *et al.* (2012) reported that in pine system that specific soil CO_2 efflux rates per g root dry weight or per g root C were significantly lower in elevated than in ambient CO_2 plots; specific respiration rate per g root N was not different between CO_2 treatments (Table 4). However, specific respiration rates per g soil N and C were significantly higher in elevated than in ambient CO_2 plots.

Table 4. Specific respiration rates (g C respired per g) for root biomass, root C and N and soil C and N (Adapted Runlon *et al.*, 2012).

CO_2	Root dry weight	Root N	Root C	Soil N	Soil C
Ambient	0.172	34.493	0.375	1.437	0.076
Elevated	0.146	32.262	0.315	1.848	0.098
ANOVA	p=0.03	P=0.32	P=0.02	P<0.01	P<0.01

Ambient $CO_2 \approx 365$ μmolmol^{-1}; elevated $CO_2 \approx 720$ μmolmol^{-1}.

Nutrient loss: Climate changes affect the hydrological cycle, thus modifying the transformation and transport characteristics of nutrients. Diffuse losses of nutrients, especially those from agricultural origins, are among the major contributors to the total load of nutrients to the river systems (Novotny and Olem, 1994). The main source of phosphorus and nitrogen transport in the water bodies is agriculture. Several studies have predicted adverse climate change impact on eutrophication problems (Bouraoui *et al.*, 2004). The recent unprecedented rainfall leading to floods in 13 districts of north Karnataka and the estimate made after the calamity revealed that nearly 287 million tonnes (mt) of top soil, 8 lakh tonnes of soil nutrients and 39 lakh tonnes of soil organic matter were washed away and nearly 57.5 mt of soil microbial biomass were lost from the region during this short period (Natarajan *et al.*, 2010). Climate changes are expected to alter significantly the hydrological behavior of catchments (Gleick, 1999; Arnell, 1999) modifying the nutrient transformation and transport characteristics (Murdoch *et al.*, 2000). An increase in diffuse source pollutant loads and nutrient cycling is among the effects to be expected (Murdoch *et al.*, 2000). A large area of the world was covered under land degradation and more sustainable to the negative effect of climate change (Table 5).

Soil resilience: It refers to the ability of a soils to resist or recover their healthy state in response to destabilizing influences. Burning of soil organic carbon due to increasing global climate change reduced the capacity of soil resilience over a period, but vice-versa it resulted in organic carbon increasing in soils. Research to increase resilience of soils against any adverse effects of climate change could be done in conjunction with that on soil resilience against direct adverse human impacts. Until site-specific management procedures have been elaborated, soil and crop (including trees and pasture) management should aim to maintain soil cover and organic matter supply to soil biota, while minimizing mechanical disturbance by heavy traffic, cultivation or excessive grazing intensity. A single management recipe would not be generally applicable in different conditions. Management methods for wetland need to be developed that make optimum use of any increased potential productivity, while minimizing secondary effects such as increased CH_4 or N_2O emission from the reduced soil. Such factors, and others, should be taken into account in designing an optimum management strategy for any specific natural and cultural environment. Resilience against soil reduction in practice depends on the drainage conditions, since most soils have sufficient organic matter for reduction to start within about a week after water-saturation. Soils most resilient against reduction in conditions of increased rainfall variability and incidence of high-intensity rainfall have similar properties as those resilient against the negative effects of other perturbations: high infiltration rate, high structural stability and a permanent heterogeneous system of tubular macropores, good external drainage (Brinkman, 1990).

Table 5 Soil edaphic conditions and land degradation that place major constraints on crop yields. Source: Food and Agriculture Organization of the United Nations (TERRASTAT, 2009).

Area	**Soil / major constraints%**	**Land degradation %**
Africa	85	77
Asia	70	74
Australia	81	61
Europe	69	91
North America	73	49
South America	81	78

Biodiversity: First effect of elevated CO_2 on soil microbial population would take place in the rhizosphere. Elevated CO_2 concentration altered environmental conditions due to climate change are likely to induce changes in plant physiology and root exudations. Rising CO_2 leads to increased C allocation to the root zone and potentially alter the composition of root exudates (Kandeler *et al.*, 2006). Root exudation enhanced by elevated atmospheric CO_2 concentration (Haase *et al.*, 2007). Under elevated CO_2, HCN produces *Pseudomonas* strains, the potential inhibitors of root parasitic fungi, was reduced and the proportion of siderophore producers and nitrate-dissimilating strains increased (Tarnawski *et al.*, 2006). It was noticed that under elevated CO_2, the population of fungi is higher as compared to bacteria (Carney *et al.*, 2007). Induced CO_2 change may generate intra and inter specific competition between microbes for scarce nutrients and habitats leading shift in microbial community and also there may be slow changes in the evolutionary pattern of the microbes. Elevated atmosphere temperature also significantly alters the structure of the AMF hyphal network and induce a shift from more vesicles in cooler soils to more extensive extra mycorrhizal hyphal networks in warmer soils (Hawkes *et al.*, 2008). This may be linked to faster carbon allocation to the rhizosphere and enhanced the respiration of the extra mycorrhizal mycelium at a high soil temperature (Heinemeyer *et al.*, 2006). Climate change may thus significantly impact the diversity and activities of such plant linked microbial communities (Drigo *et al.*, 2008). Huge studies have failed to find a significant response of microbial biomass C to elevated CO_2. Microbiological parameters are less stable. Some of the researchers found positive relation and others got negative relations. In general rhizosphere contains more microbial population than the bulk soil. The type and composition of root exudates i.e. Sugars, oligo saccharides, vitamins, nucleotides, flavones, auxins and stimulators mainly decide the diversity of microorganisms proliferating in the root zone (Dotaniya *et al.*, 2013). The decomposition of organic matter serves as energy for the growth of microorganism and also supplies C for their own cell synthesis. Under aerobic conditions microbes assimilate 20-40% of the substrate C and residual amount is released as CO_2 to the atmosphere. Factors influencing mineralization of C include temperature, O_2 supply, moisture, pH, available nutrients and the C/N ratio of the residues. Warming condition increases the release of CO_2 to the atmosphere.

Enzyme activities: Elevated CO_2 induced mainly an increase of enzyme activities (protease, xylanase, invertase, alkaline phosphatase, arylsulfatase) in the upper 5 cm of the soil and did not change microbial biomass in the soil profile. Since rhizodeposition and newly formed roots enlarged the pool of easily available substrates mainly in the upper soil layers, enzyme regulation (production and activity) rather than shifts in microbial abundance was the driving factor in higher enzyme activities in the upper soil. Repeated soil sampling during the third to fifth year of the experiment revealed an enhancement of enzyme activities which varied in the range of 20–80% (Kandeler *et al.*, 2006). Extracellular enzymes like phosphatases and dehydrogenase also participate in the decomposition of soil organic matter (Bell *et al.*, 2010). Studies revealed that elevated CO_2 can increase extracellular enzyme activities by increasing microbial demand for N and P (Dhillion *et al.*, 1996). Elevated CO_2 can increase C in soils and ultimately could affect N cycling, microbial activities, and plant N availability in many ecosystems (Hungate *et al.*, 2003; Edwards and Norby, 1999). Gross N mineralization rates have been reported to increase or remain unchanged under elevated CO_2. Rate of microbial immobilization can also increase (Mikan *et al.*, 2000) or decrease (Berntson and Bazzaz, 1997) with elevated CO_2. These unfavorable changes pose big threat to soil productivity, soil and water quality and sustainability of the production system (Emmett *et al.*, 2009; Benbi and Kaur, 2009).

Soil water dynamics: Paradigm shifting in rainfall pattern and intensity governed a set of climatic events and its direct effect on the soil system across the world. In water deficient part increasing ground water table due to changes in rainfall pattern and on another side rise in temperature increases the rate of evapotranspiration which causes soil drying and lowering of ground water table. This ultimately declines the natural resources in a particular area. High rainfall increasing the rate of soil erosion as well as nutrient loss from agricultural fields, reduces the fertility status of soil in long ways. The decrease in atmospheric precipitation in some parts will result in a decrease in water infiltration and water storage in the soil. In extreme conditions infiltration rate and water storage were observed very low due to lower precipitation. Rainfall intensity and variability both are spatial and temporal.

Salinization: Climate warming contributes to the degradation of freshwater quality and availability (IPCC, 2007b). A major way in which this happens is through salinization, an important limitation to agricultural productivity in semi-arid regions that is expected to intensify through the 21st century. Increases in demand for irrigation in semi-arid regions such as the Indo-Gangetic plain of northern India and Pakistan and parts of Africa driven by population growth and climate warming are expected to increase the extent of salinization in agriculture (Yeo, 1999). Climate warming can increase agricultural salinization by increasing the demand for irrigation and increasing rates of surface water evaporation. Warmer temperature can also increase salt accumulation in crops via increased transpiration rates (West and Taylor, 1980). In addition, rising sea levels driven by climate warming is expected to contribute to seawater intrusion of coastal aquifers (Don *et al.*, 2006; Antonellini *et al.*, 2008).

Degradation: Increased microbial activity due to higher CO_2 concentration and temperature produces greater amounts of polysaccharides and other soil stabilizers. Increases in litter or crop residues, root mass and organic matter content tend to stimulate the activity of soil macrofauna, including earthworms, with consequently improved infiltration rate and bypass flow of the greater number of stable biopores. The greater stability and the faster infiltration increase the resilience of the soil against water erosion and consequent loss of soil fertility. The increased proportion of bypass flow also decreases the nutrient loss by leaching during periods with excess rainfall. This refers to the available nutrients in the soil, including well-incorporated fertilizers or manure, but not to fertilizers broadcast on the soil surface. These are subject to loss by runoff or leaching.

Conclusions

The rising atmospheric CO_2 concentration affected vigorously, but its clear cut effects are complex due to mixed interaction of climate-soil-plant system. Although the interactions of global climate change and crop nutrition are not well understood, it is probable that the net effects of these changes will be negative for agricultural production in poor nations. Drought induced by higher temperatures and altered rainfall distribution would reduce nutrient acquisition, biological nitrogen fixation, and may disrupt nutrient cycling. Some of research experiments showed positive signs in relation to rising atmospheric CO_2, but it is only in limited areas. More intense precipitation events would reduce crop nutrition by causing short-term root hypoxia, and in the long term by accelerating soil erosion. Increased temperature will reduce soil fertility by increasing soil organic matter decomposition, and may have profound effects on crop nutrition by altering plant phenology. Since soil fertility has been already a primary constraint to food security in many developing regions, and crop production is already marginal, these stresses may be disastrous for vulnerable populations. An urgent effort is required to improve crop nutrition and soil fertility management in poor nations, integrating agro-ecological and socio-cultural aspects of the problem, to avert worsening of a situation that is already desperate. Use of modern as well as traditional techniques are required to reduce CO_2 concentration in the atmosphere and combat effects of climate change.

❑❑❑

6 Modelling the Impact of Climate Change on Soil Erosion in Stabilization and Destabilization of Soil Organic Carbon

Debashis Mandal, N.K. Sharma, Rajeev Ranjan and P.K. Mishra

Central Soil and Water Conservation Research and Training Institute, 218, Kaulagarh Road, Dehradun, India

Introduction

Soil erosion consists of the mechanisms of detachment, transport and sedimentation (Kinnell, 2004). Water erosion is one of the most active forms of soil erosion of the tropical, sub-tropical and arid areas because of adverse climatic conditions such as high rainfall intensities, combined with land use changes. Two major mechanisms of water erosion are diffuse and concentrated erosion. Of the two main mechanisms, diffuse erosion, by scouring the carbon-rich surface soil horizons, affects more severely the global C cycle (Lal, 2003). In diffuse erosion, soil particle detachment occurs mainly by raindrop impact and by overland flow (Kinnell, 2004; Ghahramani *et al.*, 2011). Raindrop sand overland flow, by breaking down soil aggregates through slaking and peeling induce the preferential erosion of soil organic components. This is because of their light nature and the absence of mineral interactions. The overall impact of human-induced erosion on the global C cycle remains controversial. Moreover, although the issue of the fate of the eroded SOC has been raised in the literature (Lal, 1995; Jacinthe and Lal, 2001), field data remain scarce and the consequences of CO_2 releases have been largely overlooked (Jacinthe and Lal, 2001). The debate on the impact of soil erosion on the global carbon (C) cycle and associated climate change might benefit from field-based studies on the impact of soil C erosion on C fluxes on particulate, dissolved and gaseous forms, and for different agricultural systems in order to set up mitigation strategies.

Soil is an important element of the global carbon (C) cycle. Soil erosion from agricultural land results in considerable losses of soil organic matter (SOM). Considering that approximately 1.6 billion ha or 13% of the earth surface is affected by human-induced erosion (GLASOD, 1990), and current annual soil loss in India are is high as 6.6 billion tons per year (Wen and Pimentel, 1998), erosion-induced C displacement may be an important factor affecting CO_2 concentration in the atmosphere (Lal, 1995). Soil erosion is a four-stage process involving detachment, breakdown, transport/ redistribution and deposition of sediments. All these processes have strong impact on SOM. The clay and soil organic carbon (SOC) is influenced during all four stages. Being a selective process, erosion preferentially removes the lighter fractions of a low density of < 1.8 Mg m^{-3}. A combination of mineralization and C export by erosion causes a severe depletion of the SOC pool on eroded compared with uneroded or slightly eroded soils. In addition, the SOC redistributed over the landscape or deposited in depressional sites may be prone to mineralization because of breakdown of aggregates leading to exposure of hitherto encapsulated C to microbial processes among other reasons. Depending on the delivery ratio or the fraction of the sediment delivered to the river system, gross erosion by water may be 75 billion Mg, of which 15–20 billion Mg are transported by the rivers into the aquatic eco-systems and eventually into the ocean. The amount of total C displaced by erosion on the earth, assuming a delivery ratio of 10% and SOC content of 2–3%, may be 4.0–6.0 Pg $year^{-1}$ (Lal, 2003). With 20% emission due to mineralization of the displaced C, erosion induced emission may be 0.8–1.2 Pg C $year^{-1}$ on the earth. Thus, soil erosion has a strong impact on the global C cycle and this component must be considered while assessing the global C budget. Adoption of conservation-effective measures may reduce the risks of C emission and sequester C in soil and biota.

Accelerated erosion by water and wind is responsible for one-half and one-quarter of all soil degradation, respectively (Daily 1995, Pimentel *et al.* 1995). Persistently high rates of soil erosion affect more than 1.1×10^9 hectares of land annually (Jacinthe and Lal 2001, Berc *et al.* 2003), redistributing on the order of 75 Pg soil per year, with sediment transport leading to silting of reservoirs and eutrophication of lakes. Soil erosion from agricultural lands alone, which accounts for two-thirds of the total soil loss, has been estimated to result in more than US$400 billion of damages annually (Pimentel *et al.* 1995). Soil erosion is the only way that otherwise stable, mineral-associated SOC can be relocated in large quantities and its decomposition rate enhanced during transport or reduced after transport (Starr *et al.* 2000, Lyons *et al.* 2002).The possibility of erosion-induced fate of C has received widespread interest from the scientific community and policymakers. Moreover, projected changes in climate are expected to stimulate the hydrologic cycle, increasing the intensity, amount, and seasonality of precipitation in many parts of the world, and thus accelerating soil erosion (Berc *et al.*, 2003).

The debate on the impact of soil erosion on the global carbon (C) cycle and associated climate change might benefit from field-based studies on the impact of soil C erosion on C fluxes on particulate, dissolved and gaseous forms, and for different agricultural systems in order to set up mitigation strategies. Assessment of erosion-induced C losses is usually based on comparative observations between disturbed (either loss or accumulation) and undisturbed sites (Harden *et al.*, 1999; Kimble *et al.*, 2001). Due to a large number of soil types, climatic and topographical conditions at the given sites and often lack of detailed site description, these relationships are difficult to generalize and extrapolate beyond the studied domain. To overcome the limitation of field observation studies, modelling of SOM loss due to water and wind erosion may be a viable option (Starr *et al.*, 2000). Assessment and prediction of erosion-induced release of CO_2 into the atmosphere requires integration of field results obtained at various scales and at different climatic and soil conditions with current C dynamics and soil erosion models.

Erosional redistribution of soil and associated soil organic carbon

Soil erosion is traditionally conceived as a three-step process involving the detachment, transport, and deposition of soil particles. Detachment exposes SOC that is physically protected within aggregates and clay domains, subsequently, finer soil particles and associated SOC are preferentially transported away from eroding slopes to different low-lying depositional sites (Gregorich *et al.*, 1998; Starr *et al.*, 2001). Following detachment and transport, burial usually is believed to protect SOC from decomposition, because there generally are enhanced and radiometrically old C stocks in the deep soils of agricultural lowlands and sedimentary basins (Stallard, 1998; Harden *et al.*, 1999, 2002). Most of the eroded topsoil (>70%; Stallard, 1998) remains within the adjacent topography and is stored in a variety of depositional basins, including wetlands, peat lands, estuaries, fluvial deltas, terrestrial depressions (hollows), and reservoirs within the same or adjacent topography. The increased wetness and reduced aeration at the depositional basins (compared with eroding slopes) can slow down decomposition (Stallard, 1998, Smith *et al.*, 2001, McCarty and Ritchie, 2002). Stallard (1998), on the basis of past data (Meade *et al.*, 1990) and model simulations, provided three major reasons why soil erosion should not necessarily represent loss of C from the terrestrial biosphere: First, soil redistribution downhill or downstream is usually accompanied by partial replacement of eroded upland C with new photosynthate. Second, a significant portion of eroded, C-rich topsoil is buried in different depositional settings, rather than flowing to the ocean. Erosion transports relatively fresh organic matter that is present at or near the soil surface (compared with deep soil organic matter). After successive erosive events, the C and nutrient-rich topsoil of the eroding slopes is buried in the depositional lowlands and becomes a subsoil horizon of the convergent slopes or plains (Fig. 1), probably reducing its rate of decomposition (compared with noneroded C on the contributing slopes). Third, the surface area for terrestrial deposition of eroded C has increased since the beginning of the industrial revolution. The estimated 10- to 100-fold acceleration of erosion rates by anthropogenic activities in recent history has not been accompanied by a concurrent and proportional increase in sediment discharge to the ocean. The discharge of sediment and C to the ocean has remained approximately constant as a result of hydrologic projects on managed floodplains. Therefore, the recent increase in the rate of soil erosion has led to increased storage of eroded C in different types of depositional basins (Stallard, 1998). Soil erosion results in drastic modifications to the structure as well as the biological and chemical properties of the soil matrix, affecting its productive capacity and ability to sequester atmospheric CO_2.

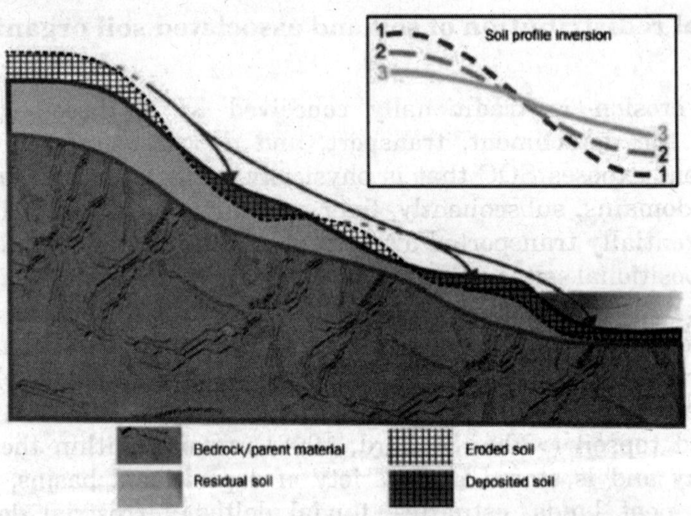

Fig. 1. Soil and soil organic carbon transport from divergent slopes to convergent or flat depositional basins and erosion-facilitated inversion of a hillslope soil profile.

Factors affecting erosion-induced soil organic matter turnover

Wide variety of views on the role of erosion on C dynamics within ecosystems indicates that the processes involved in detachment, transport, deposition and mineralization of SOM are poorly understood. Discrepancies in estimates of erosion influence on soil C may point out to differences in SOM quality at various landscape positions (Schimel *et al.*, 1985a) resulting from SOM transport and sorting by water. The SOM trapped in watershed deposits can be either sequestered or undergo increased rate of mineralization depending on local conditions.

Types of soil organic matter and its distribution in soil matrix

Accelerated soil erosion (Blaikie and Brookfield 1987) has a significant influence on the global C budget. The net amount of CO_2 released from the biosphere to the atmosphere as a result of landuse change over time is likely to be equivalent to about 75% of total fossil fuel C emissions. It is estimated that since the industrial revolution, land conversion and degradation have caused up to 200 Pg C that was originally in the biosphere to be released to the atmosphere (DeFries *et al.* 1999).The SOM consists of a great variety of organic compounds. For modelling purposes, however, it is usually classified into three pools based on the rate of mineralization and turnover (Stevenson, 1982; Parton *et al.*, 1987). Labile or easily mineralizable compounds along with microbial and fungal biomass generally comprise about 5–15% of the total SOM. This pool has turnover rate of month to years and, perhaps, is of the greatest interest for SOM erosion modelling. Slow pool with turnover time of several decades comprises 20–40% of the total SOM. Stable or recalcitrant pool has turnover time of hundreds to thousands of years and in most soils comprises remaining 60–70% of the total SOM (Rice, 2002).

Sediment enrichment

Runoff from eroding landscapes is enriched in clay sized particles (Pert and Walling, 1982), particulate organic carbon (POC), and dissolved organic carbon (DOC) (Lal, 1995). Suspended sediments in rivers will, in majority, be less than 0.062 mm diameter (Pert and Walling, 1981). The smallest size SOC fraction (less than about 0.005 mm) is thought to be the most stable size fraction in soil (Paul *et al.*, 1995) because of the physical protection afforded in soil aggregates (Kay, 1998) and the recalcitrance of humic materials (Paul *et al.*, 1995). Physical protection in aggregates is undermined by processes such as tillage and water erosion (Lal, 1997). Erosion is one of the only soil processes that can remove stable SOC in large quantities so its effects may be dramatic. The data presented by Mitchell *et al.* (1998) on modelling carbon storage in soil showed that erosion by water is the most significant factor affecting the SOC balance in the north central USA.

Enrichment ratio of SOM is the ratio between concentrations of SOM in sediment to those in undisturbed soil. Enrichment ratio >1 is the result of preferential transport of either soil or SOM. Enrichment of eroded sediment with nutrients and SOM in particular has been widely reported (Owens *et al.*, 2002). Its mechanism consists of two processes: dispersion of soil aggregates and their sorting during transport. Raindrop impact is one of the primary erosive forces acting on soil. It was shown that upon impact with soil aggregate, raindrop removes its outer layer (slaking and peeling process), thus releasing microaggregates. The mechanism of enrichment is explained by the fact that the outer layers of soil aggregates have increased concentration of sorbed chemicals including SOM compared with the inner core (Ghadiri and Rose, 1991a). Raindrop impact causes the aggregates to slake and peel. As a result of this process eroded sediment not only have finer size characteristic than the original soil, but also SOM is unevenly distributed between coarse and fine particles (Palis *et al.*, 1997). These structural factors, although difficult to quantify, need to receive more attention when modelling erosion-induced SOM dynamics. The sorting of material transported by water is caused by the differences in drag, gravitational, and cohesive forces acting on individual particles. The drag force is a function of the particle diameter and shape, and the gravity force is a function of particle mass. The enrichment ratio of SOM as high as 5 has been reported (Zobeck and Fryrear, 1986) for certain conditions. Although the amount of loose, poorly decomposed non-cohesive plant fragments in most agricultural soils is relatively small, its highly preferential transport may also have a significant impact on SOM redistribution (Ghadiri and Rose, 1991b). Enrichment ratio of SOM tends to be greater for more aggregated soils with higher concentration of clay than less aggregated and coarse-textured soils (Palis *et al.*, 1997). The SOM concentration in various sizes of aggregates tends to decrease with the severity of erosion (Bajracharya *et al.*, 1998). It also varies with rainfall duration. For example, in an experiment under various levels of erosion, it was shown that enrichment ratio of SOC in a sandy loam soil varied from 1.45 to 2.65 (Fig. 2) with severity of erosion (Mandal *et al.*, 2012).

Expected climate change impacts on soil erosion and runoff from cropland

Soil erosion rates may be expected to change in response to changes in climate for a variety of reasons, the most direct of which is the change in the erosive power of rainfall (Nearing, 2001; Pruski and Nearing, 2002). A second dominant pathway of influence by climate change on erosion rates is through change in plant biomass. The mechanism by which climate changes affect biomass and by which biomass changes runoff and erosion, are complex (Pruski and Nearing, 2002). For example, anthropogenic increases in atmospheric carbon dioxide concentrations cause increases in plant production rates and changes in plant transpiration rates (Rosenzweig and Hillel, 1998),which translate to an increase in soil surface canopy cover and more importantly, biological ground cover. On the other hand, increase in soil and air temperature and moisture will likely cause faster rates of residue decomposition due to an increase in microbial activity. More precipitation may also lead to an increase in biomass production. Higher temperatures may translate to higher evaporation rates, while more rainfall would tend to lead to higher soil moisture levels.

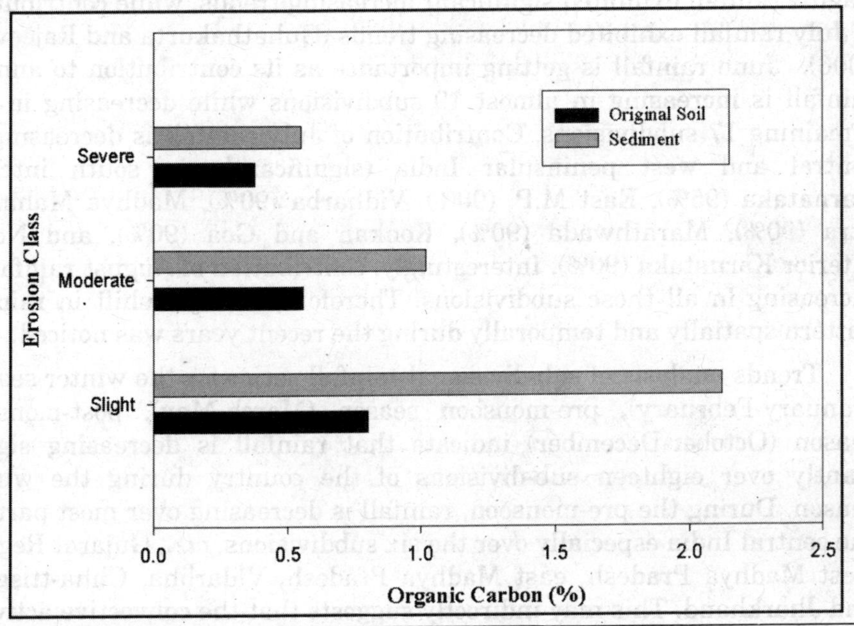

Fig. 2. Concentration of organic C (%) in soil and sediment under different erosion levels.

Analysis of long-term rainfall data for over 1100 stations across India show pockets of deficit rainfall over eastern Madhya Pradesh, Chhattisgarh and Northeast region in Central and Eastern India (Rao *et al.*, 2007) especially around Jharkhand to Chhattisgarh. In contrast, increasing trends (+10 to 12%) in rainfall are observed along the west coast, northern Andhra Pradesh and parts of NW India (National Action Plan on Climate Change, 2008). However, there are indications of shifts in monthly rainfall patterns in southern peninsular region. Comparison of the mean annual rainfall since 1991 onwards with the earlier normal (1961-90) rainfall at Bangalore showed that the peak monthly rainfall had shifted from September to October. In general, the shift seems to be of about 20-25 days. Similar patterns are also observed at Anantapur and Bijapur. The moisture condition prior to rainfall largely controls hydrological parameters including infiltration and runoff generation. However, the surface condition and surface cover are also very important controlling factors of infiltration. In relatively wetter regions including the west coast, northern Andhra Pradesh and parts of NW India, runoff would be primarily controlled by rainfall amount and is consistent with the saturation or storage control model of runoff generation. In such regions, runoff should, therefore, respond positively for increasing precipitation.

Trend analyses has been carried out on monthly basis for all the 36 subdivisions of the country and suggest that contribution of June and August rainfall exhibited significant increasing trends, while contribution of July rainfall exhibited decreasing trends (Guhathakurta and Rajeevan, 2006). June rainfall is getting importance as its contribution to annual rainfall is increasing in almost 19 subdivisions while decreasing in the remaining 17 subdivisions. Contribution of July rainfall is decreasing in central and west peninsular India (significantly in south interior Karnataka (95%), East M.P. (90%), Vidharba (90%), Madhya Maharashtra (90%), Marathwada (90%), Konkan and Goa (90%), and North interior Karnataka (90%). Interestingly, contribution of August rainfall is increasing in all these subdivisions. Therefore, a major shift in rainfall pattern spatially and temporally during the recent years was noticed.

Trends analysis of sub-divisional rainfall series for the winter season (January-February), pre-monsoon season (March-May), post-monsoon season (October-December) indicate that rainfall is decreasing significantly over eighteen sub-divisions of the country during the winter season. During the pre-monsoon, rainfall is decreasing over most parts of the central India especially over the six subdivisions, *viz*., Gujarat Region, west Madhya Pradesh, east Madhya Pradesh, Vidarbha, Chha-ttisgarh and Jharkhand. This may indirectly suggests that the convective activity, which is the main cause for the rainfall activities during the pre-monsoon season, is decreasing over the central parts of the country. However, during the post-monsoon season, rainfall is increasing signific-antly in the sub-divisions, *viz*., Saurashtra and Kutch, Marathwada and Rayalseema.

In case of rainfall, the observed southwest monsoon seasonal rainfall at all India level does not show any significant trend (Rupa Kumar *et al*, 1992). However at regional level, variations in rainfall patterns have been observed.

Analysis of rainfall events with intensities of 10, 100 and above 100 mm (Goswami *et al*, 2006) indicates that in the recent period the frequency of rain events with intensities of more than 100 mm have increased (Fig. 2) while there had been a significant decrease in the frequency of moderate rain events over central India from 1951 to 2000. Such extreme precipitation events can lead to enhanced soil erosion thereby reducing the productive capacity of the soils (Mandal *et al.*, 2010).

In drier region, possible changes in soil water regime should alter runoff production much more indirectly through vegetation cover and soil faunal activity than directly. So, any increase in precipitation should result in enhanced activity and protective vegetative cover which in turn produces low runoff. Under natural conditions, at regional scale the effect of annual rainfall on runoff coefficient (runoff depth divided by runoff depth expressed in percentage terms) is totally offset by the vegetative cover and litter fall and by the destruction of surface crust by faunal activity. Therefore, the drier the climate, higher is the runoff coefficient.

Temperature changes also affect biomass production levels and rates in complex ways. Corn biomass production may increase with increasing temperature, particularly if the growing season is extended, but then may decrease because of temperature stresses as the temperature becomes too high (Rosenzweig and Hillel, 1998). Again, biomass changes impact soil surface cover, which greatly impacts erosion. Another potential impact of climate change is associated with changes from snowfall to rainfall. If decreased days of snowfall translate correspondingly to increases in days of rainfall, erosion by storm runoff is liable to increase. Even changes in soil surface conditions, such as surface roughness, sealing and crusting, may change with shifts in climate, hence impacting erosion rates.

The spatial patterns of mean annual temperature trends shows significant positive trend over most parts of the country except over the parts of Rajasthan, Gujarat and Bihar where significant negative trends were observed. Season-wise temperature trends indicated that maximum increase in temperature was observed in post-monsoon season (0.7°C) followed by winter season (0.67°C), pre-monsoon season (0.5°C) and monsoon season (0.3°C).

Soil moisture is sensitive to rainfall inputs and evapo-transpiration rates, which are sensitive to air temperature. Infiltration rates are impacted by residue cover and soil consolidation. Canopy and ground cover are modelled as dynamic variables, and influence infiltration, runoff and erosion in different and interactive ways. The model (WEPP) is well suited for studying the complex interactions involved in assessing climate change impacts on erosion.

Potential change in rainfall erosivity (R-factor)

Estimation of future climate change provided by Global Circulation Models (GCMs) do not provide the type of detailed storm information needed to directly calculate predicted R-factor must be used to analyze the GCM output relative to erosivity changes (Nearing, 2001).

Renald and Freimund (1994) evaluated erosivity at 155 locations within the continental United States and developed statistical relationships between the R-factor and both total annual precipitation at the location and a modified Fournier coefficient (Fournier, 1960; Arnoldus, 1977), F, calculated from monthly rainfall amounts as:

$$F = \frac{\sum_{i=}^{12} p_i^2}{P} \qquad (1)$$

Where, p_i(mm) = average monthly precipitation and P(mm) = average annual precipitation.

The derived relationships between R-factor and P developed by Renald and Freimund (1994) were:

$$\text{R-factor} = 0.04830P^{1.610} \quad (r^2 = 0.81) \tag{2}$$

$$\text{R-factor} = 587.8\text{-}1.219P+0.004105P^2 \quad (r^2= 0.73) \tag{3}$$

and the relationships between R-factor and F were

$$\text{R-factor} = 0.7397F^{1.847} \quad (r^2 = 0.81) \tag{4}$$

$$\text{R-factor} = 95.77\text{-}6.081F+0.04770F^2 \quad (r^2 = 0.75) \tag{5}$$

Where, the R-factor is in units of MJ mm $ha^{-1}year^{-1}$.

Using these relationships, precipitation output from GCMs can be analyzed for trends in R-factor changes. Equations 2 and 4 provided a better fit on the lower end of the data range and Equations 3 and 5 fit better on the upper end; therefore, Renald and Freimund (1994) recommended using Equation 2 when P was less than 850 mm and Equation 3 when P was more than 850 mm. Likewise, they recommended using Equation 4 when F was less than 55 mm and Equation 5 when F was more than 55 mm.

Using these relationships, precipitation output from GCMs can be analyzed for trends in R-factor changes. Nearing (2001) used output from two coupled Atmospheric Ocean Global Climate Models, one developed by the U.K. Meteorological Office's Hadley Centre (Gordon *et al.*, 2000; Pope *et al.*, 2000; and Wood *et al.*, 1999) and the other from the Canadian Centre for Climate Modelling and Analysis (Boer *et al.*, 2000; Reader and Boer, 1998; McFarlane *et al.*, 1992).

Changes in rainfall erosivity for the two models were computed for two time intervals. In the first case, erosivity value from 2040 to 2059 were compared with those from 2000 to 2019 and in the second case erosivity values from 2080 to 2099 were compared with those from 2000 to 2019. Erosivity changes were computed in two ways: (i) as a function of change in average annual precipitation for the twenty year periods using Eqn. 2 and 3, and (ii) as a function of the Fournier coefficient for the twenty year periods using Eqn. 4 and 5.

Runoff and erosion at selected locations

The complex interactions of climate change impacts on runoff and erosion rates become evident when looking at the modelling results using specific examples of climate change scenarios. Pruski and Nearing (2002) investigated the changes expected in runoff and erosion as a function of the climate changes estimated for the 21st century using output from the Hadley Centre model (Gordon *et al.*, 2000; Pope *et al.* 2000; and Wood *et al.*, 1999) under corn and wheat management systems in the United States. Data spanning eleven decades, from 1990 to 2099, were obtained from Global Circulation Models developed by the U.K. Meteorological Office's Hadley Centre (Gordon *et al.*, 2000; Pope *et al.*, 2000; and Wood *et al.*, 1999). Soils were chosen from the most common at each location and the simulated crops were chisel plow and no-till winter wheat.

The pathways modelled runoff and erosion rates were affected by the climate change input were quite complicated. The basic results from the study of Pruski and Nearing, 2002 are shown in Table 1. In terms of annual values, there are eight (2×2×2) possible combinations (Groups) of trends in changes for precipitation, runoff and erosion. Five of the eight possible combinations (Groups) were observed in the modelling results.

Table 1. Precipitation, runoff and erosion estimated for 1990 and changes (Δ) estimated for the period of 1990-2099 (from Pruski and Nearing, 2002).

Location	Crop	Soil	Prec 1990 (mm)	Runoff 1990 (mm)	Erosion 1990 (mm)	Δ Precip-itation	Δ Runoff	Δ Erosion (%)	Group
Syracuse	Corn	Pawnee	1040.5	248.9	20.90	-99.6 mm	-5.5(b)	42.4(e)	4
NE		Shaps-burg		181.4	24.76	-9.6%	-2.9(b)	36.5(e)	4
		Wymore		224.8	19.92	(e)	-5.7(b)	47.4(e)	4
	Wheat	Pawnee		218.5	1.81		-24.5(e)	-13.1(d)	2
		Shaps-burg		144.7	4.12		-16.8(d)	-13.9(d)	2
		Wymore		197.5	2.18		-24.3(e)	-9.7(c)	2
Temple TX	Corn	Houston	847.5	332.1	31.84	-60.3 mm	-2.9(a)	5.5(a)	4
		Branyon		310.4	30.69	-7.1%	-3.1(a)	6.2(a)	4
		Tarrent		189.8	29.54	(b)	-7.4(b)	-5.2(a)	2
	Wheat	Houston		250.7	3.04		23.7(c)	65.4(e)	8
		Branyon		220.0	1.93		23.0(c)	153.2(e)	8
		Tarrent		147.9	3.91		17.1(c)	41.4(d)	8
W Lafayette	Corn	Drum-mer	1343.5	265.8	15.32	34.8 mm	-2.7(a)	77.6(e)	5
IN		Crosby		200.9	19.69	2.6%	6.1(b)	67.6(e)	1
		Starks		243.1	16.72	(b)	1.7(a)	71.1(e)	1
		Toronto		251.3	16.68		-0.6(a)	73.9(e)	5
	Wheat	Drum-mer		231.2	2.46		-16.3(d)	16.5(d)	5

Location	Crop	Soil	Prec 1990 (mm)	Runoff 1990 (mm)	Erosion 1990 (mm)	Δ Precipitation	Δ Runoff	Δ Erosion (%)	Group
		Crosby		160.0	2.59		-7.1(b)	15.2(d)	5
		Starks		233.9	2.80		-13.3(d)	16.2(d)	5
		Toronto		229.5	2.80		-15.0(d)	15.4(d)	5

Group	Precipitation	Runoff	Erosion	Significance levels
1	↑	↑	↑	(a) 0-24.99%
2	↓	↓	↓	(b) 25-49.9%
3	↑	↑	↓	(c) 50-74.9%
4	↓	↓	↑	(d) 75-89.9%
5	↑	↓	↑	(e) 90-100%
6	↓	↑	↓	
7	↑	↓	↓	
8	↓	↑	↑	

Change in climate and CO_2 can alter the rates of soil erosion by wind or water (Table 2). Soil erosion can be directly affected by change in quantity and quality of erosive forces, for instance by the amount and intensity of precipitation (erosion by water), and wind speed and direction (erosion by wind) on an event basis. Change in climate and CO_2 can indirectly affect erosion through effects on the degree and timing of crop cover, and the production and decay of residue. Change in soil water content, affected by changes in the ratio of precipitation to evapotranspiration, can also influence erosion. Generally water erosion increases (Wischmeier and Smith, 1978) and soil erosion decreases as soil become wetter. Water erosion and wind erosion both tend to be dominated by extreme events which might occur only rarely.

Table 2. Potential interactive effects of changes in climate and CO_2 on Net Primary Production (NPP), soil erosion and soil organic carbon

Potential Effects on NPP

Direct effects:

- Temperature
- CO_2
- Insolation

Indirect effects:

- Soil water content

Soil nutrient availability

Soil erosion

Potential effects of soil erosion

Direct effects:

Precipitation amount, Intensity, Wind speed

Indirect effects:

Plant cover

Residue on surface

Soil water content

Potential effects on soil organic carbon

Indirect effects:

Residue production

Residue, SOC decay

Soil temperature

Soil water content

Residue, SOC quality

Soil erosion

Sediment transport and sediment yield

Criterion for an erosion-induced terrestrial carbon sink

Soil erosion results in a C sink when C uptake rates outpace C release rates across landscapes in a given situation. Whether the rates of C uptake are higher than the rates of C loss is determined by several factors: (1) the rate of C input from plant production at the eroded site; (2) the inherent recalcitrance of SOC; (3) the fraction of plant C that enters the SOC pool rather than being mineralized to CO_2 (the humification coefficient, or *he*); (4) how much, and to what extent, eroded C is stored in more protected forms at the sites of deposition; and (5) the energy (C) cost of irrigation- and fertilization-supplemented production. In other words, what determines whether soil erosion and deposition constitute a C sink is the disequilibrium created by the replacement and stabilization of C in eroding and depositional parts of the watershed, respectively.

Organic carbon pool in the soil moderates physical, chemical and biological soil processes. The balance between various pools of carbon is commonly estimated by considering the differences between vertical inputs and outputs of carbon while ignoring lateral components. The vertical inputs include net primary production (NPP), application of organic manures and amendments, and the vertical outputs include carbon emission and leaching. The lateral components of carbon loss include carbon removal by water, wind and tillage erosion. Although the largest fluxes of carbon occur vertically, however in some soils lateral flux of carbon through soil erosion can also be a significant component.

Soil erosion alters the carbon fluxes because it removes and redistributes the C enriched sediment and accelerates the process of mineralization. In fact, each process of soil erosion including detachment, transportation and deposition affects carbon dynamics. The amount of carbon erosion depends on the magnitude of sediment removal, soil properties, surface cover condition and soil organic carbon content are some of the factors which affect the magnitude of carbon loss. Because of the deposition of carbon enriched sediments in the lower landscape position and continued removal of soil from upper positions, soil carbon concentration decreases in upper landscape position.

The sediment transported in the depositional area contains fine organic and clay particles due to preferential removal of such particles by surface runoff as they are low in density. In addition, the organic carbon is mostly concentrated in upper few centimeters of soil surface, which is the region of active perturbation by erosional processes. The process of soil erosion in altering C-dynamics entails three distinct stages: (i) detachment, (ii) transportation/ redistribution and (iii) deposition. The first process by which erosion initiates the losses of C-pools is the breaking of soil aggregates. The erosive forces of raindrops and sheering forces of runoff lead to breakdown of structural aggregates and exposure of the hitherto encapsulated SOC within the aggregate to microbial processes which in turn increase in mineralization of CO_2. The removal of carbon by erosional process is redistributed along the landscape (Fig. 1). Some of it is deposited in depressions and transported offsites to rivers, lakes and eventually to the oceans. During the depositional stage, the rate of aerobic decomposition may be low and reduces the emission of CO_2. However, prevalence of anaerobic conditions accentuates methanogenesis and denitrification leading to efflux of CH_4 and N_2O (Jacinthe *et al.*, 2001) with the relative global warming potential respectively, of 21 and 310 times that of CO_2.

Magnitude of the erosion-induced sink

For a given watershed, assuming that NPP is maintained at eroding sites, the amount of C that is stabilized, *Cs* (the sink term in units of kg C per m² per year), relative to the noneroding or steady-state system is calculated as:

$$C_s = S_e \bullet f_{c,s} \bullet f_{e,d}\,(1 - f_{d,o}),$$

where S_e is the amount of eroded soil (kg soil per m² per year), $f_{c,s}$ is the fraction of C in the eroded sediments (kg C per kg soil), $f_{e,d}$ is the fraction of eroded C deposited in local depositional basins, and $f_{d,o}$ is the fraction of deposited C that is oxidized or decomposed after it reaches the depositional basin. Above equation is formulated as such because the amount of eroded C (*Ce*, in kg C per year) is a product of S_e and $f_{c,s}$, while $C_d = C_e \bullet f_{e,d}$ and $Cs = Cd \bullet fd,o$. Here, *Cd* is the net amount of SOC deposited (kg C per year), not transported into oceans or oxidized during the detachment and transport of eroded soil particles, while *Cs* is the amount of C that is stabilized, indicating long-term C storage. It is estimated that erosion redistributes about 75 Pg soil annually worldwide (Pimentel *et al.* 1995).Typically, *fc,s* ranges between 1.5% and 5% (Lal 2001). Accordingly, 1.1 to 3.7 Pg SOC is redistributed by erosion annually. Moreover, it is generally accepted that 70% to 90% of eroded sediments are deposited within the same or an adjacent watershed (Stallard 1998). For lack of a better estimate and for simplicity, here we use an average value of 0.8 for *fe,d*. Thus, the SOC deposited within the same watershed or adjacent watersheds (*Cd*) can be 0.8 to 3.4 Pg. At the depositional site, the fraction of deposited SOC lost to decomposition and oxidation (*fd,o*) is estimated to be between 20% and 50% (Jacinthe *et al.* 2001). Taking the lowest value for *fd,o* of 0.2, it is possible that between 0.4 and 2.8 Pg C can be stabilized in a variety of depositional basins.

Lal (2003) reported that about 60-70% of eroded carbon gets redistributed over the landscape and about 10% is transported to the ocean. He further reported that about 20-30% of carbon transported by erosion is emitted to the atmosphere. The fate of carbon transported by erosion is shown in Fig. 3.

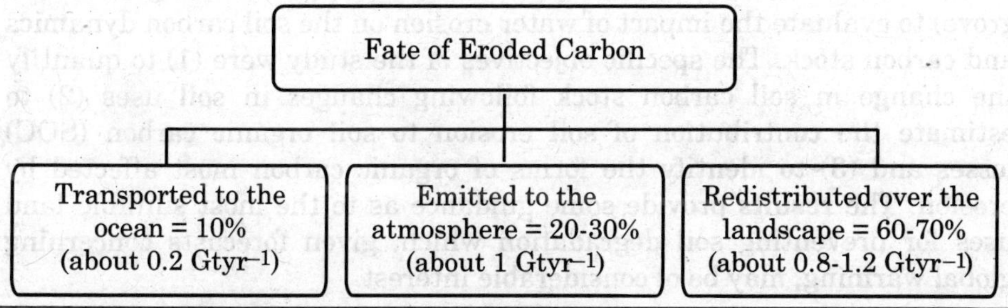

Fig. 3. Distribution of C transported by erosion

Modelling

Modelling is a useful tool for estimating soil carbon dynamics under various topographic and management condition. The most common C models are EPIC, CENTURY, APEX, ECOSYS, Roth-C, WEPP and CQUESTER. The ability of models to simulate C dynamics is a function of input parameters which reflect soil management. Pedotransfer functions (PDFs) and principal component analysis (PCA) are useful tools for studying site-specific relationships between carbon concentration and other soil properties. Data on soil properties from long term experiment are important input parameters for process based models and PDFs to predict various C-pools under different soil use and management.

Among the different pathways by which carbon is lost from agricultural fields is in sediments. Erosion reduces the amount of soil C by (1) promoting soil degradation and so reducing productivity and (2) removing C from one site and depositing it elsewhere (Gregorich *et al.*, 1998). According to Jacinthe *et al.* (2001) the organic carbon content in sediments and eroded material is more than twice that of the original soils, the same authors estimating that 5.7 Pg SOC is removed annually by water erosion in terrestrial ecosystems. While some authors (Smith *et al.*, 2001; Berhe *et al.*, 2007) contend that retention of the eroded C in terrestrial and aquatic systems could lead to carbon sequestration, others (Anderson *et al.*, 1986; Beyer *et al.*, 1993; Zhang *et al.*, 2006) indicate that these materials contain labile C fractions which under aerobic conditions, could undergo mineralization. Little work has been done on the fate of the organic carbon lost through water erosion and its contribution to soil losses after changes in soil use in semiarid Mediterranean areas, where erosion is one of the main causes of soil degradation. Furthermore, the assessment of C losses has frequently formed part of nutrient loss studies in surface runoff plots or part of erosion plots studies (Owens *et al.*, 2002). However, in very few (even in semiarid areas) was the loss of C the central focus of the study and the different carbon pools lost by erosion have hardly been analysed. We present the results of an experiment developed in a semiarid area of South-East Spain under three different land uses (forest, abandoned agricultural field and non-irrigated olive grove) to evaluate the impact of water erosion on the soil carbon dynamics and carbon stock. The specific objectives of the study were (1) to quantify the change in soil carbon stock following changes in soil uses (2) to estimate the contribution of soil erosion to soil organic carbon (SOC) losses and (3) to identify the forms of organic carbon most affected by erosion. The results provide some guidance as to the most suitable land uses for preventing soil degradation which, given forecasts concerning global warming, may be of considerable interest.

General structure of soil organic matter models

Little is known of the properties and fate of SOC that is translocated from erosion to deposition points on the landscape. The physics of particle settling would tend to suggest that the extent of erosional translocation is negatively correlated with the size and density of detached soil aggregates and primary particles. Thus, soil erosion is a highly selective process that preferentially removes the smallest and lowest density components of soil and transport them great distances (Lal, 1995). It is important to develop diagnostic models that will improve our understanding of the underlying mechanisms and processes affecting erosion induced losses of SOC. Available data on erosion and deposition of SOC are very limited, so our approaches to diagnostic modelling are based on the information on the transport of soil and drawing inferences concerning the closely related transport of SOC.

The empirical model fits the observed data from individual events better than the linear parametric model, particularly in the region of the graph showing critical large events. Because the empirical relationship is a power law, it is not as easy to draw broader inferences from the results as with the linear parametric model. For instance, calculation of cumulative C losses from data on cumulative soil losses is not trivial. The general logarithmic linearity seen in these data are, essentially, the same functional forms as the parametric model discussed above with a log-linear relation between enrichment ratio and soil loss remarkably similar to what has been observed by previous researchers (Massey and Jackson, 1952). The proposed empirical relationship is of the form:

$$\text{SOC loss} =. a. (\text{soil loss}).^{b} \qquad (6)$$

Where a and b are statistical constants. This empirical formulation does not require an explicit expression for ER, an advantage for watershed scales where source materials of different SOC contents contribute to a composite mixture in runoff. A study by Collins, *et al.*, (1997) showed the complexity of relating eroded sediment to source material in intermediate sized watersheds (10 km scale). In this study ER ranged from 0.5 to greater than 5 depending on stream bank, forest, pasture, or cultivated soil source material. This empirical model only predicts C losses from these watersheds for events with given soil loss. The two watersheds do, however, exhibit nearly identical equations and it would be an interesting course of future research to add more data to this graph from different areas and different scales in an effort to understand the underlying processes affecting the relationship between soil and SOC loss.

Each of the modelling strategies has associated advantages and disadvantages but these strengths and weaknesses generally complement one another. One disadvantage common to all these models is that they do not assess indirect effects such as reduction in SOC inputs because of crop productivity decline (Gregorich, *et al.*, 1998) and increased carbon dioxide losses of translocated SOC (Lal, 1995). Empirical models are helpful for understanding the relationships between soil and SOC losses in specific watersheds and statistical estimates of accuracy may be established once the models are calibrated. However, there is a need for long-term monitoring to obtain calibration data, the derived relationships only apply to specific watersheds, and this approach requires input data for specific runoff events. Simulation models play an increasingly important role in C pools assessment at different spatial scales as well as in understanding of processes underlying C fluxes in ecosystems (Izaurralde *et al.*, 1998). Models are also essential tools for devising and evaluating management practices intended to balance global C fluxes. The Soil Organic Matter Network (SOMNET) database (Molina and Smith, 1998) identified 33 SOM dynamics models available for use today and the database is being continuously updated. A great variety of models designed for different spatial scales and time steps can be classified into four groups according to the conceptual approach to SOM turnover in soil: (i) process-based (single or multicompartment), (ii) cohort, (iii) food-web chain, and (iv) combined. The majority of models are process-based multi-compartment models. The characteristic features of these models are: (i) subdivision of the SOM into several "homogeneous" pools each with its unique decomposition rate, (ii) assumption that decomposition of SOM follows first-order kinetics, (iii) defined relationship between the dynamics of C and N pools (Paustian, 1994). Different C pools of different

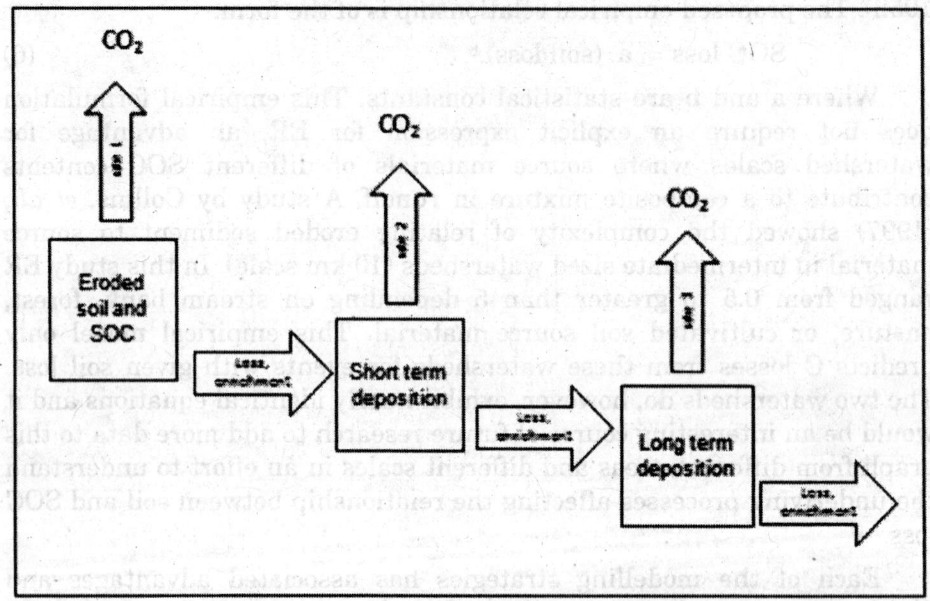

Fig. 4. Soil organic matter fluxes as influenced by soil erosion.

properties combined with flows of C between the pools represent the structure of process-based models (Molina and Smith, 1998). From plants C is successively transferred into microbial biomass, SOM pools of various stability or CO_2. The output from a component of the SOM system may be split or have a back loop to account for the process microbial succession (Molina and Smith, 1998). Modular structure of SOM models allows flexibility and ability to expand the model structure to accommodate new processes and flows as empirical data such as connection between microbial community and soil structure, structural effects on decomposition rates and effect of water, wind and tillage erosion becomes available. Each module occupies specific position in the C flow hierarchy, and is characterized by a decay rate. Most of SOM models do not explicitly specify C flows due to erosion, but allow including these flows if necessary. A general concept of SOM dynamics as affected by soil water erosion is presented in Fig. 4.

Principal governing equation of erosion component in soil organic matter models

A simple model of organic carbon balance at landscape scale can be represented by the following mass balance equation

$$\Delta\ SOC = (SOC_a + A) - (E + L + M)$$

Where SOCa represents antecedent SOC pool, A represents addition of C through inputs, E represents C transported by erosion, L represents C leaching and M represents mineralization of SOC.

After the inclusion of soil erosion as an explicit term in the mass balance of soil carbon, the change in the soil carbon stock is modelled as

$$\frac{dC}{dt} = I - (k_0 - k_e)C$$

Where, C= Soil C stock (kgm^{-2}), I= zero order introduction of soil carbon by plant growth ($kgm^{-2}yr^{-1}$), k_o= first order loss of slow carbon by oxidation/decomposition (yr^{-1}), k_e= first order loss or addition of slow carbon by erosion (negative k_e) or deposition (positive k_e) (yr^{-1}) (Stallard, 1998). Leaching is usually not included in the equation because its magnitude is believed to be small compared to the other fluxes.Three major processes involved in erosion-induced SOM turnover can be defined as: (i) physical removal of SOM from slopes and convex landforms; (ii) deposition of SOM in depressions and on concave landforms; (iii) change in the rate of mineralization of displaced C (Fig. 1). Loss of C by SOM models is usually estimated in relation to soil loss (Gregorich *et al.*, 1998). One of the most common approaches is a linear relationship (Starr *et al.*, 2000):

$$C_{loss} = A \cdot C_{SOIL} \cdot Er \quad (7)$$

where A= spatial average soil loss ($t\ ha^{-1}\ year^{-1}$);

C_{SOIL}= concentration of organic C in soil (%); Er = enrichment ratio of eroded sediment relative to the original soil (dimensionless).

Soil loss is included as a component in several SOM models such as Century or EPIC (Goss *et al.*, 2001). Century model employs the Universal Soil Loss Equation (USLE) (Wischmeier and Smith, 1978), which depends on rainfall to determine erosive energy. The USLE uses six erosion factors to compute the average annual soil loss on field slope as (Eq. 7):

$$A = RKLSCP \tag{8}$$

where R = rainfall erosivity factor, which includes runoff factor from snowmelt; K = soil erodibility factor, which is correlated with soil particle size distribution, OM content, permeability, etc.; L= slope length factor; S = slope steepness factor; C = cover and management factor, which describes the type of soil cultivation and the degree of soil protection by plant canopy; P = support practice factor related to the type of erosion protection practice such as terracing, strip cropping, etc.

Another erosion prediction tool, MUSLE, is used in EPIC model (Williams, 1995). The limitation of the described soil loss prediction equations is their inability to account for soil deposition. The SOM turnover process in depositional areas is a key component of C balance in terrestrial ecosystems (Bajracharya *et al.*, 1998). Although SOM mineralization in soil as affected by soil structure and degree of disturbance has been studied (Vanveen and Kuikman, 1990; Parton *et al.*, 1993), attempts to incorporate the contribution of displaced sediment into CO_2 flux were limited (Harden *et al.*, 1999; West and Wali, 2002).

The principal governing equation of C flux between SOM model components is first-order relationship of the form Eq. (9):

$$dCs/ dt = -kmpC_{SOIL} + h \tag{9}$$

Where, C_{SOIL} = concentration of organic C in soil; t = time; k = first-order decomposition coefficient; m, p = correction factors for soil temperature and moisture; h = additional rate independent of decomposition rate, such as erosion, deposition or net primary production.

One of the most significant changes, which occur with soil during transport, is change of texture of sediment relative to the soil from which it originates due to sorting of material. It has been shown that soil texture affects the mineralization of labile fraction of SOM as well as has an influence on transformation of labile SOM into slow and recalcitrant SOM (Vanveen *et al.*, 1984).

The SOM decomposition process in soil depends on soil depth. Temperature and moisture reduction coefficients (Voroney *et al.*, 1981) are included in SOM transformation equation (Eq. 4) to accommodate for differences in these variables with soil depth. Majority of SOM dynamics models treat soil profile as a combination of layers with different decomposition rates (Molina and Smith, 1998). This subdivision on layers has an effect on how erosion component of the model operates. A scheme was proposed (Schimel *et al.*, 1985b; Bouwman, 1989) in which the upper soil layer, which becomes thinner due to the loss of soil, is compensated from the second layer. An amount of soil equal to the one lost to erosion is transferred from the second (lower) soil layer into the first (upper) As a result, the SOM in the upper layer becomes diluted.

Enrichment ratio of SOM in sediment is usually logarithmically related to sediment loss (Massey and Jackson, 1952; Ghadiri and Rose, 1991b)

$$E_r = bA^d \ldots \qquad (10)$$

Where E_r = enrichment ratio; n, b = coefficients (dimensionless);

A= soil loss (t ha^{-1}).

Coefficients n and b vary widely with type and texture of soil, which complicates the incorporation of the logarithmic equation into erosion component of SOM dynamics models. Many models use constant values for enrichment ratio (Voroney *et al.*, 1981; Bouwman, 1989), because sediment delivery is estimated as an average annual sediment yield (such as by USLE) while for the logarithmic relationship to be employed sediment yield data for a specific event is needed. Logarithmic relationship though is used in models such as CREAMS (Silburn and Loch, 1989) and EPIC (Williams, 1975) which have more powerful hydrological component capable of generating stochastic events. Because rare large storms are responsible for most of the soil loss from watersheds (Edwards and Owens, 1991), logarithmic expression for E_r needs to be incorporated into SOM dynamics models. Enrichment ratio also tends to decrease with the rainfall duration (Palis *et al.*, 1997) in a logarithmic fashion.

The change in C budget over time on eroding site can be generalized as follows (Harden *et al.*, 1999) (Eq. 11):

$$dC_S /dt = C_{NPP} - k_s C_{SL} - k_E C_{EROS} + C_{LHZ} \qquad (11)$$

Where C_{NPP} = net primary production (input) of C; CSL= amount of C in the top layer of soil; C_{EROS} = amount of C eroded from the site; C_{LHZ} = amount of C, which is incorporated into upper layer from the lower layer of soil; ks and k_E = decomposition rates of soil and sediment C. Monreal *et al.* (1997) demonstrated using Century that erosion rate on Chernozemic and Gray Luvisolic soils was linearly correlated with SOM change. Similar trend was also observed in field measurements of SOM and soil losses for two watersheds in Ohio (Starr *et al.*, 2000). Monreal *et al.* (1997) estimated that depending on management practice 12–46 years was required to achieve steady-state level of SOM when erosion was included into simulation, which was 5–50% longer than when erosion component was not included. Simulated SOM losses were in agreement with those obtained directly assuming 1.18 SOM enrichment factor. Gregorich *et al.* (1998) used Century model to simulate the effect of erosion on soil C on a black chernozem under a wheat-fallow rotation and a thin black chernozem in Canada. The soil loss estimated from the model was verified using ^{137}Cs technique (Ritchie and McHenry, 1990). Out of 67% of total C lost from the cultivated field after 71 years of agricultural activity (comparing to undisturbed prairie), 47% was attributed to loss by soil erosion.

Policy implications

Regardless of how well deposition sequesters large amounts of C in some settings, erosion is not a sensible management strategy for C sequestration. The seriousness of the soil erosion problem in different parts of the world is well documented in the literature. Since soil erosion reduces soil quality, its unexpected contribution to C sequestration should not be interpreted as advantageous. Advocating that marginal lands be kept marginal because they are far from C saturation and potentially can remain at their highest sequestration potential makes no ecological sense. Despite massive global efforts to reclaim them, it is still very hard to rehabilitate degraded lands; they stand as testimony to the limits of soil development, in terms of replacing fertility and corresponding C on the timescale of accelerated erosion. Proper management of erosional and depositional landscapes is crucial to C sequestration and to protecting these lands for future use. In the policy arena, under the provisions of the Kyoto Protocol, accountable emission mitigation measures are limited to provable and accepted changes in C storage resulting from afforestation, reforestation, and deforestation (IPCC 2000). As it is currently formulated, this policy makes it difficult to realize the sequestration potential of protecting marginal lands from disturbance. The contribution of managing marginal lands should be considered, because tillage of C-rich alluvial soils, draining of floodplains for agriculture, and dredging of reservoirs all release large amounts of CO_2 that could otherwise be protected in land preserves.

Conclusions

Soil erosion by water is one of the path way of C loss. Water erosion on sloping land contribute significantly to the losses of C. soil aggregates disintegration and dispersion, preferential removal of organic colloids, redistribution of transported C or mineralization of deposited C are some of the mechanization by which soil C is lost through emission and some portion is buried promoting long term C sequestration. The fate of eroded C is very complex as it can be oxidized at the eroding site during transport, depositional site and aquatic systems. Thus, soil erosion is probably a source rather sink of C.

Reclamation of the degraded lands through various conservation measures is a potential alternative for increasing C pools in soil system. Crop rotation, cover crops, crop residue, manure application, agro forestry and high biomass producing bio energy crops are some of the practices that increase soil carbon while conserving soil and water.

The annual transport of particular organic C to the ocean is estimated at 0.09 to .57 Pg C per year. Sedimentologist extrapolate these projection further and surmise that so called burried sink in the aquatic systems is a good thing with regard to C sequestration. However, soil scientist argue that the magnitude of oxidation of SOC during the various stages of soil erosion is significant. The magnitude of oxidation of eroded material may depend on the composition of particulate organic material though some of the material reaggregated and protected against mineralization. The proportion of eroded SOC that is mineralized during the transportation of eroded particles may be as high as 30%.

Erosion vs climate studies suggest that increased rainfall amount and intensity will lead to grater rate of erosion therefore it will definitely increase C loss unless amelioration meseasures are taken. In terms of processes, rainfall amount and intensity are certainly the most direct and important factors controlling erosional changes under climate change scenario. Change in climate and increase CO_2 concentration in the atmosphere can cause changes in soil erosion which in turn affect the long term productivity and sustainability of agricultural system.

❑❑❑

7 Simulating Crop Yield, Soil Processes, Greenhouse Gas Emission and Climate Change Impacts with InfoCrop Model

K.K. Bandyopadhyay, B. Chakrabarti and H. Pathak*

Division of Agricultural Physics, Indian Agricultural Research Institute, Pusa, New Delhi 110 012, India

*CESCA, Indian Agricultural Research Institute, Pusa, New Delhi - 110 017, India

Introduction

Crop growth simulation models provide a means to quantify the effects of climate, soil and management on crop growth, productivity and sustainability of agricultural production system. InfoCrop is a generic model to simulate growth, development, water use, nitrogen balance and crop-pest interactions for major annual crops (Aggarwal *et al.*, 2006a). It is a user-friendly system that is targeted for increasing applications of crop modelling in research and development. Its general structure is based primarily on SUCROS (van Laar *et al.*, 1997) and further supported by MACROS (Penning de Vries *et al.*, 1989), WTGROWS (Aggarwal *et al.*, 1994), and ORYZA1 (Kropff *et al.*, 1996) models. The model is written in Fortran Simulation Translator programming language (FST/FSE; Graduate School of Production Ecology, Wageningen, The Netherlands; van Kraalingan, 1995). Another window based user interface version of the model has been developed to facilitate its greater applications in agricultural research and development by the stakeholders not familiar with programming. The user-interface of this software has been written using Microsoft.Net framework while in the back-end it has FSE models and databases in MS-Access.

Basic structure of the InfoCrop model

In the InfoCrop model, different processes affecting crop growth and development are considered *viz.*, phenology, photosynthesis, partitioning of biomass, leaf area growth, storage organ numbers, source: sink balance, transpiration, uptake, allocation and redistribution of nitrogen etc. InfoCrop also simulates the impact of pest infestation on crop yield. The model is able to simulate impact of water and nitrogen on crop growth and development. Soil carbon module of the model simulates soil organic carbon dynamics by simulating carbon mineralisation and immobilization. The carbon module is linked with the nitrogen module which simulates different processes of N transformation in soil like N uptake, nitrification, denitrification, volatilization and leaching. Emission of GHGs such as carbon dioxide (CO_2), methane (CH_4) and nitrous oxide (N_2O) is simulated by the C and N module. The model can also simulate the impact of future climate change on growth and yield of crops. Based on the simulation results the crop management options like sowing date, irrigation and fertilizer application can be optimized for achieving higher yield.

Thus InfoCrop considers following processes of growth and development, soil water, nitrogen and carbon, and crop-pest interactions:

- Crop growth: photosynthesis, respiration, partitioning, leaf area growth, source: sink balance, phenology, transpiration, N uptake, allocation and redistribution, effects of stresses
- Crop-pest interactions: damage mechanisms of insects, diseases and weeds
- Soil water balance: root water uptake, inter-layer movement, drainage, evaporation, runoff, ponding
- Soil nitrogen balance: mineralisation, uptake, fixation, nitrification, volatilization, inter-layer movement,denitrification, leaching
- Soil organic carbon dynamics: mineralisation and immobilization
- Emissions of Green house gases: CO_2, CH_4, N_2O
- The relational diagram showing different processes in InfoCrop model has been depicted in Fig. 1.

The model requires the following simple and easily measurable inputs.

Weather data: Daily weather data of any location is required for simulating crop growth and development. Weather parameters like maximum and minimum temperature, rainfall, solar radiation and wind speed is required for running the InfoCrop model.

Crop data: Crop name, variety and different varietal coefficients (thermal time of different growth stages, base temperature, specific leaf area) are required by the model.

Soil data: Soil texture, thickness of soil layer, bulk density, saturated hydraulic conductivity, soil organic carbon, soil moisture status, pH, EC, ammonical and nitrate nitrogen content of soil is needed. The InfoCrop considers 3 soil layers and thickness of each layer is to be defined by the user prior to simulation.

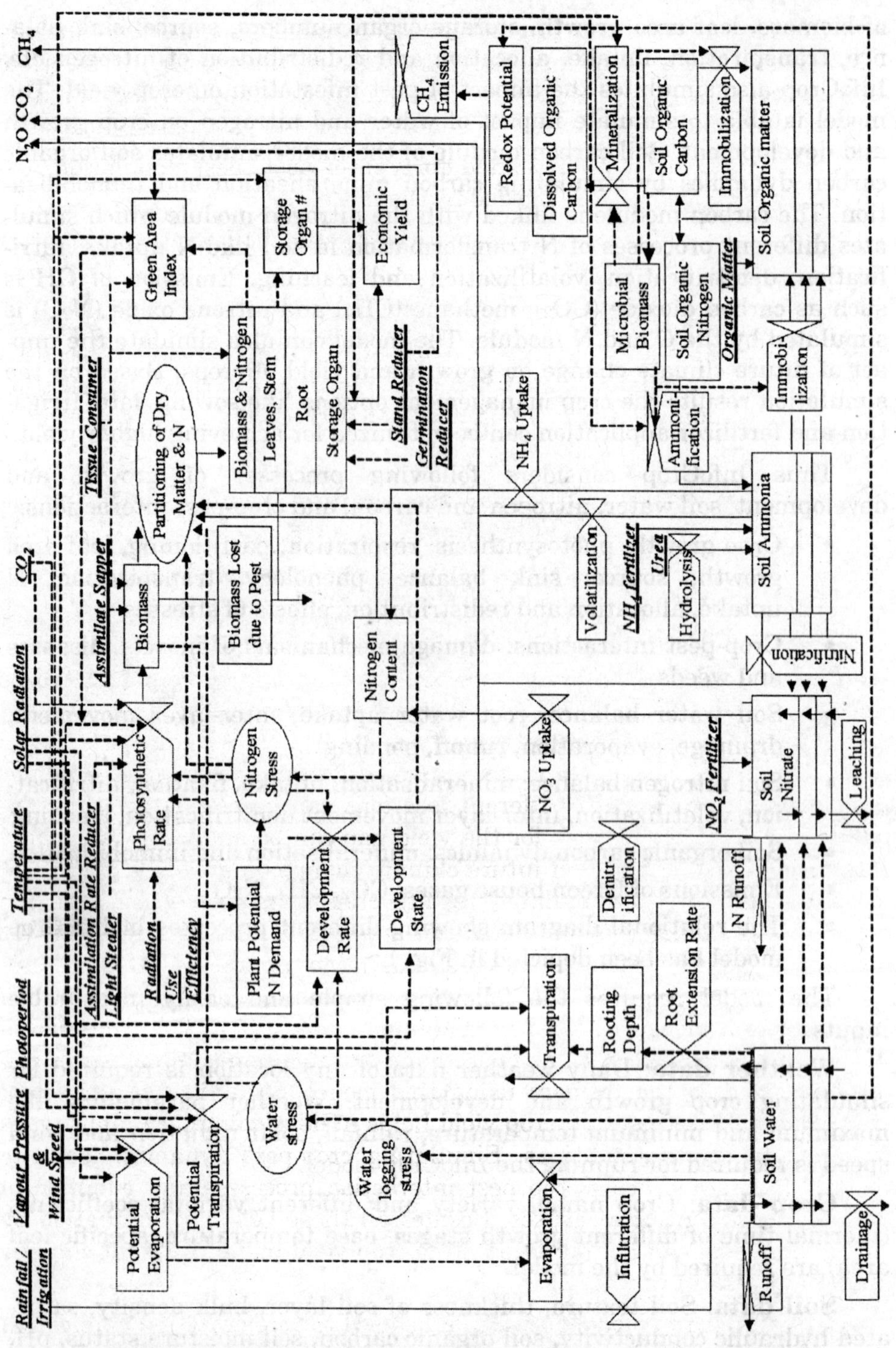

Fig. 1. Relational diagram showing the different processes in the InfoCrop model (Source: Aggarwal *et al.*, 2006a).

Management data: Inputs on crop management practices include date of start of simulation, sowing/transplanting time, seed rate, depth of sowing, seedling age (in case of rice), irrigation date and amount of irrigation applied, date, amount and type of fertilizer and manure application.

Pest data: Pest data related to type of pest, pest population and their time of appearance is needed by the model.

The InfoCrop model is sensitive to the following parameters (Aggarwal *et al.*, 2004):

- Variety: its physiology
- Agronomic management: dates of planting/ transplanting, seed rates, amount and time of irrigation and N fertilization (including organic) in different soil depths
- Soil: pH, depth, texture, fertility
- Environment: radiation, temperature, rainfall, wind speed, vapor pressure, flooding, frost
- Pests: population/severity of insects, diseases and weeds

The InfoCrop model outputs include duration of crop, crop biomass and yield, soil organic carbon status, water use by crop, N uptake, ammonia (NH_3) volatilization, nitrate leaching and emission of GHGs including CH_4, CO_2 and N_2O (Fig. 2). The simulation results help the end users in analyzing and interpreting their experimental data. Potential as well as actual yields of crops under different water and nutrient management practices at different locations can be estimated using the model and possible reasons for the yield gap can be analyzed. The model can also simulate impact of future climate change on growth and yield of crops. Based on the simulated results crop management options like sowing date, irrigation and fertilizer application can be optimized for achieving higher yield. Impact of cultivation practices on long term changes in soil organic carbon can be assessed and suitable carbon sequestration measures can be formulated using this model. Emission of GHGs and N losses can be estimated and suitable remedial measures can be suggested to the farmers using this model. Studying the impact of various abiotic stresses on crop yield helps in developing new plant types tolerant to stress conditions. Simulating crop-pest dynamics leads to better understanding of crop pest interaction processes and adoption of suitable pest control measures. Besides all these benefits, simulation by InfoCrop model can help in forecasting crop yield under different climatic and management conditions.

Thus InfoCrop model output can be used to accelerate the application of available knowledge at field, farm and regional level for the following purpose (Aggarwal *et al.*, 2004):

- Analyzing and interpreting experimental data

- Estimating potential yields
- Estimating yield gaps: principal causes and their contribution
- Assessing impacts of climatic variability and climatic change
- Optimizing management- Dates of planting, variety, irrigation and nitrogen fertilizer
- Quantifying change in soil organic carbon with land use
- Assessing environmental impact- percolation and N losses, greenhouse gases (CO_2, methane and nitrous oxide) emissions
- Designing and evaluating plant types
- Understanding genotype by environment and management by pests interactions
- Forecasting yield and production
- Assessment of yield loss due to pests and its control
- Increasing the efficiency of multi-environment experiments
- Assisting in crop insurance schemes management

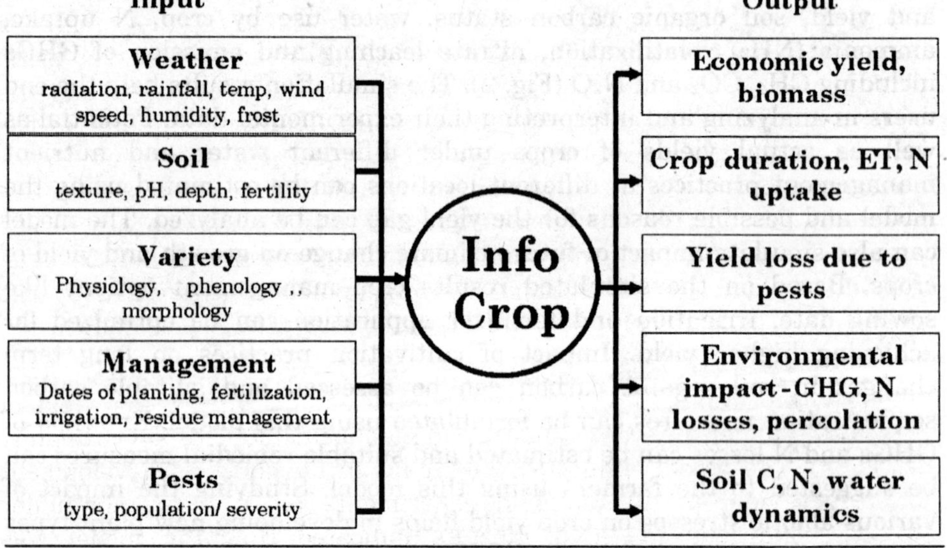

Fig. 2. Major inputs and outputs of the InfoCrop model (Source: Aggarwal *et al.*, 2006a).

Modelling crop yield using InfoCrop

The InfoCrop-rice and InfoCrop-wheat models were developed and validated under contrasting agro-environments of tropics. Results showed that the model was able to simulate satisfactorily dry matter and grain yield of rice and wheat crop in diverse agro-environments (Fig. 3) (Aggarwal *et al.*, 2006 b).

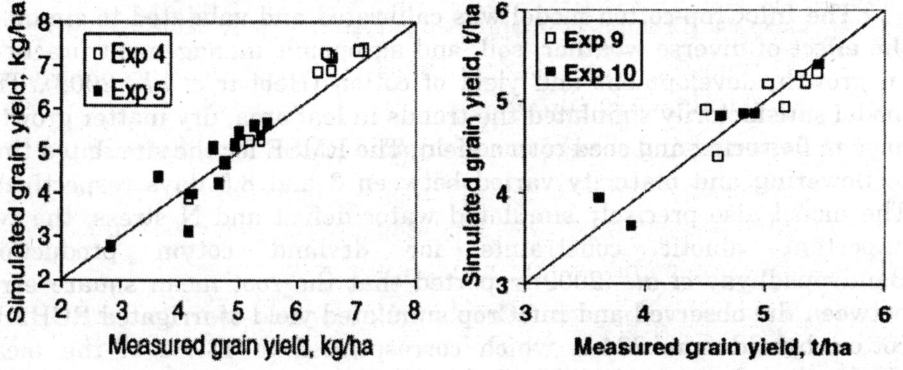

Fig. 3. Measured and simulated grain yield of (1) rice and (2) wheat crop (Source: Aggarwal et. al., 2006b)

Bandyopadhyay *et al.* (2012) reported that InfoCrop model could account for 76% variation in the observed grain yield of wheat under different irrigation and nitrogen management practices in a sandy loam soil. The root mean square error between the observed and predicted grain yield of wheat was to the tune of 19.2% of the mean observed grain yield of wheat. The index of agreement between the observed and the predicted grain yield of wheat (D) was 0.99. The predictability of the model was better for irrigated treatment than rainfed treatment. It was also observed that the global warming potential during wheat growth (i.e. the emission of green house gases expressed in CO_2 equivalent) increased with the increase in the water application. There was a trade-off between the wheat production and global warming potential estimated by the model. This calls for efficient use of water and nitrogen to reduce the global warming potential.

The InfoCrop-potato model was developed for simulating yield of potato crop under different climatic and management conditions (Singh *et al.*, 2005). The model simulated the trends in phenological development and tuber yield in diverse field experiments fairly accurately under tropical conditions in India. Results indicated that the model was adequate to simulate the effects of agro-climate and various management of inputs on potato yield resulting in increased efficiency of agronomic experiments.

The InfoCrop-SORGHUM model (Srivastava *et al.*, 2010) was validated for monsoon and winter sorghum crop in different locations viz., Hyderabad, Indore, Coimbatore, Nagpur, Parbhani, Udaipur and Dharwad, which represent major sorghum growing areas in India. The validation was carried out for eight varieties of sorghum crop. The model satisfactorily simulated the phenology and grain yield of sorghum cultivars in different locations.

The InfoCrop-cotton model was calibrated and validated to simulate the effect of diverse weather, soil, and agronomic management practices on growth, development and yield of cotton (Hebbar *et al.*, 2008). The model satisfactorily simulated the trends in leaf area, dry matter growth, days to flowering and seed cotton yield. The RMSE for the simulated time to flowering and maturity varied between 3 and 8.5 days respectively. The model also precisely simulated water deficit and N stress, the two important abiotic constraints for dryland cotton production. Bandyopadhyay *et al.* (2009) reported that the root mean square error between the observed and InfoCrop simulated yield of irrigated RCH2 Bt cotton hybrid was 284.1, which corresponded to 13.7% of the mean observed seed cotton yield. The index of agreement (D index) between the observed and simulated seed cotton yield was 0.67. There was significant reduction in water use efficiency of cotton but significant increase in global warming potential and carbon efficiency (g CO_2 eq. GHGs emission kg^{-1} seed cotton yield) with the increase in the level of irrigation.

The InfoCrop-coconut model was developed for simulating yield of perennial crop like coconut (Kumar *et al.*, 2008). The InfoCrop-coconut model was calibrated and validated with data compiled from published studies comprising many physiological, agronomical and nutritional experiments in diverse geographic locations throughout India. Simulated trends in phenological development, total dry mass and its partitioning, and nut yield agreed closely with observed values. For a range of agro-climatic zones, simulated potential yields varied from 26 to 30 Mg ha^{-1} $year^{-1}$ and potential annual dry mass production varied from 52 to 62 Mg ha^{-1}, depending on environment.

Modelling long-term soil organic carbon dynamics

The InfoCrop model was used to assess long term dynamics of soil organic carbon (Aggarwal *et al.*, 2006b). Published data of a long-term rice–wheat cropping system (Bhandari *et al.*, 2002) conducted at Ludhiana in the Indo-Gangetic plains for 13 years (1985–1997) was used for this purpose. The publication reported occasional measurements of soil organic carbon for the surface soil layer in few years.

However, the InfoCrop is yet to have the capability to automatically simulate cropping systems in a multi-year time horizon. Therefore, at the beginning of each crop season, crop and soil parameters were initialized. The soil parameters at the harvest of rice were physically noted and used as the initial conditions for the following wheat crop. Similarly the soil parameters noted at the end of wheat season were used for the following rice crop. This sequence was followed every year. The control treatment showed that the soil organic carbon was maintained at around 0.4% in all years.

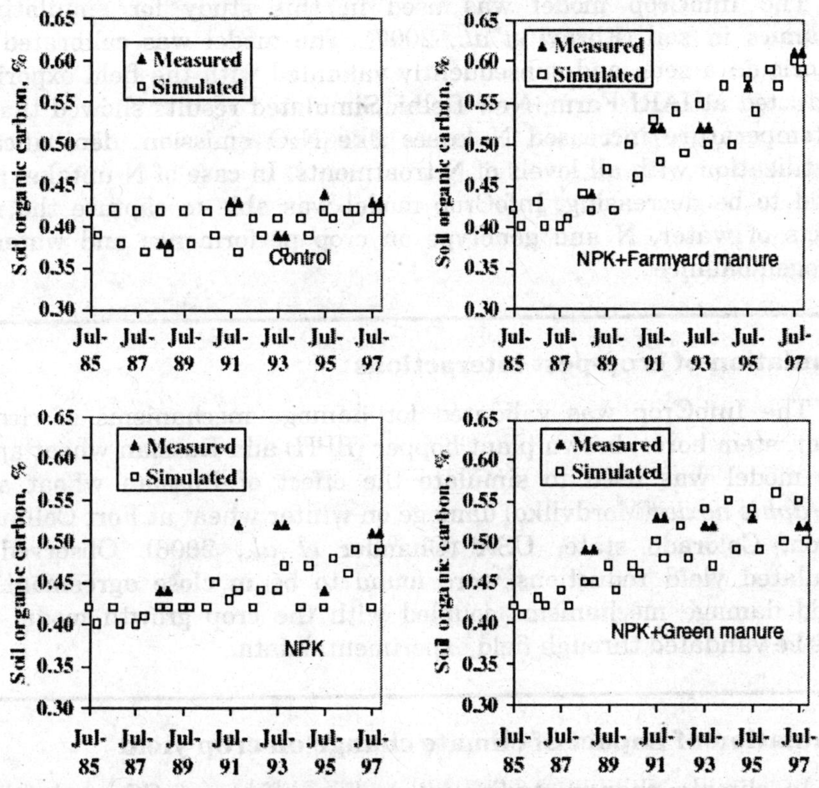

Fig. 4. Observed and simulated changes in soil organic carbon with time in a rice–wheat system at Ludhiana in north-western India for treatments contrasting in nutrient management. NPK refers to an application of 100% recommended dose of the ferilizers (Source: Aggarwal *et al.*, 2006b).

This gradually increased to 0.5% when the recommended dose of NPK was added and the build-up was further enhanced over time by the addition of organic matter. At the end of the 13 years of the experiment, measured soil organic carbon was 0.51% and the simulated value was only marginally lower (0.49%) in 100% NPK treatment (Fig 4).When green manure or farmyard manure was added, the model satisfactorily simulated the trend of change in soil organic carbon. Considering the limited availability of soil organic carbon data for calibration of the model, it can be concluded that despite these discrepancies, the model simulated the trend of change in organic carbon in all four treatments reasonably well.

Modelling soil N dynamics

The InfoCrop model was used in this study for simulating N dynamics in soil (Ebrayi *et al.,* 2007). The model was calibrated with historic data sets, and subsequently validated with the field experiment conducted at IARI Farm, New Delhi. Simulated results showed that rise in temperature increased N losses like N_2O emission, denitrification, volatilization with all levels of N treatments. In case of N uptake, it was found to be decreasing. InfoCrop model was able to capture the major effects of water, N and genotype on crop performance and water and nitrogen balance.

Simulation of crop-pest interactions

The InfoCrop was validated for damage mechanisms of rice leaf folder, stem borer, brown plant hopper (BPH) and Russian wheat aphids. The model was used to simulate the effect of Russian wheat aphid, *Diuraphis noxia* (Mordvilko) damage on winter wheat at Fort Collins and Akron, Colorado state, USA (Chander *et al.*, 2006). Observed and simulated yield reductions were found to be in close agreement. The aphid damage mechanisms coupled with the crop growth model could thus be validated through field experimental data.

Simulation of impact of climate change on crop yield

In climate change studies interaction effects of CO_2, rainfall and temperature can be best studied using crop growth simulation models. Many scientists have used various crop growth models to quantify the yield loss due to climatic variation. The output provided by the simulation models can be used to make appropriate crop management decisions and to provide farmers and others with alternative options for their farming system.

The InfoCrop model was extensively used for studying impact of climate change on crops. Sensitivity of crop yield to temperature increase was studied by Aggarwal *et al.* (2006 b) using InfoCrop model. The study revealed that 1°C increase in temperature resulted in only a small decrease in grain yield of rice and wheat, while subsequent increase in temperature caused considerable reduction in yield. This decrease was related with a reduction in leaf area index and the grain number with increasing temperature. In case of wheat this response was, as expected, in a normal planted crop with optimal temperature throughout the growth period. Increase in temperature beyond 1°C resulted in a gradual decrease in yield of wheat reaching 30% at +5°C. This was due to reduction in crop duration, especially during the grain-filling phase.

Increase in thermal time for vegetative phase resulted in increased crop duration in rice as well as in wheat. Any change in crop duration from the optimal period resulted in a decrease in grain yield in both the crops.

According to Aggarwal and Rani (2009), climate change is likely to cause significant reduction in wheat production in India even in near future. Simulated impact of climate change on wheat crop done with InfoCrop model showed that temperature rise by 1°C, 3°C and 5°C with no associated CO_2 increase will lead to a decrease of 6, 19 and 27.5 million tons of wheat. Temperature rise if accompanied by increase in CO_2 level to 450 ppm will reduce yield loss by 4 to 5 million tons.

Mustard crop was also found to be sensitive to changes in carbon dioxide and temperature (Boomiraj *et al.*, 2010). Future climate change scenario analysis using the InfoCrop-mustard model showed that mustard yields are likely to reduce in both irrigated and rainfed conditions in India. Impact of climate change on mustard in terms of its yield reduction would be maximum in eastern India, where projected rise in temperature for 2080 scenario is highest, followed by central and northern India. Lower temperature during the winter season in northern India, caused less yield reduction of irrigated mustard, while projected reduction in rainfall would make rainfed crop more susceptible to climate change.

Climate change impact study on maize crop was carried out by Byjesh *et al.* (2010) using the InfoCrop-MAIZE model under Indian subcontinent conditions. Results indicated a reduction in productivity of maize in mid Indo-Gangetic plains and southern plateau regions of India during both monsoon and winter seasons. Maize yields in monsoon season are projected to be adversely affected due to rise in atmospheric temperature; but increased rainfall will offset those losses in upper Indo-Gangetic plains. Insignificant change in yield of monsoon maize and increased yield in winter maize due to climate change in upper Indo-Gangetic plains showed the possibility of expansion of maize area in northern parts of the country.

Modelling greenhouse gas emission

The InfoCrop model was also used to estimate GHGs emission from rice and wheat field (Aggarwal *et al.*, 2006 b). Field experiment was conducted with rice (variety Pusa 44) and wheat (variety HD 2329). Emissions of methane (CH_4), carbondioxide (CO_2) and nitrous oxide (N_2O) were measured frequently from the plots following the standard methodologies (Pathak *et al.*, 2003). The temporal variation in measured methane emission in 2000 and 2003 crop seasons varied from almost nil to 1.5 kg ha^{-1} day^{-1}. This was due to changes in soil carbon and water management. The simulation model was in general able to explain these variations throughout the crop season satisfactorily. The model showed no significant methane emission immediately after transplanting of rice for the first few days. This could have been due to inadequate characterization of initial soil organic carbon in the field and the possibility of unsatisfactory simulation of surface ponding of water by the model.

In rice the measured emission of N_2O varied from 1.0 to 18 g ha^{-1} day^{-1} in two years of sampling (Fig. 5). There were three/four peaks in the measurements coinciding with fertilizer application. InfoCrop also simulated these peaks at the right time and in right quantities. The simulated peak of emission was almost double of the measured peak.

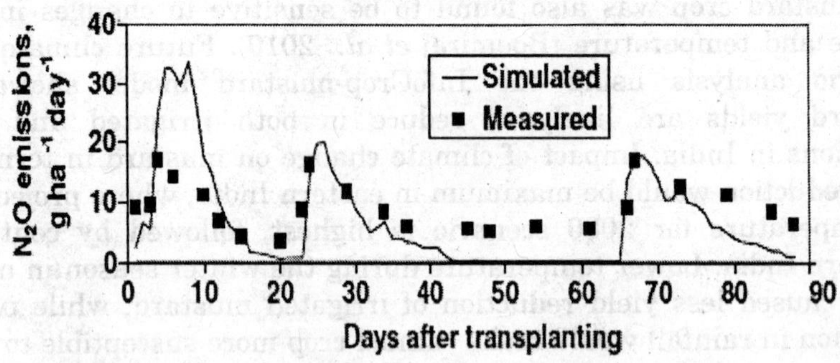

Fig. 5. Measured and simulated emissions of nitrous oxide from rice crop (Source: Aggarwal *et al.*, 2006b).

The measured emission of CO_2 varied from 12 to 17 kg ha^{-1} day^{-1}. The model also showed similar variation in CO_2 emission at all times except towards maturity where there were small dips in emission. In wheat crop, the measured CO_2 emission varied from 10 to 22 kg ha^{-1} day^{-1} whereas the average emission was 15 kg ha^{-1} day^{-1}. InfoCrop was in general able to simulate much of this variation across times. In wheat, N_2O emission showed four peaks in the measurements, which varied from 2 to 13 g ha^{-1} day^{-1}. InfoCrop was able to simulate these patterns reasonably well except during the first 15 days (Fig. 6).

Spatial inventory of greenhouse gas emissions from rice and wheat soils using InfoCrop model

The InfoCrop simulation model was evaluated to calculate methane (CH_4), nitrous oxide (N_2O), and carbon dioxide (CO_2) emissions from soils under rice and wheat (Bhatia *et al.*, 2012). Indian rice fields covering 42.21 million ha (Mha) emitted 2.07, 0.02, and 72.9 Tg of CH_4-C, N_2O-N and CO_2-C, respectively, with a global warming potential (GWP) of 88.5 Tg CO_2-C eq. Annual GHG emission from 28.08 Mha of wheat-growing areas was 0.017and 43.2 Tg of N_2O-N and CO_2-C, respectively, with a GWP of 44.6 Tg CO_2-C eq. Intermittent irrigation in rice reduced methane emissions by 40%. However, application of farmyard manure in rice increased the GWP of 18.54 Mha irrigated rice by 41%.

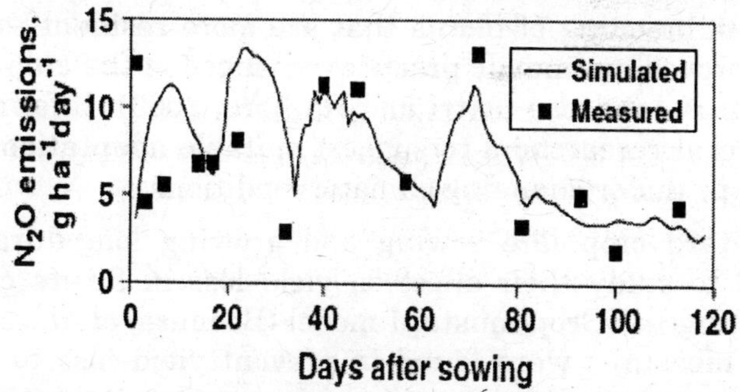

Fig. 6. Measured and simulated emissions of nitrous oxide from wheat crop (Source: Aggarwal *et al.*, 2006b).

Emission of CH_4 ranged from <1 Gg to 80 Gg CH_4 per district. High emission values, observed in some of the districts of West Bengal and Orissa in eastern India, were due to higher soil organic C content, continuous submergence of the fields, and larger area under rice. Emission of N_2O-N from various districts of India ranged from <1 Mg to 500 Mg N_2O-N. Emission of N_2O-N was higher from the south-eastern (Andhra Pradesh) and northern (Punjab) states of the country. Few districts of Andhra Pradesh emitted more than 1.5 kg ha^{-1} because of higher doses of N input. The high emissions of N_2O-N were due to larger area under rice, and more use of N fertilizers. Emission of CO_2-C varied between <1 and >2000 Gg CO_2-C per district. Eastern and southeastern India showed higher CO_2-C emission, because of high SOC content in these regions. The Emission of N_2O-N from the wheat growing areas of India ranged from <0.2 kg N_2O-N ha^{-1} to >1.0 kg N_2O-N ha^{-1}. Some district of South eastern India emitted high N_2O because of high doses of N application in irrigated wheat. The emission of CO_2-C in wheat varied between <1000 to >1800 kg CO_2-C ha^{-1}. The GWP of the wheat growing regions of the country was <4000 to >6000 kg CO_2 equivalent ha^{-1} per year. District wise the GWP ranged from <0.001 to > 4 Tg CO_2 equivalent per district per year. The highest GWP was observed in the districts lying in the Indo Gangetic plains (IGP) of India as this is the major wheat growing belt of the country. The preparation of this spatial inventory will be helpful in removing the uncertainties and identifying the hotspots of GHG emissions and implementing management practices for mitigating the emissions.

Adaptation and mitigation strategies

Adaptation strategy is a set of actions that lead to a specific response to reduce the vulnerability of change (Boomiraj *et al.*, 2010). The identification of suitable response strategies is a key to sustainable agriculture. The important mitigation and adaptation strategies required to cope with anticipated climate change impacts include adjustment in sowing dates, breeding of plants that are more resilient to variability of climate, improved agronomic practices, changes in the cropping mixtures, and agriculture land use (Attri and Rathore, 2003). InfoCrop model was used by several researchers to suggest suitable adaptation strategies for different crops under changing climatic conditions.

In mustard crop late sowing and growing long-duration varieties were tested to study their effect on yield loss in future climate change scenarios using InfoCrop-mustard model (Boomiraj *et al.*, 2010). Both the adaptation measures were found to prevent yield loss to certain extent across the country. But these adaptation practices would be more beneficial for central and northern India. Benefit of late sowing was attributed to lower temperature during anthesis of mustard crop which prevented yield loss. Shortening of crop growth period due to temperature rise was less in long duration varieties as compared to existing ones which helped in preventing yield loss of mustard crop. Simulation study using InfoCrop-maize model showed that in maize crop, development of new cultivars with growth pattern in changed climate scenarios similar to that of current varieties in present conditions and sowing them on time could be an advantageous adaptation strategy for minimizing the vulnerability of maize production to climate change in India (Byjesh *et al.*, 2010). InfoCrop-rice and InfoCrop-wheat model were used to simulate the impact of future climate change on productivity of rice and wheat crops and suggest suitable adaptation measures. Simulated results showed that with improved management of N and water, productivity of rice and wheat crop can be improved in 2050 climate change scenario (Bhatia *et al.*, 2010).

Conclusions

The InfoCrop, an indigenously developed generic crop growth simulation model, can be successfully used to simulate the yield of a diverse group of crops viz., rice, wheat, maize, sorghum, mustard, soybean, potato, cotton and coconut etc., simulate soil organic carbon and nitrogen dynamics, soil water balance, crop-pest interactions and GHGs emission. Using remote sensing and GIS technologies, the yield predictions of these crops can also be upscaled to regional level. This model can also be used to simulate the impact of future climate change on crop yield and GHG emissions. This information can be used to develop adaptation and mitigation strategies for the future climate change.

❑❑❑

8 CropSyst Model and its Application in Natural Resource Management

Sangeeta Lenka and N.K. Lenka

Division of Soil Physics, Indian Institute of Soil Science,
Nabibagh, Berasia Road, Bhopal - 462 038, India

Introduction

Use of crop growth simulation models has been a potential and convenient research tool for quantitatively understanding the effect of climatic, edaphic and agronomic management factors and their interaction on crop growth and productivity. Models are means to capture, condense and organize knowledge. Models have been developed, validated and evaluated for several crops under different agro-climatic conditions to quantify the effect of nutrient and water interaction on crop growth and yield and also for predicting nutrient and water balance in soil system. These models attempt to integrate numerous factors that affect crop growth and development such as plant available soil water, nitrogen, temperature, wind, genetics, management choices, and pest infestations. The strength of these models as research tools resides in their ability to capture the soil–environment–plant interactions, but their initialization generally requires a number of physiological and pedological parameters that are not easily acquired. Both simple and more complex approaches have important limitations especially when yield assessment must be carried out on regional scale. Here we discuss the potential application and uses of CropSyst model in India and across globe.

A number of opportunities are now available for the use of crop simulation models for quantifying the effect of various factors including weather in agriculture. Accurate and real-time estimation of crop yield on regional, national and international scales is becoming increasingly important in both developing and developed countries. In particular, crop yield estimation may play a fundamental role in supporting policy formulation (e.g. new EU Common Agriculture Policy) and decision-making in agriculture (e.g. management of food shortage). Among the possible approaches that may be adopted for yield estimation at large spatial scales, the integration of crop simulation model and satellite data seems to be one of the most appropriate quantitative analysis methodologies. CropSyst has been successfully used to assess the potential impact of climate change and elevated CO_2 on regional, national and global scale. The best adaptation strategies would be easily identified and validated by use of calibrated CropSyst model.

An overview of model

CropSyst, as any other model attempting to predict crop responses to the environment, is not a universal model. It requires some field data for calibration so as to represent a particular crop or cultivar. CropSyst is a multi-year crop, daily time step cropping system simulation model developed to serve an analytical tool to study the effect of climate, soils, and management on cropping systems productivity and the environment. The model simulates the soil water budget, soil–plant nitrogen budget, crop canopy and root growth, dry matter production, grain yield, residue accumulation and decomposition and erosion. Management options include cultivar selection, crop rotation, irrigation, nitrogen fertilization, tillage operations and residue management. The development of CropSyst started in the early 1990s, evolving to a suite of programmes.

CropSyst and associated programmes can be downloaded free of charge over the Internet. One key feature of CropSyst is the implementation of a generic crop rotation via single set of parameters. Simulations can last a fraction of a year to hundreds of years. CropSyst is a suite of programs designed to work co-operatively, providing users with a set of tools to analyze the productivity and the environmental impact of crop rotations and cropping systems management at various temporal and spatial scales. The main components of the CropSyst suite are: CropSyst parameter editor, a cropping systems simulator (CropSyst model), a weather generator (ClimGen), a GIS-CropSyst simulation co-operator (ArcCS), a watershed analysis tool (CropSyst Watershed), and several utility programmes. A description of the main processes in CropSyst is given in the following sections.

Water budget

The water budget in the model includes precipitation, irrigation, runoff, interception, water, infiltration, water redistribution in the soil profile, deep percolation, crop transpiration, and evaporation. Water redistribution in the soil can be simulated by a simple cascading approach or a numerical solution of the Richard's soil flow equation (Campbell, 1985; Ross and Bristow, 1990). CropSyst offers two options to calculate reference crop ET (ET_0): the Penman /Monteith model (Monteith, 1965) and the Priestley/Taylor model (Priestley and Taylor, 1972). The implementation of the Penman/Monteith model follows the methodology suggested by FAO (Allen *et al.*, 1998). This option requires daily maximum and minimum temperature, solar radiation, maximum and minimum relative humidity (or dew-point temperature), and wind speed. The Priestley/Taylor model requires only temperature and radiation data, but the user must provide an appropriate value of the Priestley/Taylor constant.

Nitrogen budget

The mineral N budget in CropSyst includes separate budgets for nitrate and ammonium. Processes include N transformations, ammonium sorption, symbiotic N fixation, crop N demand and crop N uptake. Nitrogen transformations (net mineralization, nitrification, and denitrification) and ammonium sorption follow the approach presented by Stockle and Campbell (1989) while symbiotic N fixation is based on Bouniols *et al.* (1991). Crop N uptake was modelled by adapting the approach presented by Godwin and Jones (1991), where N uptake is determined as the minimum of crop nitrogen demand and potential nitrogen uptake. Crop nitrogen demand is the amount of nitrogen the crop needs to meet growth requirements plus its deficiency demand. The deficiency demand is the difference between the crop maximum and actual nitrogen concentration.

Crop phenology

The simulation of crop development is based on thermal time, which is the required daily accumulation of average air temperature above a base temperature and below a cutoff temperature to reach given growth stages. The accumulation of thermal time may be accelerated by water stress. This can be conceptualized as a response to increased crop temperature. Relations between air and crop temperatures for stressed and unstressed crops, expressed as a function of the vapor pressure deficit of the atmosphere, can be found in the infrared thermometry literature (e.g., Jackson, 1982).

Biomass accumulation

The core of these calculations is the determination of unstressed (potential) biomass growth based on crop potential transpiration and on crop intercepted PAR. This potential growth is then corrected by water and nitrogen limitations, if any, to determine actual daily biomass gain.

Given the common pathway in leaves for carbon and vapor exchange, there is a conservative relationship between crop transpiration and biomass production. Thus, the potential daily biomass production can be calculated (Tanner and Sinclair, 1983).

Leaf area development

The increase of leaf area during the vegetative period, expressed as leaf area per unit soil area (leaf area index, LAI), is calculated as a function of biomass accumulation:

$$LAI = \frac{SLA\,B}{1 + pB}$$

where LAI is in $m^2\ m^{-2}$, B is accumulated aboveground biomass (kg m^{-2}), SLA is the specific leaf area ($m^2\ kg^{-1}$), and p is a partition coefficient ($m^2\ kg^{-1}$) controlling the fraction of biomass apportioned to leaves (a value of zero apportions all biomass to leaves).

Yield

Yield simulation depends on total biomass accumulated at physiological maturity (B_{PM}) and the harvest index (HI=harvestable yield/ aboveground biomass):

$$Y = B_{PM}\,HI$$

where Y is yield and B_{PM} both expressed in kg m^{-2}. The harvest index is determined using as base an unstressed harvest index modified according to stress intensity (water and nitrogen) and crop sensitivity to stress during flowering and grain filling.

Crop growth response to elevated atmospheric CO_2

Two parameters in CropSyst define potential crop growth in response to water use and radiation capture. In order to establish the effect of elevated atmospheric CO_2 on growth, values of these parameters under specified atmospheric CO_2 concentrations are required. Determining the rate of change of these parameters in response to changes in CO_2 concentration requires complex carbon assimilation models that are not suitable for application at the cropping systems scale. Therefore, the implementation in CropSyst relies on experimental evidence of crop growth responses to CO_2. These experiments report the percent increase of growth under a specified atmospheric CO_2 concentration compared to growth at a baseline concentration. This information is processed using a modified version of the approach introduced by Stockle *et al.* (1992).

Crop rotations

CropSyst simulations are performed using a daily time step within a period specified by start and ending dates. During this period, state variables such as residue biomass (surface and incorporated into the soil), evapotranspiration, soil water content, soil N content (nitrate and ammonium), soil organic matter and others are updated daily regardless of a growing crop is present or not (fallow). Crops and their corresponding management are initiated during this period according to a sequence given by a crop rotation template.

Data requirements

Five input data files are required to run CropSyst: Simulation Control, Location, Soil, Crop, and Management files. Separation of files allows for an easier link of CropSyst simulations with GIS software. Definitions, usage, and range of variation of all parameters required by CropSyst are given in the User's Manual (Stockle and Nelson, 2000), and are also available in the Help facility of the model interface.

The Simulation Control file combines the different types of input files as desired to produce specific simulation runs. It specifies the start and ending day of the simulation and the crop rotation to be simulated, and sets the values of all parameters requiring initialization. Also, inputs to this file allow users to switch on/off the simulation of soil erosion, soil salinity, nitrogen and CO_2 effects on crop growth, and to select soil water redistribution and runoff models (Stockle *et al.*, 2003).

The Location file includes information such as latitude, weather file code name and directories, rainfall intensity parameters, selection of ET models (Penman-Monteith or Priestley-Taylor) and associated parameters, and generalized information on wind for locations where daily wind data are not available.

The Soil file includes surface soil cation exchange capacity and pH for the estimation of ammonia volatilization, parameters for the SCS curve number approach for runoff calculation, and parameters for the Revised Soil Loss Equation for erosion calculation. For each soil layer, thickness and texture must be specified. Based on this information, pedotransfer functions (Saxton *et al.*, 1986) are used to calculate bulk density, volumetric water content at water potentials of -33 kPa (Field Capacity) and -1500 kPa (Wilting Point), and air entry potential and Campbell b value for the relationship between volumetric water content and water potential (Campbell, 1985). Whenever available, actual measurements instead of values estimated from texture can be used.

The Management file includes scheduled and automatic management events. Management events can be scheduled using actual date, relative date (relative to year of planting), or using synchronization with phenological events (e.g., number of days after flowering). Scheduled events include irrigation (application date, amount, and salinity concentration), nitrogen fertilization (application date, amount, source, and application mode), tillage operations, and residue management. The automatic event manager (irrigation and nitrogen fertilization) checks continuously the soil water and nitrogen content and it can be specified to provide management for maximum growth or to implement deficit strategies.

The Crop file allows users access to a common set of parameters to represent different crops and crop cultivars. This is a key feature of CropSyst. The file is structured in the following sections: phenology (thermal time requirements to reach specific growth stages), morphology (Maximum LAI, root depth, specific leaf area, leaf area duration, and other parameters defining canopy and root characteristics), growth (transpiration biomass coefficient, radiation-use efficiency, stress response parameters, etc.), residue (decomposition and shading parameters for crop residues), nitrogen (defining crop N demand and root uptake), harvest index (unstressed harvest index and stress sensitivity parameters), salinity tolerance, and CO_2-elevation response. Crop parameters are the only input data that require calibration within a narrow range to properly represent specific crops and cultivars. However, those parameters defining the bulk of the crop response to the environment and management can be determined through field experiments. These parameters are the transpiration-biomass coefficient (Tanner and Sinclair, 1983), the radiation-use efficiency (Monteith, 1977), the specific leaf area, the stem/leaf partitioning coefficient, and the leaf area duration. In addition, thermal time requirements for different crop development stages can also be recorded from field observations. The basic experimental data set must include growing season evolution of biomass (leaves and stem), LAI, intercepted PAR, soil water, seasonal daily weather, total biomass at harvest and yield.

Model validation and examples of applications

Model evaluation is conventionally made by comparing simulation outputs with data collected from the 'real world' system represented by the model. However, such evaluations can be limited by several factors, making it somewhat difficult to establish the true performance of models. Detailed information on the initial conditions of the system is required to conduct these comparisons, information that it is not always available or is confounded by significant spatial variability under typical field situations. Model evaluation is increasingly difficult as the system under consideration becomes more complex, as is the case with crop rotations and cropping systems evaluated over several years. In such a case, many types of data are required to test the various processes simulated by the model. Furthermore, not all model outputs of interest can be evaluated because the corresponding measurements are difficult or unfeasible to obtain. Another problem in model evaluation is the choice of quantitative indices used to evaluate model performance. These statistical indices normally rely on one-to-one (simulated vs. measured) comparisons, neglecting measurement errors and other sources of variation inherent to field experiments. Moreover, discrepancies between simulated and measured values in time series are not properly evaluated, greatly penalizing simulation outputs that have even a modest time shift with respect to measured data.

The CropSyst model has been evaluated in many parts of the world by comparing model estimates to data collected in field experiments. CropSyst has been applied to perform risk and economic analyses of scenarios involving different cropping systems, management option and soil and climatic conditions.

Water balance studies

Irrigation water management and saving is essential to conserve this scarce key input of agriculture. Therefore, it is of prime importance to save water and enhance crop water productivity. Water saving can be achieved through reduction in (i) the unproductive water loss from crop fields by soil water evaporation (E) and (ii) evapotranspiration (ET). While evaluating a water saving technology, it is important that one should know the nature of water saving and the components of water balance. The later are not easily measured in the field and vary widely with soil type, location and season. Under such situations, combination of field experimentation and modelling is the powerful approach that can give the conclusive results. Badini *et al.*, (1997) made the application of CropSyst model for agroclimatic assessment of improved millet in Burkina Faso. The model was used for simulating the soil water budget components and millet production both spatially and temporally. Pannkuk *et al.* (1997) using CropSyst model observed that the root mean square error (RMSE) for ET was 7% of the mean and 14% for the wheat yield. Linear regressions between simulated and observed values, forced to zero intercept gave slopes, which were statistically not different. Donatelli *et al.* (1997) from an evaluation study of CropSyst reported reasonably accurate model simulations for maize, soybean and barley growth. Jalota *et al.* (2009) successfully used CropSyst model to study the integrated effect of different water saving technology in rice at Ludhiana.

They further used the model to simulate crop biomass, water balance components, crop water product-ivity and rice yield in relation to different management interventions.

Nutrient and water management

Nutrient and water management in crop field is the key to increase resource use efficiency. CropSyst model has been successfully used to study the interactive effect of these two resources in different agro-ecological conditions. Based on preliminary validation, CropSyst appears to be a promising tool to analyze best management practices for water and nitrogen (Stockle *et al.*, 1993). Pala *et al.* (1996) using the CropSyst model for simulation of durum wheat growth under different water and nitrogen regimes in a Mediterranean environment, reported index of agreement between predicted and observed values fluctuating from 0.90 to 0.98 and magnitude of overall RMSE about 10% of the observed means. CropSyst model was chosen to study the interactive effect of water and nitrogen on crop yield and water productivity in maize-wheat cropping system at New Delhi (Lenka *et al.*, 2011). The treatments were three levels of water regimes and five nitrogen levels in split plot design. Three levels of water regime, namely W_1, W_2 and W_3 referred to limited, medium and maximum irrigation. The five nitrogen levels were: T_1 to T_5 (T_1, 0% N; T_2, 75% N; T_3, 100% N; T_4, 150% N; T_5, 100% N from organic source). CropSyst requires several parameters for calibration and validation. These parameters were set based on typical field observations, or taken from CropSyst manual or from literature pertaining to site-specific Indian conditions. A few parameters that tend to fluctuate among cultivars were calibrated. The model was initialized each time prior to maize sowing in 2002. Crop parameters used for this simulation are presented in Table 1. The calibrated parameters were adjusted using three points in the data set (no N/high water, high N/no water, high N/high water i.e. W3T1, W1T3, and W3T3) for both the crops maize and wheat. The data set of first year maize 2002 and wheat 2002-03 were used for calibration. These data included grain yield, above ground biomass, LAI, actual ET, Total N uptake. The maize and wheat cultivars simulated were KH-101 and HD 2687, which are the varieties largely used in the region.

Table 1. Crop parameters for CropSyst simulations of maize and wheat growth and yield*

Parameters	**Maize 2002**	**Maize 2003**	**Wheat 2002-03**	**Wheat 2003-04**	**Source**
Degree-days emergence (°C-day)	150	58	54	97	O
Degree-days peak LAI (°C-day)	1331	1094	602	484	O

Parameters	Maize 2002	Maize 2003	Wheat 2002-03	Wheat 2003-04	Source
Degree-days flowering (°C-day)	1358	1123	633	503	O
Degree-days maximum grain filling (°C-day)	1465	1207	760	603	O
Degree-days maturity (°C-day)	2135	1507	1642	1418	O
Base temperature (°C)	10	10	6	6	L
Cutoff temperature (°C)	35	35	30	30	L
Optimum mean daily temperature (°C)	27	27	20	20	L
Maximum root depth (m)	1.8	1.8	1.5	1.5	L
Maximum expected leaf area index (LAI)	4	4	6	6	O
Specific leaf area (m^2 kg^{-1})	25	25	24	24	C
Stem/leaf partition coefficient	3	3	2.5	2.5	C
Leaf duration (°C-day)	1000	1000	800	800	C
ET crop coefficient	1.05	1.05	0.8	0.8	C
Maximum water uptake rate (mm per day)	16	16	13	13	M
Critical canopy water potential (J Kg^{-1})	-1000	-1000	-1600	-1600	C
Wilting canopy water potential (J Kg^{-1})	-1800	-1800	-2200	-2200	C
Biomass/transpiration coefficient (KPa)	8.5	8.5	7.5	7.5	C
Light to above ground biomass conversion (g MJ^{-1})	3	3	3	3	M
Maximum Harvest Index	0.4	0.4	0.4	0.4	O
N uptake adjustment (0-2)	0.2	0.2	0.2	0.2	O
Maximum nitrogen concentration at emergence (kg kg^{-1})	0.018	0.018	0.02	0.02	O
Maximum nitrogen concentration at maturity (kg kg^{-1})	0.011	0.011	0.011	0.011	O
Minimum nitrogen concentration at maturity (kg kg^{-1})	0.005	0.005	0.009	0.009	O

* Parameters were set as observed from field data (O), extracted from CropSyst manual (M), site specific datas from literature (L), or set by calibration (C).

Source: (Lenka *et al.*, 2011).

After the calibration of model for maize and wheat using W3T1 (maximum water, no nitrogen), W1T3 (no water and maximum nitrogen), and W3T3 (maximum water and maximum nitrogen), the model was validated against other nitrogen and irrigation levels in maize and wheat. The model validation was done for grain yield, above ground biomass, LAI, actual ET, total N uptake and soil moisture of next year maize 2003 and wheat 2003-04. The evaluation of model performance was carried out using statistical tools, *viz.*, mean biased error (MBE), mean absolute error (MAE), root mean square error (RMSE), R^2 (Pearson's correlation coefficient) and d (Willmott's index of agreement). The calibrated model was validated on independent dataset observed in the year 2002-03 and 2003-04. Comparison of experimental (O) and simulated (P) results with respect to grain yield across irrigations and nitrogen levels is given in fig. 1 and 2. The model responded well to different levels of irrigation with significant R^2 values and index of agreement (d) between observed and predicted. Significant correlation was observed for biomass and grain yield but at higher N-rates, the model underestimated the grain yield. Scatter plot of simulated and observed above ground biomass is shown in Fig. 3. There was a good match between simulated and observed data with higher coefficient of determination (0.81-0.97).

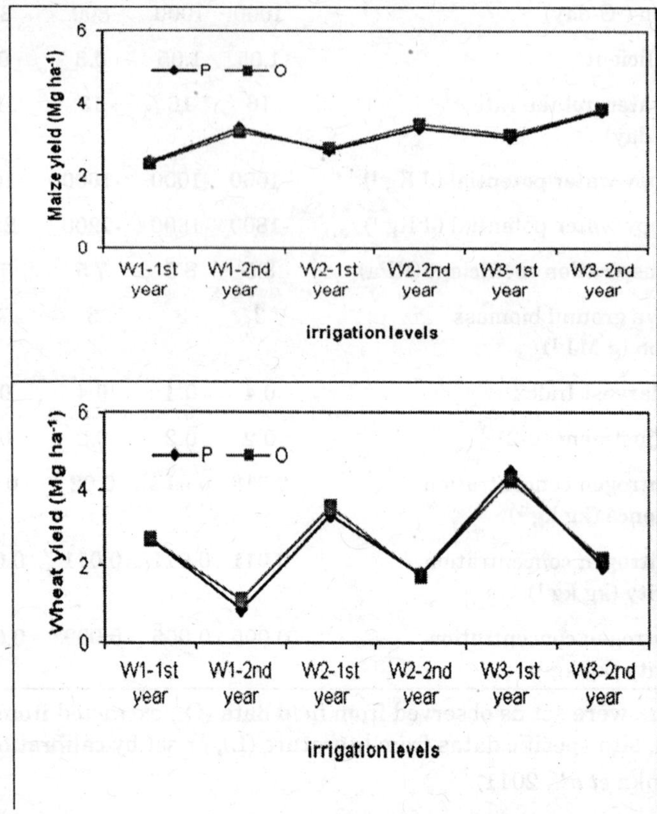

Fig. 1. Effect of irrigation levels on predicted and observed grain yield (Mg ha^{-1}) across nitrogen levels in maize and wheat crops (Lenka *et al.*, 2011)

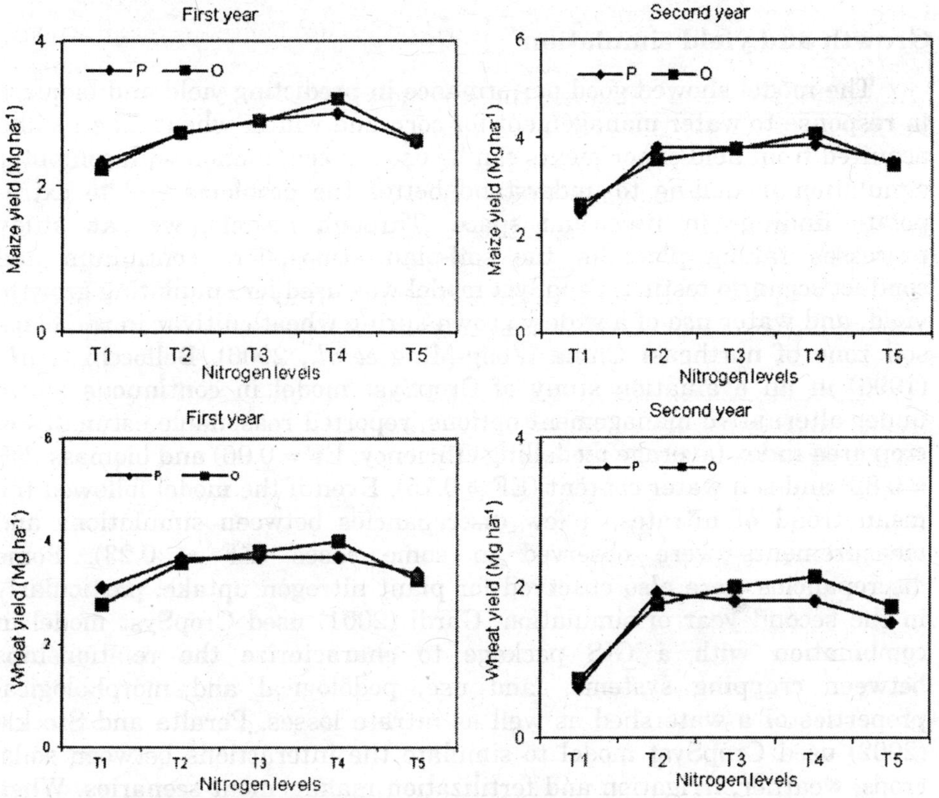

Fig. 2. Effect of nitrogen levels on predicted and observed grain yield (Mg ha^{-1}) across irrigation levels in first and second year of cropping sequence. (Lenka *et al.*, 2011)

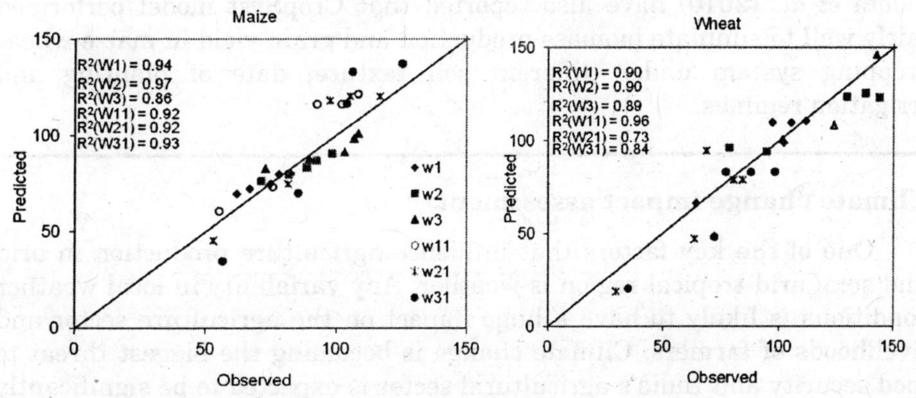

Fig. 3. Observed and predicted nitrogen uptake (kg ha^{-1}) for maize and wheat in 2002-03 and 2003-04 under minimum (W_1, W_{11}), medium (W_2, W_{21}) and maximum irrigation (W_3, W_{31}) (Lenka *et al.*, 2011)

Growth and yield simulation

The model showed good performance in predicting yield and biomass in response to water management for corn and winter wheat. The results acquired from field experiments can be used in conjunction with computer simulation modelling to understand better the problems and to extrapolate findings in time and space. Through models, we can study processes taking place in the soil-plant-atmosphere continuum and conduct scenario testing. CropSyst model was used for simulating growth, yield, and water use of a widely grown spring wheat cultivar in the black soil zone of northeast China (Zong-Ming *et al.*, 2006). Bellocchi *et al.*, (1996) in an evaluation study of CropSyst model in continuous maize under alternative management options, reported reasonable estimates of crop area index (average modelling efficiency, EF = 0.96) and biomass (EF = 0.82) and soil water content (EF = 0.75). Even if the model followed the mean trend of nitrates, a few discrepancies between simulations and measurements were observed in some cases (EF = 0.23). Some discrepancies were also observed for plant nitrogen uptake, particularly in the second year of simulation. Gardi (2001) used CropSyst model in combination with a GIS package to characterize the relationships between cropping systems, land use, pedological and morphological properties of a watershed as well as nitrate losses. Peralta and Stockle (2002) used CropSyst model to simulate the interactions between soils, crops, weather, irrigation and fertilization management scenarios. When scenarios with fertilization rates above current recommended rates were simulated, potato had the largest nitrate leaching than that during maize growth. Excess irrigation did not significantly affect the bulk of simulated nitrate leaching, although it enhanced leaching during crop periods. Jalota *et al.* (2010) have also reported that CropSyst model performed fairly well to simulate biomass production and grain yield in maize-wheat cropping system under different soil texture, date of planting and irrigation regimes.

Climate change impact assessment

One of the key factors that influence agriculture production in arid and semi arid tropical region is weather. Any variability in local weather conditions is likely to have a huge impact on the agriculture sector and livelihoods of farmers. Climate change is becoming the biggest threat to food security and India's agricultural sector is expected to be significantly influenced by the concurrent processes of climate change and globalization. Climate change could lead to sea level rise, increased weather variability, more droughts and spread of infectious diseases (Balasubramanian and Palaniappan, 1995). Appreciable changes in temporal and spatial evolutions in temperature and rainfall are expected to cause effect on both productivity and quality of agricultural produces (Allen, 1990). According to the latest assessment report of the United Nations Intergovernmental Panel on Climate Change (IPCC), there is unequivocal evidence that the earth's climate system is warming due to anthropogenic greenhouse gas (GHG) emissions. Atmospheric carbon dioxide concentration has been increasing exponentially in the last 250 years (Keeling and Whorf, 2005). Climate models project that the global temperature will rise by about 1.4 to 5.8°C by the year 2100. Since agricultural practices are climate-dependent and yields change from year to year depending on climate variability, the agricultural sector is particularly exposed to changes in climate.

For climate change impact assessment, crop growth models have been widely used to evaluate crop responses (development, growth and yield) by combining future climate conditions, obtained from General or Regional Circulation Models (GCMs and RCMs respectively), with the simulation of CO_2 physiological effects, derived from crop experiments. Many of these impact studies were aimed at assessing crop development shifts and yield variations under changes in mean climate conditions. These analyses showed that increasing temperatures generally shortened the growing period of commercial crops (e.g. Bindi *et al.* 1996; Singh *et al.* 1998; Guerena *et al.* 2001; Giannakopoulos *et al.* 2009), resulting in a shorter time for biomass accumulation. On the other hand, changes in yields were not homogeneous and dependent on crop phenology (e.g. summer and winter crops), crop type (e.g. C_3 and C_4 plants) or environmental conditions (water and nutrient availability) (Ainsworth and Long 2005; Brassard and Singh 2008; Giannakopoulos *et al.* 2009). Other studies stressed that changes in climate variability, as can be expected in a warmer climate, may have a more profound effect on yield than changes in mean climate (Mearns *et al.*, 1996; Southworth *et al.*, 2000; Porter and Semenov, 2005). As such, policy analysis should not rely on scenarios of future climate involving only changes in means (Katz and Brown 1992; Semenov and Barrow 1997). Furthermore, the changes in the frequency of extreme climatic events during the more sensitive growth stages have been recognized as a major yield-determining factor for some regions in the future (e.g. Easterling and Apps 2005; Schneider *et al.* 2007). Accordingly, both changes in mean climate and climate variability (including extreme events) should be considered for a reliable climate change impact assessment in agriculture. For this purpose, Moriondo *et al.* (2011) used a modified version of the CropSyst growth model (Stockle *et al.* 2003) to account for heat wave impacts on crop fertility and yield. Following, this modified version of CropSyst was coupled to the outputs of a RCM for both the baseline (1961–1990) and future SRES scenarios A2 and B2 (2070–2100). They have also explored the possible spin-off of the proposed modelling approach for selecting feasible adaptation strategies. CropSyst had also been applied for the assessment of potential impacts related to climate change which may be an indication for a proven reliability of the model to realistically capture the yield limiting (stress) factors water and heat. Sadras (2002) used CropSyst to study the interaction between rainfall and nitrogen fertilisation of wheat in environments prone to terminal drought in Australia. He observed simulations are to be consistent with known physiological principles, e.g. reduction in harvest index, rather than in shoot biomass, accounting for yield reduction of crops under severe terminal drought in response to high nitrogen supply.

Conservation agriculture (CA) and residue management

CA has also emerged a major way forward from the existing unsustainable conventional agriculture, to protect the soil from degradation processes and make agriculture sustainable. Empirical evidences have been accumulating to show that zero/minimum tillage based agriculture along with crop residue retention and adoption of suitable crop rotations can be highly productive, provided farmers participate fully in all stages of technology development and extension (FAO, 2001). Among the major crop models that consider the most important features of conservation agriculture, the Cropping System Simulation Model, CropSyst (Stöckle *et al.*, 2003), has been applied successfully under a range of climatic conditions and for a variety of crops, such as maize, barley, rice, sorghum, potato, alfalfa and cotton. Crop modelling of wheat accounts for most of the published simulation studies with CropSyst. This comprises studies on durum wheat in northern Syria (Pala *et al.*, 1996), winter and spring wheat in Washington (USA) in response to different fallow and tillage management practices (including no-till; Pannkuk *et al.*, 1998), as well as winter wheat in Italy (Bechini *et al.*, 2006), the Turkish Central Anatolia Plateau (Benli *et al.*, 2007) and northwest Uzbekistan (Djumaniyazova *et al.*, 2010).

Conclusion

CropSyst model has emerged and evolved as good cropping system simulation model since its inception. CropSyst has been developed and evaluated at field regional, national and global scale for assessing large agricultural impacts and adaptation to climate variability. It can be successfully used to quantify the effect of different factors (soil, climate, water, nutrient etc.) on crop growth and performance. Accurate and proper input data generation over time is essential for the model to realistically simulate given agro-ecological condition. The real challenge under climate change conditions is to use adaptation strategies, which are improved agricultural management practices, to reduce the damage caused by climate change on the yield of the growing crops. CropSyst has also performed satisfactorily in climate change and impact assessment studies.

❑❑❑

9 Climate Change and Crop Production System: Assessing the Consequences for Food Security

Dillip Kumar Swain* and Sushree Sagarika Satapathy

Agricultural and Food Engineering Department,
Indian Institute of Technology, Kharagpur – 721 302, India
*Corresponding author: swain@agfe.iitkgp.ernet.in

Introduction

The growth in human population over the past two centuries and the economic development with an increased consumption of resources has led to marked changes on global climate. The climate change is evident through a range of interacting factors such as increasing concentrations of greenhouse gases in the atmosphere, climate variation/change, rising sea level, loss of biodiversity, etc. Besides the observational documentation on the reality of climate change, other research has increased our knowledge of the causes of the climate change. A topic of particular societal concern is that of climate change and global food security. The climate change has affected the agricultural sector through direct and indirect effects of changes in temperature and precipitation in the biological and physical environment (Williams, 1995; Brown and Rosenberg, 1997). Climate change associated with increasing levels of CO_2 may bring benefits to agricultural production in some parts of the world, especially northern latitudes above about 55^0, while the places particularly tropical and sub-tropical countries (developing world) will be adversely affected (Parry *et al.*, 2004; Stern, 2006; Hadley- Centre, 2006). Many parts of the developing world are anticipated to be exposed to significant changes in temperature and rainfall pattern, thereby increasing the frequency and intensity of extreme events such as droughts and floods (IPCC, 2001; Tyson *et al.*, 2002). The developing economies are particularly sensitive to the direct impacts of climate change given their often heavy dependence on agriculture and ecosystems, and because of their high poverty levels and geographic exposure (Stern, 2006). If the climate change effects dominate, world crop yields are likely to be decreased significantly (Parry *et al.*, 2004) as projected using climate change scenarios of Intergovernmental Panel on Climate Change (IPCC): Special Report on Emission Scenarios (SRES) driven experiments conducted using the UK Hadley center's third generation coupled atmosphere-ocean global climate model (HadCM3) (Johns *et al.*, 2003).

To compound the forecasted negative impacts of climate change on crop production, overall demands for food will increase as global population continues to rise from six billion people today to an anticipated nine billion by 2050 (UN-DESA, 2004). The overall crop production will need to continue to increase by 50% over the next few decades to meet this anticipated demand, although predicting future global food production is complex (Do¨o¨s, 2002). There is a need to redesign the agricultural system that is more responsive to climate change adaptation, environmental preservation, regional and national food security, and poverty and hunger alleviation.

As optimum crop production becomes more complex, involving fertilizer, pest control, genotype, environment, and cultural practices, conducting trials that take all these factors into account becomes increasingly complex and expensive. A suitably validated crop simulation model could be used to test hundreds of such combinations in a brief time at limited expense. Such simulations can adequately describe relative trends in yields caused by environment variation (Penning de Vries *et al.*, 1989). Well-evaluated crop simulation models can be used to assist the decision-making processes in agriculture for climate change adaptation/ mitigation. There are gaps in currently available knowledge regarding some aspects of mitigation of climate change, especially in developing countries. Appropriate adaptations can greatly reduce the magnitude of climate change impacts on agriculture. It is the need of time to assess the climate change impacts and evaluate the agro-adaptation measures for increased crop production to ensure global food security.

Climate change

Climate change on a global scale refers to changes in the climate of the earth as a whole, including temperature increases (global warming) or decreases, rainfall variability, shifts in wind, and rise in sea level. Climate change is one of the greatest environmental, social and economic threats being faced by the planet in present scenario. Regional changes in climate, particularly increase in temperature, have already affected a diverse set of physical and biological systems in many parts of the world. According to NASA, the year 2005 was the warmest year in over a century. Four among the last five years (2002-2006) rank within the five warmest years (2005, 1998, 2002, 2003, and 2006) in the instrumental records of global surface temperature since year 1850. Over the past 30 years, the earth has warmed by 0.6°C while over the past 100 years it has warmed by 0.8°C. The trend of increase in temperature since 1950 for global and lower latitude is shown in Fig. 1.

With observational studies documenting the reality of climate change, scientists speak of 'forcing' to denote the external factors that give rise to climate change. These include natural forcing, such as volcanism and changes in the sun, as well as anthropogenic forcing, such as adding to the amounts of atmospheric gases that contribute to the greenhouse effect. In addition, climate change can occur without forcing, simply because of internal variability of the climate itself.

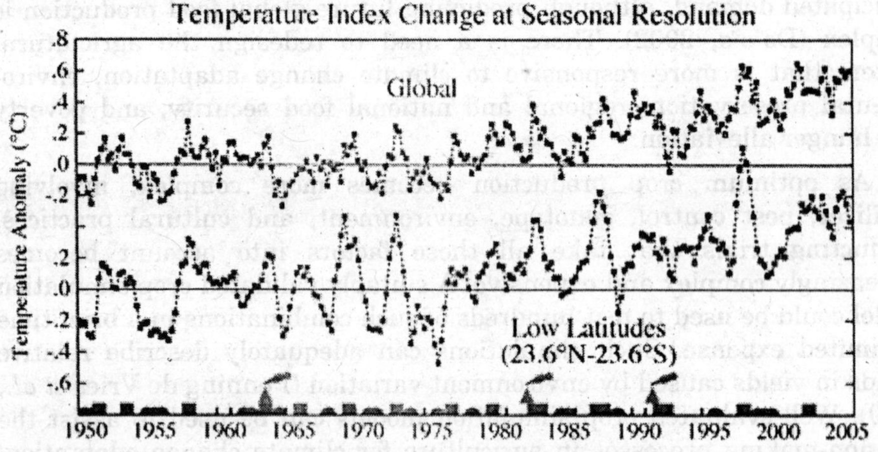

Fig. 1. Temperature changes since 1950 (source NASA, 2007)

Carbon dioxide as one of the major contribution to greenhouse effect of the atmosphere is increasing steadily in recent years. During the last 20 years of the twentieth century, about three-quarters of the anthropogenic emission of CO_2 into the atmosphere is due to fossil fuel burning and the rest resulted from land-use changes, predominantly deforestation. Other gases contribute to greenhouse effect such as methane, nitrous oxide, chlorofluorocarbons, etc which are also increasing in the atmosphere due to human activities, including fossil fuel, cattle, rice agriculture, chemical farming, landfills, etc. Until recently, about half of the contribution of human activities to increasing the greenhouse effect could be attributed to carbon dioxide, and the combined effect of all the other anthropogenic emission of greenhouse gases made up the other half. Since pre industrial times global emission of greenhouse gases has increased by 70% i.e. from 28.7 Gt to 49 Gt of CO_2 equivalents in between 1970 and 2004. The largest growth in greenhouse gases emission has come from energy supply sector (an increase of 145%), followed by other major contributors like transportation (an increase of 120%), industry (65%), land use and land cover change and forestry (40%), agriculture

(27%) and buildings (26%). Also global population growth was 69% and global income growth was 77% in the period of 1970 to 2004, which are the main drivers of increasing energy related CO_2 emissions (IPCC, 2007a)

The climate system is made up of the global atmosphere, oceans, land surface, ice and snow, and many of the interactions among these components, including some biological and chemical processes. Computer simulation plays a central role to understand the range of possibilities for climate change in coming decades. The key tool in this type of research is the global climate model or general circulation model (GCM). This class of models has been developed to incorporate the most important features of the climate systems. These models are continuously under development and are revised as improved knowledge of specific aspects of climate becomes available. The modern GCMs are effectively a synthesis of much of our present understanding of the climate system. The most realistic twentieth-century climate simulations are obtained using various combinations of natural and anthropogenic forcing as inputs to GCMs.

According to most of the climate change scenarios, the greenhouse gas concentration will increase through 2100, with a continued increase in average global temperature. The extent of temperature increase and the time period to reach a particular increase is still uncertain and thus it creates uncertainty about future greenhouse gas, aerosol emissions and the earth's response to changing conditions. Natural influences such as changes in the sun and volcanic activity may also affect the future climate changes, and effects of these are also uncertain (IPCC, 2007b)

The Special Report on Emission Scenarios (non-mitigation) of IPCC projects an increase of 9.7 Gt CO_2 equivalents to 36.7 Gt CO_2 equivalents i.e. (25% - 90%) between year 2000 and 2030 in the amount of global greenhouse gas emissions. Also CO_2 emission is projected to grow in the range of 45% to 110% (IPCC, 2000). The IPCC fourth assessment report published in 2007 states: 'Warming of the climate system is unequivocal, as is now evident from observations of increase in global average air and ocean temperatures, wide spread melting of snow and ice, and rising global average sea level'.

IPCC projections of future climate change (Somerville and Jouzel, 2008) are as follows:

- During the next two decades, we expect about 0.2ºC per decade of further warming. This anticipated future warming continues observed recent trends, which are themselves consistent with earlier IPCC projections.
- Tropical cyclone will become more intense.
- Heat wave and heavy precipitation will become more frequent.
- Precipitation will tend to increase in high latitudes.

- Precipitation will tend to decrease in subtropical latitudes.
- The Atlantic meridional overturning circulation will become slow.
- Sea level will rise by about 0.2 to 0.6 m in the twenty-first century.
- Warming and sea level rise will continue for centuries.
- Snow cover and ice will continue to shrink.

Crop production system

Crop production is the exploitation of plant morphological and physiological responses within a soil and atmospheric environmental system with optimum agronomical management practices to attain potential production per unit of land, water and other natural resources. The major components of crop production system are crop system, soil system and climate system.

Crop system: The main principle in crop production system is harvesting as much solar energy as possible and converting the solar energy into chemical energy by photosynthesis. Photosynthesis links inorganic and organic world by converting the inorganic compounds like water and CO_2 to the organic compounds like sucrose, starch, etc. The raw materials produced through photosynthesis result in crop growth and development of storage organs such as leaves, stems, fruits, tubers, grains, etc. During photosynthesis, some of the carbon in the sucrose (CH_2O) is combined with other elements (N, P, K, S, etc) and retained in plant tissue, and some is used to provide energy for synthesizing tissue, releasing CO_2 in the process. Penning de Vries *et al.* (1989) described a sound quantitative basis for computing the rates of tissue synthesis based on the photosynthetic rate and on the composition of tissue being synthesized. They computed the net product weights of protein, carbohydrate, lipid, lignin, organic acid, or mineral if 1 g of CH_2O was used to provide both the carbon in the final product and the energy for product synthesis. Thus, depending on the composition of plant tissue, conversion efficiency (E) can be computed and used to convert photosynthesis rate into dry matter accumulation rate. For vegetative tissue, this conversion efficiency is in the range of 0.65-0.75 (g tissue)/ (g CH_2O).

Respiration is the loss of CO_2 from plants in growth and maintenance processes. Growth respiration is accounted for in the conversion efficiency and maintenance respiration is the loss of CO_2 due to the breakdown and resynthesis of existing tissue and depends on temperature.

The equation for crop dry weight growth rate is

$$\frac{DW}{dt} = E\ (Pg - RmW)$$

Where

$\frac{DW}{dt}$ = rate of dry weight growth of crop, (g tissue)/($m^2 \cdot h$)

W = total plant dry weight, g/ m^2

Rm = maintenance respiration rate, (g CH_2O) / [(g tissue). h]

E = conversion efficiency of CH_2O to plant tissue, (g tissue)/(g CH_2O)

Pg = canopy gross photosynthesis rate, (g CH_2O) / (m^2. h)

Total biomass growth rate is partitioned into canopy (Wc) and root (Wr) biomass with the function fc(N).

$$\frac{dWc}{dt} = \left(\frac{dW}{dt}\right) fc\,(N)$$

and

$$\frac{dWr}{dt} = \left(\frac{dW}{dt}\right) [1.0 - fc(N)]$$

Where, fc(N) is the function of stage of development, which is the fraction of the total biomass partitioned to the canopy i.e. above ground biomass.

Crop 'development' refers to the progression of the crop through its life stage, which may be measured by various vegetative and reproductive plant characteristics such as number of leaves on the plant or the appearance of first flower or first fruit. Development rate is usually expressed as the inverse of time between events, such as inverse of hours between successive leaves on the main stem or of days from plant emergence to flowering. Development rate in plants is very sensitive to temperature and sometimes to day-length but are usually insensitive to light, CO_2, water, nutrient stress, etc unless the stresses are severe.

Soil system: Soils are made up of four components: minerals, air, water, and organic matter. In most soils minerals represent around 45% of the total volume, water and air about 25% each, and organic matter from 2-5%. The mineral portion consists of three distinct particle sizes classified as sand, silt, and clay. Sand is the largest size particle in soil, which is largely the mineral quartz, though other minerals are also present. Since quartz contains no plant nutrients, sand is the lowest contributor to soil fertility of the three soil particle sizes. Silt particles are much smaller than sand but, like sand, silt is mostly quartz. The smallest of all the soil particles is clay that contains appreciable amounts of plant nutrients. The textural designation of a soil is derived from the relative portions of sand, silt, and clay. The suitability of soil to particular crop depends on texture in addition to soil depth, depth of water table, salinity and alkalinity. Though a particular crop can be grown in various type of soil, the most suitable soil for each crop differs. For example, rice, cotton, sugarcane, sorghum, etc prefer heavy textured soil like clay loam, silty clay loam, silty clay and clay. Medium textured soils like loams, silty loam, silt and sandy loam are suitable for most of the crops. Sandy soils and loamy sands are light textured soils and these are suitable for growing peanut, potato, tobacco, root crops and fodder crops.

Another soil characteristic, 'soil structure' is different from soil texture. Structure refers to the combination or 'aggregation' of sand, silt and clay particles into larger secondary clusters. Good structure is apparent when the sand, silt, and clay particles are aggregated into granules or crumbs. Both texture and structure determine pore space for air and water circulation, erosion resistance, looseness, ease of tillage, and crop root penetration. However, while texture is an innate property of the native soil and does not change with agricultural activities, structure can be improved or destroyed readily through our choice and timing of farm practices.

Crop production system is influenced by soil physical, chemical and biological environments. The favorable physical environment for proper crop growth is deep soil without hard layer, high available water holding capacity, good aeration, and less soil strength or mechanical resistance. The favorable chemical environment is of steady nutrient availability for crop plants. The biological environment is composed of favorable organism belonging to both plant and animal kingdom, which facilitates nutrient release pattern in soil and improves soil structure, thereby crop growth and development.

Climate system: The most important climatic factors that influence growth and development of the crops are solar radiation, temperature, and rainfall. Solar energy provides light required for seed germination, leaf extension, growth of stem and shoot, flowering, fruiting, and thermal condition necessary for physiological function of plant. Specifically, solar radiation influences the production of enzymes useful in photosynthesis, development of photosynthetic apparatus, growth, yield formation and yield of crop plants. Temperature influences the growth and development of crop through changes in biochemical reactions, uptake of carbon dioxide, enzymatic activity and photosynthesis. For each plant species, there are maximum and minimum limits of temperature at which growth is nil or negligible and optimum temperature at which growth is maximum. The optimum temperature for cool season crops like wheat, barley, potato, etc, is 20 to 28°C and for warm season crops like rice, maize, sugarcane, peanut, etc is 30 to 38°C. Rainfall amount and its distribution help in selection of crops and cropping system of a particular region. The decisions on agricultural operation such as land preparation, sowing time, irrigation scheduling, intercultural operations, harvesting time, etc are influenced by rainfall pattern. Water deficit affect the normal physiological processes in the plants leading to reduced cell growth, stomatal closure, and decreased CO_2 intake and photosynthesis.

Effects of climate change on crop production system

Increased intensity and frequency of storms, drought and flooding, altered hydrological cycles and precipitation variance have implications for future food production. The net effect of a changing climate on crop production system is complicated due to the interactions of several simultaneous biophysical processes. The climate change may be beneficial in some cases and detrimental in others. In beneficial cases, increasing level of CO_2 has shown to increase photosynthesis rates and to increase stomatal resistance in crops, leading to increased water use efficiency (Kimball *et al.*, 2002). Another beneficial impact would be prolonged crop growing season in areas limited by cold temperature i.e. at high altitudes and high elevation. Semiarid regions may get benefit from increased precipitation, because warmer atmosphere can hold more water vapours.

A warmer and more variable climate is likely to have negative as well as positive effects on agricultural region around the world. Potential negative effects include frequent drought and floods, heat stress, increased outbreak of disease and pests, shortening of crop growth periods and in coastal region; increased flooding and salination due to sea-level rise and impeded drainage. An important negative effect that warming has on crops is the shortening of their growth period. Warmer temperature speeds crops through their growth cycle, especially the grain filling stage. Total yield is a product of the rate and duration of the grain filling, which is determined by accumulated temperature. Since higher temperature shortens the duration of grain filling, higher temper-ature tends to exert a negative effect on yield of most annual crops.

Crop yield: Considerable progress has been made internationally in both the measurement and modelling of the effects of climate and atmospheric CO_2 change on the growth of crops. Despite general uncertainties about the rate and magnitude of climate change, regional and global studies have consistently shown that agricultural production systems in the mid and high latitudes are more likely to benefit in the near term, while production system in low latitudes are more likely to decline (IPCC, 2001). In biophysical terms, rising temperature will create more atmospheric demand of water leading to greater water stress, and likely to push many crops beyond their limits of optimal growth and yield, especially in arid and semi-arid region. Since most of the developing countries are located in lower latitudes (arid and semi-arid region), while most developed countries located in the mid- to high-latitudes, this finding suggests a divergence of vulnerability between these groups of nations, with far reaching implications for future world food security (Table 1).

Table 1. Effect of climate change on global food production (Percent change in yield) under various scenarios (Parry *et al.*, 2004)

Scenarios	HadCM3 2080s							HadCM2 2080s	
	A1F1	**A2a**	**A2b**	**A2c**	**B1a**	**B2a**	**B2b**	**S550**	**S750**
CO_2, ppm	810	709	709	709	527	561	561	498	577
World	-5	0	0	-1	-3	-1	-2	-1	1
Devd. countries	3	8	6	7	3	6	5	5	7
Devg. countries	-7	-2	-2	-3	-4	-3	-5	-2	-1
Difference in Devd. & Devg. countries	10.4	9.8	8.4	10.2	7.0	8.7	9.3	6.6	7.7

Devd. – Developed, Devg. – Developing

Experiments at Maricopa, Arizona, USA showed that increased atmospheric CO_2 tends to increase growth and yield of wheat crops and that the relative size of this effect is bigger with drought crops than those given unlimited water supply. Crop models have been used together with outputs from GCMs to assess the impact of climate change on wheat production in Australia. The relative benefits of higher CO_2 are likely to be greater in Australia than elsewhere because wheat is often grown under conditions of limited water supply. The yield response ranges from +15 to +25% as mean temperature increases from 0 to 4°C with a doubling of atmospheric CO_2 to 700 ppm. However, this response decreases to a range of +4 to -8% if rainfall simultaneously decreases by 20% (Gregory and Ingram, 2000). In another study, Ludwig *et al.* (2009) reported that in Western Australia, the growing season rainfall (May to October) decreased by an average of 11% and the rainfall in June and July decreased by 20%, which is the main factor limiting production. Based on historical weather data, the wheat yield simulated using Agricultural Production Systems Simulator (APSIM-N wheat) model did not fall, despite the large decline in rainfall. This finding will have significant implications for estimates of future climate change impacts in this region with changes in rainfall causing non-proportional impacts on production.

The adverse effect of climate change on crop yield has been simulated in lower latitude region. Rice yield of eastern India was simulated using crop simulation models (INFOCROP and ORYZA1) for three GCM scenarios i.e. Geophysical Fluid Dynamic laboratory (GFDL), Goddard

Institute of Space Studies (GISS), and United Kingdom Meteorological Office (UKMO). The model ORYZA1 predicted the yield changes of -8%, -9% and -16%t, while INFOCROP predicted changes of -9%, -12% and -21% for the GFDL, GISS and UKMO scenarios, respectively (Table 2) (Krishnan *et al.*, 2007). An average yield change of -20% per every 1°C increase in temperature up to 5°C was noted at the current level of CO_2 (380 ppm), while elevated CO_2 simulated higher yield at each degree temperature increment (Table 3).

Table 2. Estimated changes in the rice yield (in percent) predicted by INFOCROP and ORYZA1 models at different sites of eastern India for three GCM scenarios

Sites	GFDL	GISS	UKMO
INFOCROP			
Bhubaneswer	-24	-27	-37
Chinsurah	-7	-7	-8
Cuttack	-25	-28	-41
Faizabad	-14	-18	-28
Jabalpur	-11	-14	-26
Jorhat	13	12	8
Kalyani	-9	-12	-22
Pusa	-4	-4	-5
Raipur	-2	-5	-18
Ranchi	-9	-12	-35
Average	-9	-12	-21
ORYZA 1			
Bhubaneswer	-17	-20	-27
Chinsurah	-8	-9	-10
Cuttack	-20	-20	-31
Faizabad	-9	-11	-19
Jabalpur	-11	-14	-21
Jorhat	12	13	8
Kalyani	-8	-10	-16
Pusa	-5	-6	-7
Raipur	-3	-5	-10
Ranchi	-8	-10	-26
Average	-8	-9	-16

Table 3. Mean predicted change (%) in potential rice yield of eastern India simulated through INFOCROP and ORYZA 1 models under the fixed temperature and CO_2 scenarios

CO_2 concentration in ppm	Temperature increments(°C)						Average
	0	+1	+2	+3	+4	+5	
INFOCROP							
380	0	2.68	-18.04	-28.15	-36.36	-42.49	-20.39
400	5.23	-2.90	-12.03	-23.21	-33.05	-40.32	-18.12
500	25.92	13.90	3.60	-12.32	-25.08	-35.11	-4.85
600	40.61	29.39	16.74	1.04	-15.80	-29.17	7.14
700	56.37	41.21	27.96	9.85	-7.67	-23.15	17.43
Average	25.14	16.86	3.65	-10.56	-23.59	-34.05	-3.76
ORYZA 1							
380	0	-4.27	-15.32	-26.78	-29.45	-38.96	-19.13
400	2.78	0.93	-8.44	-19.74	-23.67	-33.65	-13.22
500	18.57	11.73	0.24	-11.75	-15.43	-26.57	-3.87
600	26.8	18.54	5.03	-6.86	-10.87	-21.84	1.80
700	30.73	23.77	8.67	-3.68	-7.32	-19.04	5.52
Average	16.27	10.14	-1.96	-13.76	-17.35	-28.01	-5.78

Processing quality: Rising atmospheric CO_2 is the primary driver of global warming, but as the principal substrate for photosynthesis it also directly affects the yield and quality of crops. Elevated CO_2 leads to numerous physiological changes in cereal crops i.e. rice, wheat, barley, such as changes in the photosynthesis and assimilate translocation, nutrient uptake and translocation, water relation, and altered gene expression and enzyme activity. These altered processes are very likely to affect the chemical and physical characteristics of grains.

In order to investigate the interactive effects of CO_2 and N supply on rice quality, Yang *et al.* (2007) conducted a free-air CO_2 enrichment (FACE) experiment at Wuxi, Jiangsu, China, in 2001-2003. A long-duration rice japonica with large panicle (cv.Wuxiangging 14) was grown at ambient or elevated (ca. 200 mmol mol^{-1} above ambient) CO_2 under three levels of N: low (LN, 15 g N m^{-2}), medium (MN, 25 g N m^{-2}) and high N (HN, 35 g N m^{-2} (2002, 2003)). Across both CO_2 levels, brown rice percentage (BRP) increased significantly with increased amount of N fertilizer (Fig. 2a). The FACE significantly reduced milled rice percentage (MRP) by 2.0% across the three seasons, with averages of 1.1%, 2.4% and 2.2% in 2001, 2002, and 2003, respectively (Fig. 2b). The head rice percentage (HRP) of the FACE plants was dramatically reduced by 23.4% across the 3 years (Fig. 2c). Across both CO_2 levels, both MRP and HRP increased apparently with increasing N supply (Fig. 2c). These results indicate a higher fraction of removed outer layers (i.e., pericarp and aleurone layer) and more broken or damaged grains during milling in rice from the high CO_2 plots. The outer layer of endosperm cells transformed into aleurone layer, which absorb nutrients from maternal tissues for endosperm development, meanwhile act as storage for wastes produced from the grain filling processes. The decrease of head rice percentage under elevated CO_2 might attribute to substantial amount of assimilates retained in the alureone layer, which was removed during milling process.

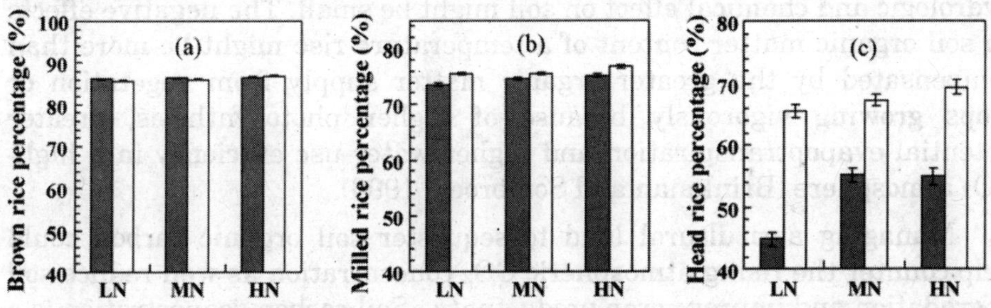

Fig. 2. Brown rice percentage (a), milling rice percentage (b) and head rice percentage (c), of rice plants subjected to ambient CO_2 (AMB; unfilled square) and free-air CO_2 enrichment (FACE, 200 mmol mol^{-1} above AMB, filled squares) under low (LN, 15 g N m^{-2}), medium (MN, 25 g N m^{-2}) and high (HN, 35 g N m^{-2}) levels of N application over three cropping seasons (2001–2003). Data are average values across 3 years with ±1 S.E. (vertical bars)

Soil carbon: Carbon is a key ingredient in soil organic matter, which is created by the cycling of organic compounds in plants, animals, and microorganisms into the soil. Well-decomposed organic matter forms humus, a dark brown, porous, spongy material that provides a carbon and energy source for soil microbes and plants.

Plants grown under elevated CO_2 generally increase the allocation of photosynthates to roots, which increases the capacity and/or activity of belowground carbon sinks (Canadell *et al.*, 1996), enhancing root turnover (Pregitzer *et al.*, 1995), rhizodeposition (Cardon, 1996), and mycorrhizal development in some but not all systems. These effects would be in synergy with better nutrient uptake by the intensive root system due to higher atmospheric CO_2 concentration. The increased production of root materials tends to raise soil organic matter content. Increased organic matter dynamics and microbial activity in soil are positive for the soil-plant system when CO_2 concentration rises gradually over decades.

In humid tropics and monsoon climates, increased intensities of rainfall events and increased rainfall totals would increase leaching rates in well-drained soil with high infiltration rates and would cause temporary flooding and water saturation, hence reduced organic matter deposition. In subtropical or semi-arid regions, the increased productivity due to higher CO_2 would tend to increase ground cover counteracting the effects of higher temperature. If there would be locally much less rainfall, and increasing intra- and inter-annual variability, these could lead to less dry matter production and hence in due course lower soil organic matter content. In temperate climates, minor increases in rainfall total, would be expected to be largely taken up by increased evapotranspiration of vegetation or crops at the expected higher temperature, so that the net hydrologic and chemical effect on soil might be small. The negative effects on soil organic matter content of a temperature rise might be more than compensated by the greater organic matter supply from vegetation or crops growing vigorously because of higher photosynthesis, greater potential evapoptranspiration and higher water use efficiency in a high-CO_2 atmosphere (Brinkman and Sombroek, 1996).

Managing agricultural land to sequester soil organic carbon could help counter the rising atmospheric CO_2 concentration as well reduce soil degradation and improve crop productivity. Soil carbon sequestration is a land-based option to reduce green house warming potential of the atmosphere. Soil aggregation is the primary process that protects organic matter from biological oxidation to form CO_2. Tillage practices lead to disruption of larger aggregates, partly by mechanical forcing, and partly because of exposing the soil to the harsh environment of soil surface where wet-dry and/or freeze-thaw cycles are most frequent. The loss of organic carbon associated with tillage-induced destruction of soil aggregates in agricultural soil can be considered as a conversion of organic carbon from a protected pool to a labile pool that is then mineralized to CO_2. The loss of organic carbon from the soil can be stopped by minimizing the tillage operation as minimum or no tillage. Tillage practices for rainfed maize production in United States indicated soil carbon sequestration of 337 kg C/ha/year in no-tillage system over conventional tillage (Duxbury, 2005).

Agricultural pests: A gradual, continuing rise in atmospheric CO_2 will affect pest species directly (i.e. the CO_2 fertilization effect) and indirectly (via interactions with other environmental variables). However, individual species responses to elevated CO_2 differ; consumption rates of insect herbivores generally increase, but this does not necessarily compensate fully for reduced leaf nitrogen. In general, leaf chewers (lepidoptera) tend to perform poorly (Joutei *et al.*, 2000), whereas suckers (aphids) tend to show large population increases (Bezemer and Jones, 1998)– indicating that pest outbreaks may be less severe for some species but worse for others under high CO_2. Some experiments under elevated CO_2 have suggested that aphids may become more serious pests, although other studies have discerned no significant effects on sap-feeding homopterans. Most studies have concluded that insect pests will generally become more abundant as temperatures increase, through a number of inter-related processes, including range extensions and phenological changes, as well as increased rates of population development, growth, migration and over-wintering.

Temperature directly affects virtually all biological rate processes in insects and population may change dramatically in response to climate change (Coxwell and Bock, 1995). Shifts may favour increased food consumption and digestion, more rapid development and faster growth rate, increased survival, or higher fecundity (Woods and Kingsolver, 1999; Peterson *et al.*, 2000). The small size of insect makes it more important to recognize microclimate rather than macroclimate when assessing response to global climate change (Lactin and Johnson, 1998). Operative thermal environments relevant for insects may change in ways that are not obvious from general description of local weather conditions. Insects are ectotherms of generally small size, and rely on external heat source and sink to control body temperature. Often, insects control body temperature using a variety of anatomical and behavioral means called thermoregu-lation to keep body temperature around 38ºC for as much of time as possible. As insects have so much control over body temperature and microclimates have so much impact, broad-scale prediction of impact of temperature changes must be tempered and scaled to the finer level.

Mitigation and adaptation

Agriculture is a source as well as victim of greenhouse gas emission. The agricultural activities such as ploughing land, shifting cultivation for agricultural expansion, chemical farming, and livestock release these greenhouse gases into the atmosphere. According to IPPC, the annual amount of greenhouse gases emitted by the agricultural sector is estimated between 5.1 and 6.1 Gt CO_2 equivalents, which is approximately 10–12% of total greenhouse gas emissions in 2005 (Barker *et al.*, 2007). Among the agricultural sector, the greenhouse gas emission is maximum from soil (38%), followed by enteric fermentation (32%), biomass burning (12%), rice production (11%), and manure (7%). However, by sequestering carbon dioxide in the soil, agriculture has the potential to be a considerable CO_2 sink to largely reduce the greenhouse gas emission to mitigate their effect on crop production and on the livelihoods of farmers.

The IPCC has suggested a range of measures for mitigation of greenhouse gas emission from agricultural system (Smith *et al.*, 2007).

The prominent mitigation options are improved crop land management including nutrient management, tillage/residue management, and water management; improved grazing land management; and restoration of degraded soil. Organic agriculture has the potential to reduce greenhouse gas emission by sequestering CO_2 in soil. The global warming potential of organic farming is considerably smaller as compared to conventional farming (chemical farming). In two long term comparison experiments with arable crops in Switzerland, the global warming potential of all crops was reduced by 18% in organic plots (Nemecek *et al.*, 2005). Additionally, under dry condition, organic agriculture may outperform the conventional agriculture.

Farmers can adopt coping mechanisms that withstand climate variability through activities such as the use of drought- or salt-resistant crop varieties; more efficient use of water resources; changing crop management: crop diversification, sowing time, nutrient management, tillage method, etc; and improved pest management practices. Modification sowing/planting time could be an adaptation strategy to counteract the adverse effect of climate change. Adjustment of planting time may help to offset any detrimental effects of climate change on rice production. From an adaptation study in rice production of eastern India, Krishnan *et al.* (2007) simulated an average yield changes of +6.6, +4.1 and -9.8% during July 15 sowing, which was considerably higher than the yield changes of -19.7, -20.3 and -30.7% during June 15 sowing for the GFDL, GISS and UKMO scenarios, respectively at Cuttack (Fig. 3). Likewise, the Jorhat site showed +27.1, +24.3 and +13.4% yield changes for the sowing on July 1 under the GFDL, GISS and UKMO scenarios, respectively; these changes were considerably higher than those of +12.1, +12.6 and +8.3% for the June 15 sowing for the corresponding scenarios (Fig. 4). Further postponement in sowing did not improve the grain yield, probably due to low incident solar radiation and temperature. There were striking differences in the predicted yield changes among the three scenarios. The large differences were mainly due to the sensitivity of spikelet sterility to temperature. Adaptation practices require extensive high quality data and information on climate, and on agricultural, environmental and social systems affected by climate, with a view to carrying out realistic vulnerability assessments and looking towards the near future.

Swain and Thomas (2009), using CERES-Rice simulated rice grain yield of two popular cultivars namely 'Lalat'(medium duration type, 120d) and 'Swarna' (long duration type, 145d) for climate change scenarios and to evaluate agro-adaptations at three locations (Kharagpur, Dumdum and Purulia) in eastern India. An advanced planting time of 10 to 30 days from current farmers' planting time (15 July) is expected to mitigate the adverse effect of climate change on the rice yield. Cultivar selection and plating time adjustment could be suitably managed for better adaptation to minimize the risk of yield loss due to climate change. At Kharagpur and Dumdum, among the different planting dates, the 15 June planting simulated highest grain yield whereas planting on 15 August simulated the lowest grain yield of both the cultivars. Unlike Kharagpur and Dumdum, at Purulia, among the different planting dates, the planting on 5 July simulated the highest grain yield of both the cultivars, where 15 August planting continued to simulate the lowest grain yield.

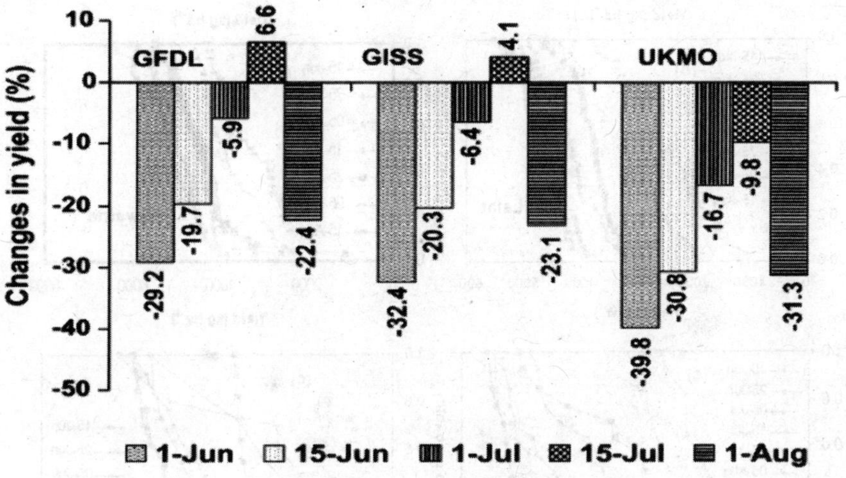

Fig. 3. Changes in yield (%) under the GCMs scenarios of the rice variety IR 36 under different sowing dates grown during the *kharif* (summar) season at Cuttack

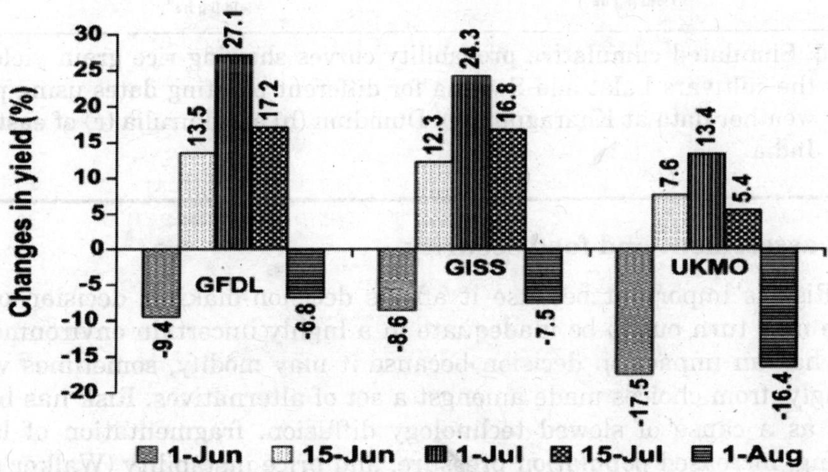

Fig. 4. Changes in yield (%) under the GCMs scenarios of the rice variety IR 36 under different sowing dates grown during the *kharif* (summar) season at Jorhat

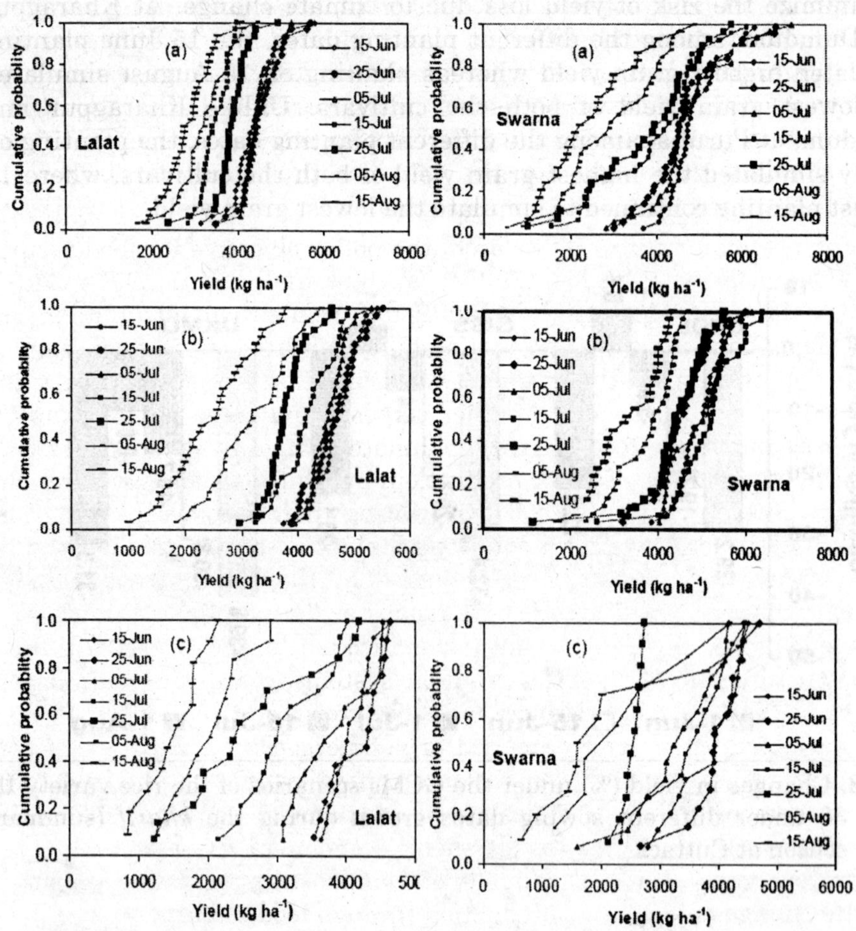

Fig. 5. Simulated cumulative probability curves showing rice grain yield of the cultivars Lalat and Swarna for different planting dates using past weather data at Kharagpur (a), Dumdum (b) and Purulia (c) of eastern India

Risk assessment and food security

Risk is important because it affects decision-making; decision once made may turn out to be inadequate in a highly uncertain environment. Risk has an impact on decision because it may modify, sometimes very strongly, from choices made amongst a set of alternatives. Risk has been cited as a cause of slowed technology diffusion, fragmentation of land holding, increased population pressure, and price instability (Walker and Ryan, 1990). Indeed, agricultural production has to continue with management of risk, with impacts on food security at regional, national and world level. Informed decision making under risk involves a marriage of two strands of information – the decision maker's expectation about the future happening, and the decision maker's preference. Many researchers often have difficulties in formalizing their expectation concerning the future, even when based on extensive past experience. Expectation has to be derived in terms of probability distribution into the future (mean yield with standard deviation) but not as point prediction, for better risk management.

In view of agricultural research, the policy objective for food security is synonymous with yield stability (Walker and Ryan, 1990). At the household level, food security for family may relate to an adequate income level, dependent not only on yield stability but also on off-farm employment opportunities. The year-to-year variation in food production, whether caused by variable weather or pests and diseases, for example, has direct impact on food security. Climate change is likely to affect the physical availability of food production by shifts in temperature and rainfall, people's access to food by lowering their incomes from agricultural industry, or lowering a country's foreign exchange earnings by the destruction of its export crops because of the rising frequency and intensity of tropical cyclones. Some groups are particularly vulnerable to climate change: low-income groups in drought-prone areas with poor infrastructure and market distribution systems, low to medium-income groups in flood-prone areas who may lose stored food or assets, farmers who may have their land damaged or submerged by a rise in sea-level, and fishers who may lose their catch to shifted water currents or through flooded spawning areas. However, it is thought that by 2030 more countries will have improved their economies, infrastructure and institutions, and may be capable of compensating for the impact of climate change on domestic production through food imports.

Climate variability is already the major cause of year-to-year fluctuations in production in both developed and developing countries. The largest reduction in crop production will occur in developing countries. These impacts would require adaptation efforts that in many cases will be hardly affordable for people living with little access to basic resources. Some of the impacts of climate change on food production, which are already visible and seem to be advancing at a higher rate than previously anticipated include:

- Increased heat stress to crop, e.g. higher night-time temperatures, which could adversely affect grain formation and other aspects of crop development
- Possible decline in precipitation in some food-insecure areas such as southern Africa and the northern region of Latin America
- Increased evapotranspiration rates caused by higher temperatures, and lower soil moisture levels

- Concentration of rainfall into a smaller number of rainy events with high intensity, increasing erosion and flood risks
- Changes in seasonal distribution of rainfall, with less amount in the main crop growing season
- Sea level rise, aggravated by subsidence in parts of some densely populated flood-prone countries, displacing millions
- Food production and supply disruption through more frequent of severe extreme events

Conclusion

Climate is changing at a tremendous rate and is influencing the agricultural sector to a great extent. The climate change associated with increasing levels of CO_2 may be beneficial to agricultural production in northern latitudes above about 55°, while the lower latitudes (developing world) will be adversely affected. The global average temperature is expected to increase by 0.2°C per decade for the next two decades. Many parts of the developing world are anticipated to face significant changes in temperature and rainfall pattern with increasing the frequency and intensity of extreme events such as droughts and floods. If the climate change effects dominate, world crop yields are likely to be decreased significantly, which will bring the food security issue into ever sharper focus in the future. Providing information to decision makers that can help, even in a limited way, to combat risk and ensure food security is thus a research area that warrants intense activity. At regional and national levels, a wide variety of biophysical crop models and techniques are applied for yield forecasting and evaluation of alternate agro-management practices for climate change adaptations/mitigations. Also, the crop models are coupled with socioeconomic model for food security assessment. As per all models and applications, the integrity of information produced by crop models to study risk and food security depends on sound models, sound validation procedure, and sound input data.

❑❑❑

10 A Biophysical Model to Analyze Climate Change Impacts on Rainfed Rice Productivity in the Mid-hills of Northeast India

B.U. Choudhury*, Anup Das, K. P. Mohapatra, Taznin A. Choudhury and S.V. Ngachan
ICAR Research Complex for NEH Region, Umiam, Meghalaya
*Corresponding author: burhan3i@yahoo.com

Introduction

In the recent past, water scarcity has been emerged as the potential threat to the sustainability of rice production *vis-à-vis* food and livelihood security for the growing populations of Northeastern Region (NER) of India. Rice being the staple food and major crop, occupies 3.51 M ha of the total 4.13 M ha net sown area. Rainfed lowland rice production system alone occupies more than 80% of the total rice area of NER (Munda *et al.*, 2009) and contributes more than 85% of the total 6.17 million ton food grain production. However, productivity of rice in the region excluding the state Assam known as Northeastern Hill Region (NEH) of India, is much less (<1.7 t ha^{-1}) than the national average. As a result, the region is still in deficit of about 1 million tonnes of rice grain annually (Munda *et al.*, 2009) and 12% deficit in food grain requirement for a population of 45 million. The deficit in rice as well food grain requirement is likely to increase by another 10% in the next 8 years (2020).

Despite receiving more than 2000 mm annual average rainfall across NEH, water scarcity is an important constraint to crop productivity because of erratic and uneven rainfall distribution affecting timeliness of rice crop establishment, and resulting in dry spells during the monsoon season and terminal drought stress. This situation prevails in vast areas where farmers are solely or largely dependent on rainfall for the rice crop. Unproductive (deep percolation , under bund seepages, over bund runoff etc.) water loss in farmers field accounted for 65-73% of rainfall during rice growing season and thus, water productivity with respect to rainfall received during crop growth periods is very low (< 0.16 g grain kg^{-1} water). Farmers of the region follow subsistence farming; they grow rice mostly without adequate supplementation of fertilizer inputs (NPK). In very few pockets, farmers apply only nitrogen fertilizer at blanket dose in the form of di-ammonium phosphate (Das *et al.*, 2011). Due to heavy downpours of monsoon rains during rice growing season, most of the applied N-fertilizer is lost either by runoff in sloppy lands or by over bund flow/leaching in valley lands. As a result, N-use efficiency in lowland paddy is very low (<25%). Sole dependence on erratic and uneven rainfall distribution, lack of water availability for meeting crop water requirement at critical periods, occurrence of intermittent terminal drought stresses etc. pose severe threat to the sustainability of rice ecosystem. The risk of crop failure and food insecurity has been further complicated by the lack of any sustainable water management strategies including life saving irrigation approaches in these fragile hilly ecosystems of India. Due to the impact of regional climate change, like other region, NEH is also experiencing abrupt changes in rainfall pattern and temperature (Kumar, 2011), which may further pose a serious threat to water availability and rice production *vis-à-vis* food and environmental security of the region. Study area (mid altitude Meghalaya) is also experiencing an increase in maximum temperature (@0.086°C/year), decrease in minimum temperature (@ 0.011°C/year) and decline in monsoon rainfall (@ 1.70 mm/year) with high frequency of anomalies of either deficit or excess in occurrence (Choudhury *et al.*, 2012).

For devising any immediate adaptation as well as long term mitigation strategies under changing climate scenario for this fragile hilly ecosystem, adequate information showing the effect of climate change particularly, temperature and carbon-dioxide (CO_2) on rice productivity is the major bottleneck. No such systematic intensive studies have been conducted to quantify the impact of climate change on rice productivity in the region. Although, a host of factors, both environmental (rainfall, temperature, solar radiation, CO_2 concentration etc.) and management (soil, fertiliser, irrigation, pest-diseases etc.) are responsible for crop perfor-mance, yet, many of these factors have contrary impacts on crop yields. Change in temperature (-/+) from critical limit along with water shortage may reduce the crop productivity substantially while more CO_2 concentration in the atmosphere may boost crop productivity (Jiang, 2009). Generation of such information through field experimentation is possible only under controlled environmental conditions like CO_2 and temperature growth chamber (CTGC), FATE, FACE etc. which are very expensive and highly skill oriented.

The best possible alternative is the use of location specific well calibrated and validated simulation models, since they are the most power tool for such studies to have the conclusive results (Jalota *et al*,

2006). Quantification and simulation of climate change's impact on crop yields can help in understanding as to what extent climate risk can be managed through management strategies. Agricultural production systems simulator (APSIM) is a biophysical model developed to simulate biophysical processes in farming systems, particularly in relation to the economic and ecological outcomes of management practices faced with climate risk. APSIM has the ability to simulate crop and soil biophysical process under climate change scenarios; researchers have done enormous work in investigating climate change impacts on wheat crops (Reyenga *et al.*, 2001; Luo *et al.*, 2005; Jiang, 2009; Gaydon *et al.*, 2012) APSIM has been listed as a primary investigating tool in climate change by UNFCCC (2002). In the present simulation study using a calibrated and validated APSIM model, an attempt has been made to quantify the effect of projected climate change (elevated CO_2 and modification of temperatures) on rainfed lowland rice productivity under different rates of N-fertilization in the NEH region of India.

Status of water resources of Northeastern region

Northeastern regions (NER) of India alone account for nearly 40% of the country's renewable fresh water resources, mostly due to the occurrence of annual rainfall of 641 km³ at an annual average of 2474 mm. This region is endowed with world's highest rainfall (>1000 cm annually) area i.e. Mousinram near Cherrapunji. However, due to erratic variability in rainfall (both in spatial and temporal dimensions), irony of the region is that even Mousinram, which receives the highest rainfall in the world, also suffers most often from severe water scarcity during pre and post monsoon months. The total surface water and replenishable ground water resources of NER are estimated as 629 km³ and 33.84 km³, respectively. Unfortunately, most of the surface water in the region is confined to two major river basins (Brahmaputra – Barak) which are in highly dynamic stage due to Himalayan glacial fed origin and steep gradient. As a result, only 3.8% of total surface water resources (24 km³) of the region are utilizable. Currently, the total utilizable water resource (both utilizable surface and net available ground water) of the region is 54.8 km³ for a population of over 39 million (as per 2001 census). Although, the average per capita per year total surface water (Brahma-putra and Barak basins) in NER is as high as 16589 m³, yet, the total utilizable water resources per capita per year is as low as 1404 m³. It is projected that by 2021, an additional 20 million population will be added to the region. Consequently, the already low per-capita per year total utilizable water availability (1404 m³) will further reduce to < 1000 m³ and the region will be pushed from already water stressed to water scarce zone (as per the international norms, less than 1700m³, the region is classified as water stressed and below 1000 m³, water scarce) (Kumar *et al.*, 2005). This will be further compounded by the increase in fresh water demands from present 27.4 km^3 to 43.0 km^3 by the year 2025 and agricultural sector will be the major consumer (>65%) of it. Therefore, sufficient food grain production to cater the demand of burgeoning population growth and to continue other developmental process are at peril since availability of water resources for human use in NER is at great stake.

Climate change and water balance/hydrological cycle

Glaciers play an important role in maintaining ecosystem stability as they act as buffers and regulate the runoff water supply from high mountains to the plains during both dry and wet spells. They release more water in a drought year and less water in a flood year and thus ensure water supply even during the lean years. Major river basin (Brahmaputra and sub basins) of NE region is of Himalayan origin and is fed by Himalayan glaciers. Apart from monsoon rains, the glacial melt water from the snow and ice keeps the Brahmaputra and its sub basins (Barak- Subansiri- Jia-Bharali) perennial and plays a significant role in water resource development of NE region of India. Average intensity of mountain glaciations is 1.3% for Brahmaputra and it varies from 0.4 to 4% for its tributaries.

Abrupt changes in climate will significantly modify hydrological cycle and water balance front of NE region, most likely in adverse way. A warm climate will alter rainfall distribution pattern (temporal and spatial dimensions). Intense rain may occur over fewer days and thus, the total amount of rainwater received over longer time span will be further shortened. As a result, magnitude of runoff loss of water will increase by several folds from the existing very high rates. Dominance of hilly terrain features and faulty land use practices in NER will further encourage the severity of runoff and subsequent loss of fertile soils by water erosion processes. Currently, faulty land use practices are causing an annual loss of more than 40 million tonnes of soil and 10.65, 0.37 and 6.05 thousand tonnes of available N, P_2O_5 and K_2O, respectively (Prasad and Sharma, 1993). Increase in quantum of sediment loads and siltation of river basins, reservoirs and other water storage bodies will be further accelerated. Consequently, the region will experience more frequent occurrence of floods and thus, its extension to even in non-flood prone areas. Currently, flood is affecting annually 3760 km^2 area of the region. Total flood prone areas of the region have already exceeded 35000 km^2. With abrupt change in climate, flood prone areas may further increase substantially. Frequent landslides and other forms of land degradation will be more frequent and conspicuous. All these will bring down uncertainty about the sustainability of habitats, productive land mass, water bodies and thus existence of human lives.

Excessive and rapid rate of runoff and soil erosion due to rainfall variation will cause consistent decline in infiltration rate and other hydrodynamic properties (water transmission-retention) of the soil. As a result, reduction in ground water recharge and subsequent fall in already falling water table will be more severe and prevalent across NE region. The already / existing lack of sufficient moisture in the vadose zone of the soil solum during pre and post monsoon months will be further aggravated. Evaporation is one of the major components of hydrological cycle/water balance, which alone accounts for more than 50% of total water loss from free water and soil surfaces. Increase in temperature due to climate change will accelerate the rate of water loss through evaporation. As a result, availability of soil moisture during winter season will be further reduced and the occurrence of agricultural drought will be wide spread across the NE region. Ultimately, in NER, the extreme hydrological events (flood-drought cycle) will be repeated more often inter-annually. Therefore, sustainability of ecological balance, agricultural production system, socio-economic conditions of the farming community, all-round developmental process of this fragile region will be at peril.

Description of the study area

The study area experiences a tropical monsoon climate. Seventy percent of the total rainfall is received from July to September, with average annual rainfall of 2450 mm. April is the hottest month, with average minimum and maximum temperatures of 17.3°C and 29.4°C, respectively. The coldest month is December where the average minimum and maximum temperatures are 7.6°C and 20.4°C, respectively. The average relative humidity is highest in the month of June (89.4%) while January records the lowest relative humidity of 72 percent. Daily pan evaporation rate varies from 2.04 mm day^{-1} (during December) to 4.60 mm day^{-1} (during April) with a mean value of 2.89 mm day^{-1}. Daily wind speed varies from 2.58 to 4.39 km hr^{-1}, with a mean value of 3.36 km hr^{-1}. Average sunshine hours are 5.42 hr day^{-1} (Choudhury *et al.*, 2012).

Parameterization of APSIM model

APSIM model was parameterized and validated based on two field experiments on rainfed lowland rice conducted for three consecutive years (2009-2011) at the research farm of Indian Council of Agricultural Research Complex for NEH region (25°41'N latitude, 91°55'E longitude, 1010 m above mean sea level) Meghalaya, Northeast India. The experiments were laid in Factorial Randomized Block Design with three replicates having plot size of 5 m x 4m. The rice genotype 'Shahsarang' (HYV) was grown in the wet season (June/July to November/December).

Twenty one days-old seedlings were transplanted in a puddled field with a spacing of 20 cm row to row and 15 cm from plant to plant. The crop was grown under rainfed condition with sole source of water input was rainfall. Nitrogen in the form of urea was applied at the rate of 90 kg ha^{-1}, of which 50% was applied basally, 25% at maximum tillering and the remaining 25% at panicle initiation. Other basal fertilizer applications were 60 kg ha^{-1} P in the form of single super phosphate, 40 kg ha^{-1} K in the form of *murate* of potash. The APSIM model was parameterized using relevant soil, crop and climate data and subsequently validated using field experimentation. Genetic coefficients of HYV cultivar Shahsarang were generated. Satisfactory performance was achieved for biomass production and grain yield. Simulated values were within the permissible range of ±18% deviation from measured (Fig. 1). The root mean square error (RMSE) was 0.48 t ha^{-1} and normalized RMSE was 5.6% of the observed yield.

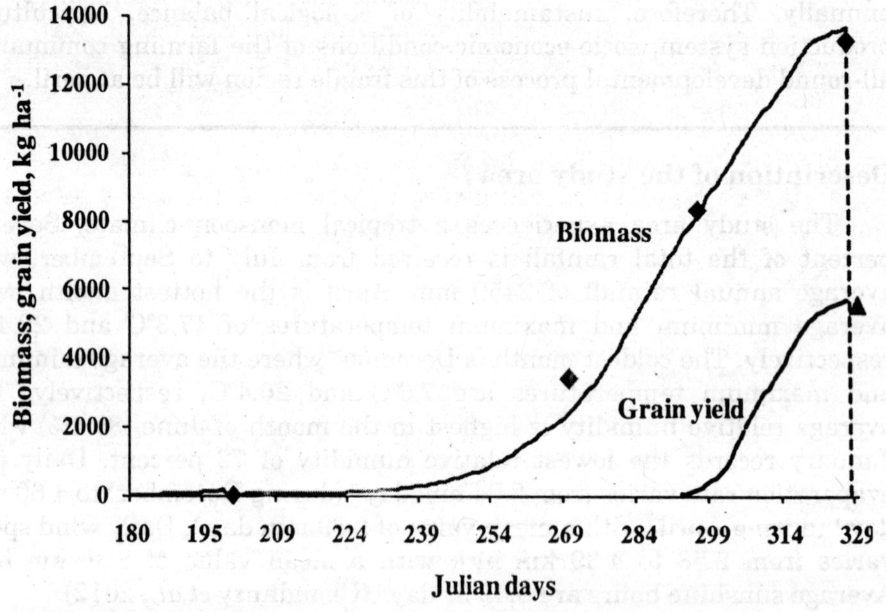

Fig. 1. Comparison of simulated (solid and dash lines) and measured biomass of the whole crop ( ), and grain yield (▲) of rice.

Climate change scenarios modelled

Scenarios considering rice growth and yield response to changes in average temperatures (-2°C, +2°C, +4°C vs. long period average temperature of 22.2°C during rice growing season) over three carbon dioxide levels (390, 550, and 700 ppm) and four N-fertilizer levels namely 0, 60, 90 and 120 kg ha^{-1} were simulated under rainfed condition using historical climate data (1987-2011) to provide a realistic measure of seasonal climatic variability for the modelled scenarios. A total of 48 scenarios consisting of three carbon dioxide levels, four temperatures and four N-fertilizer levels were simulated. The average annual rainfall for 25 years (1987–2011) was 2415 ± 373 mm, of which monsoon rains (June-September) coincides with rice growing season was 1546 ± 288 mm. The average temperature during rice growing seasons (July-November) for 25 years (1987–2011) was 22.2 ± 0.31°C (long period average, LPA). Long period average temperature of 22.2 °C temperature has been considered as ambient temperature.

Trend analysis of long period weather data and simulated grain yields

Trend analysis (+/-) of long period (1987-2011) simulated grain yields of rice under different scenario options was statistically examined in two phases. Firstly, using the non parametric Mann-Kendall test, the presence of a monotonic increasing or decreasing trend was tested based on normalized test statistics (Z) value. The trend is said to be *decreasing* if Z is negative and *increasing* if the Z is positive. In the second phase, the rate of increase or decrease in annual trend was estimated with the nonparametric Sen's slope estimator. The slope of the trend gave the annual rate and direction of change (Kendall, 1995; Helsel and Hirsch, 2002 and Salmi *et al.*, 2002).

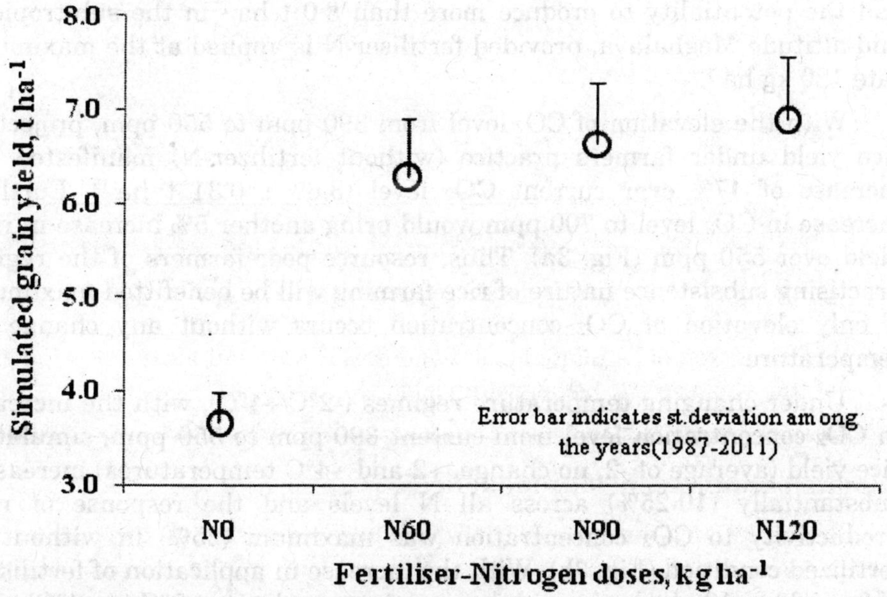

Fig. 2. Simulated rice yield (average of 1987-2011) at ambient temperature (22.2°C) and 390 ppm CO_2 concentration.

Simulated rice yield with base line CO_2 concentration and temperature

Simulation results from the scenario analysis with current CO_2 concentration of 390 ppm (IPCC, 2011) and ambient temperature of 22.2°C revealed considerable effect of fertiliser-N levels on rice productivity (Fig. 2). APSIM model was very sensitive in capturing the effect of N-fertilization. Shifting from without N-fertilization to 60 kg N ha^{-1}, projected yield (averaged of 25 years, 1987-2011) increased by 71% than without N fertilization (3.37 ± 0.31 t ha^{-1}). On further increase in N application rate to 90-120 kg ha^{-1}, consistent increase in simulated rice yield by an additional 10-17% over 60 kg N ha^{-1} was observed.

Effect of elevated CO_2 concentration levels on simulated rice yield at ambient temperature

Without changing ambient temperature of 22.2°C, elevation of only CO_2 concentration from current 390 ppm to 550 ppm, simulated rice yield increased substantially (8-14%) across all the three fertiliser-N levels (60, 90 &120 kg ha^{-1}). On further increase in CO_2 level to 700 ppm in N-fertilized condition (60-120 kg N ha^{-1}), an additional increase of 4-7% rice yield over 550 ppm was registered (Fig. 3a). Simulation studies reflected that under elevated CO_2 level of 700 ppm, HYV rice cultivar Shahsarang had the potentiality to produce more than 8.0 t ha^{-1} in the sub-tropical mid-altitude Meghalaya, provided fertiliser-N is applied at the maximum rate 120 kg ha^{-1}.

With the elevation of CO_2 level from 390 ppm to 550 ppm, projected rice yield under farmers practice (without fertilizer-N) manifested an increase of 47% over current CO_2 level (3.67 ± 0.31 t ha^{-1}). Further increase in CO_2 level to 700 ppm would bring another 5% increase in rice yield over 550 ppm (Fig. 3a). Thus, resource poor farmers of the region practising subsistence nature of rice farming will be benefitted maximum if only elevation of CO_2 concentration occurs without any change in temperature.

Under changing temperature regimes (-2°C/+4°C), with the increase in CO_2 concentration level from current 390 ppm to 550 ppm, simulated rice yield (average of -2, no change, +2 and +4°C temperatures) increased substantially (10-25%) across all N levels and the response of rice productivity to CO_2 concentration was maximum (25%) in without N fertilized condition (Fig. 3b). With the increase in application of fertiliser-N from 60 to 120 kg ha^{-1} across temperature gradients (-2°C to +4°C), rice productivity also reciprocates and registered a consistent increase of 10-19% over current level (390 ppm). Further increase in CO_2 level from 550 ppm to 700 ppm, response of simulated average rice productivity along the temperature gradients (-2°C to+4°C) reflected a marginal increase (4-7%) across all levels of fertiliser-N. Same is the case with non-fertilized condition (without N) where projected rice yield increased only by 4% over 550 ppm CO_2 level (4.58 t ha^{-1}). Addition of fertiliser-N at 60 kg ha^{-1} to farmers practice (without N), rice productivity increased by 15% at 550 ppm and 18.7% at 700 ppm of CO_2 level, respectively. Increasing the N rate further from 60 to 90 kg ha^{-1}, an additional increase of 8.6-8.9% at both the levels of CO_2 level (550 and 700 ppm) was observed (Fig. 4b). Beyond 90 kg N ha^{-1}, response of rice productivity to elevated CO_2 levels (550 and 700 ppm) was marginal (<3.5%).

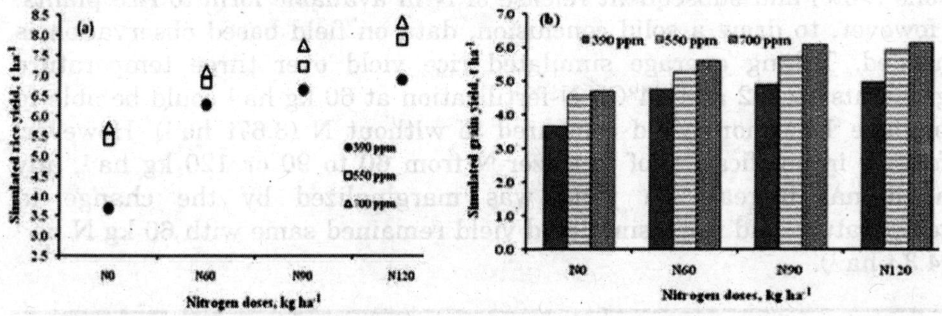

Fig. 3. Effect of elevated CO_2 concentration on simulated rice yield at (a) ambient temperature of 22.2°C and (b) averaged over temperature gradient (-2 to +4°C).

Despite changes in average temperature (-/+), elevated CO_2 level upto 550 ppm can substantially improve the rice productivity under low fertiliser -N (without N to 60 kg N ha^{-1}) intensive farming while at higher level of CO_2 concentration (700 ppm), intensive application of fertiliser-N (@ of 90 kg N ha^{-1}) can complement the rainfed rice productivity in the subtropical climate of mid-altitude Meghalaya. Therefore, fertiliser-N application manifested a synergistic effect on rice productivity under elevated CO_2 levels (550 and 700 ppm) and, thus, could be a possible adapt-ation strategies under changing climate, particularly temperature and elevated CO_2 concentration.

Effect of change in average temperatures (-/+) at basal CO_2 concentration level

With the changes in average temperature (-2°C to +4°C) over ambient temperature (22.2°C) at current CO_2 levels of 390 ppm, the simulation (1987-2011) results exhibited yield penalty of rice cultivar Shahsarang with wide variation in magnitude of reduction (22-47%) across all levels of N-fertilization (60-120 kg ha^{-1}). By decrease in temperature by 2°C from current 22.2°C, simulated rice yield declined by

18% in treatment without N- fertilization (farmers practice) while with N-fertilization (@60-120 kg ha^{-1}), corresponding reduction in projected rice yield was very high (38-42%). Similarly, increase in rate of N-fertilization (60-120 kg ha^{-1}) under elevated temperature (by 2-4°C), simulated yield declined by 22 to 25% and 36-47% at 2°C and 4°C increase, respectively. However, simulation showed that with 2°C increase in temperature under farmers practices (without N-fertilization), rice yield increased by 19% over 3.67±0.31 t grain ha^{-1} under ambient temperature (22.2°C). With 4°C increase in ambient temperature, simulated yield under farmers practices (-N fertilization) remained unaffected. Increase in temperature probably facilitated the mineralization rate of high organic carbon content of the soils (>3%) and subsequent release of N in available form to rice plants. However, to draw a solid conclusion, data on field based observation is needed. Taking average simulated rice yield over three temperature gradients (-2, +2 and +4°C), N-fertilization at 60 kg ha^{-1} could be able to produce 15% more yield compared to without N (3.67t ha^{-1}). However, further intensification of fertilizer-N from 60 to 90 or 120 kg ha^{-1}, any additional increase in yield was marginalized by the change in temperature and thus, simulated yield remained same with 60 kg N ha^{-1} (4.3 t ha^{-1}).

Effect of change in average temperatures (– / +) at elevated CO_2 concentration

At elevated CO_2 concentration levels (550-700 ppm), fall in temperature by 2°C over ambient 22.2°C, simulated rice yield declined substantially across both without -N and with N-fertilization (Fig. 4). Beneficial effect of elevated CO_2 levels (550 and 700 ppm) in increasing the simulated rice yield by 47-56% in farmers practice (without –N) and 8-20% with 60-120 kg N ha^{-1} over corresponding rice yield at CO_2 level of 390 ppm has been offset by the change in temperatures (-2 to +4°C). With the change in temperature (±2°C), the projected rice yield under different levels of N fertilization decreased by 12 to 41% and the reduction was more with the decrease compared to increase in temperature. Further increase in temperature to 4°C over 2°C, an additional decline in simulated yield by 14 to 20% was observed. The rate of decline in rice yield with the change (-/+) in temperatures was more at higher doses (90 to120 N kg ha^{-1}) compared to lower doses of N fertilization (0 to 60 N kg ha^{-1}) (Fig. 4). Thus, either shifting from without N to 60 kg N ha^{-1} or even application of higher doses of fertiliser-N (90-120 kg ha^{-1}) at elevated CO_2 levels (550-700 ppm) could not be able to compensate the decline in simulated rice yield compared to the simulated yield at ambient temperature.

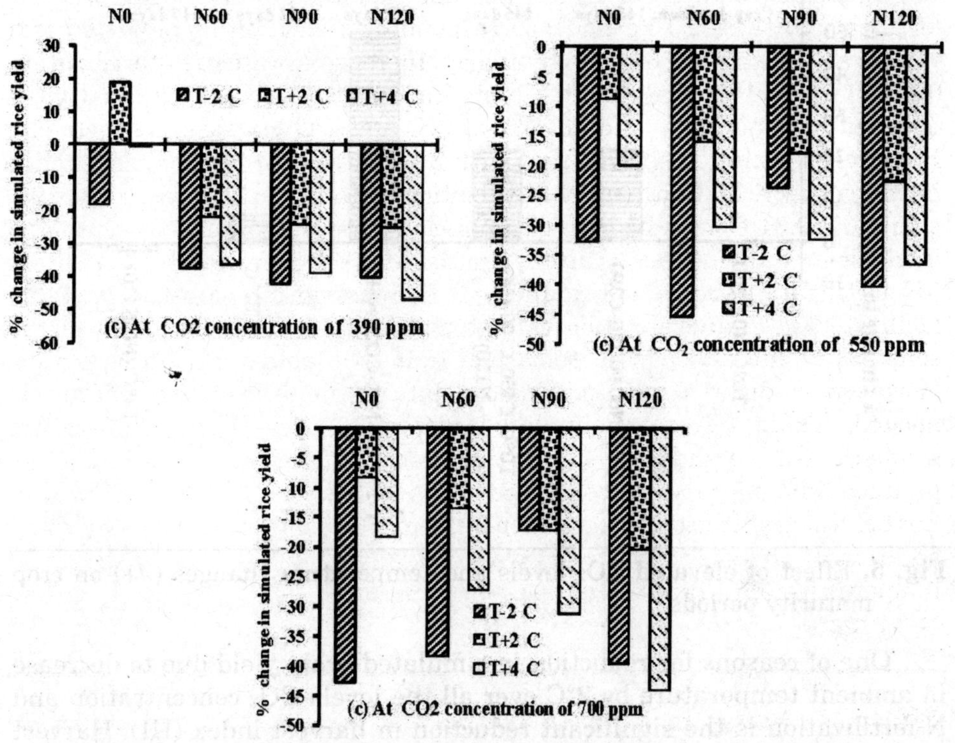

Fig. 4. Effect of change in temperature (-2 to+4°C) and elevated CO_2 levels (390 to 700ppm) on simulated rice productivity.

Effect of elevated CO_2 levels and change in temperature (–/+) on crop maturity periods and harvest index

At ambient temperature (22.2°C) and CO_2 concentration of 390 ppm, rice cultivar Shahsarang under farmers practice (without N-fertilization) matured (seed to seed) in 119 days. With the application of fertilizer-N (60-120 kg ha^{-1}), simulated maturity period increased to 145 days (22% higher than without-N) (Fig. 5). Similarly, elevated CO_2 levels (550-700 ppm) at ambient temperature also increased crop maturity period by 22% over CO_2 level of 390 ppm. Increase in CO_2 concentration might have led fertilization effect (similar to N-fertilization) on crop growth and thus, duration from seed to seed increased substantially over low level of CO_2 concentration (390 ppm). However, irrespective of CO_2 concentrations and fertilizer-N levels, decrease in temperature by 2°C over 22.2°C, crop maturity period increased by 50% (179 days) over 119 days under farmers practice (without fertiliser-N) with ambient temperature. In contrast to decrease in temperature, increase in temperature by 2 to 4°C, crop maturity period of cultivar Shahsarang shortened by 13%-20% (119-127 days).

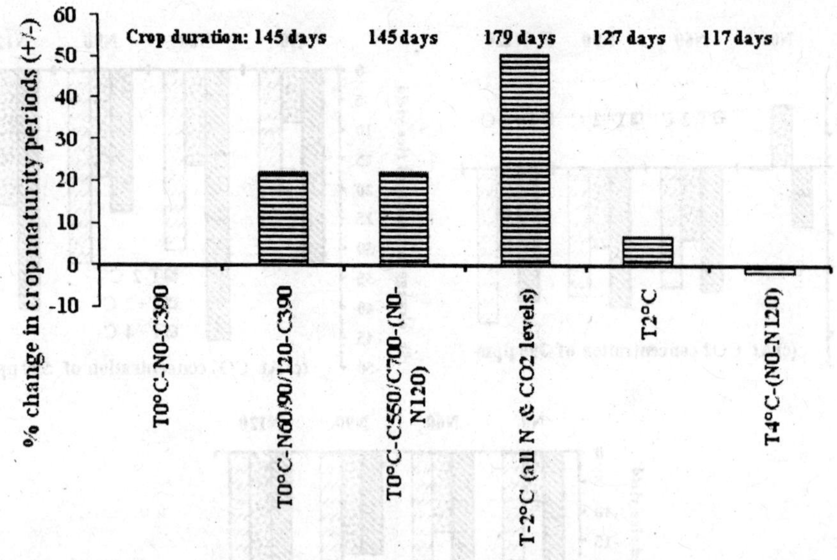

Fig. 5. Effect of elevated CO_2 levels and temperature changes (-/+) on crop maturity periods.

One of reasons for reduction in simulated grain yield due to decrease in ambient temperature by 2°C over all the levels CO_2 concentration and N-fertilization is the significant reduction in harvest index (HI). Harvest index takes into account the fraction of above ground total biomass (stem, leaf and panicle) which crop transforms into only productive spikelet/ grain formation. Low temperature did not promote in conversion of unproductive biomass (stem, leaf and panicle) to productive grain formation and as a result, HI was 52% less across all fertiliser N-levels compared to ambient temperature of 22.2°C at 390 ppm CO_2 level (HI: 0.45). Under elevated CO_2 levels (550-700 ppm), low temperature (-2°C) also reduced HI by 35-37% over ambient temperature but fertilization effect of CO_2 at elevated levels (550-700 ppm) could be able to compensate 15-17% yield recovery over low level of 390 ppm CO_2 concentration.

Besides crop duration and harvest index, elevated temperature (by 2-4°C) also substantially increased the soil evaporation losses (13-24%), and increased soil moisture stress. The drought stress factor causing the reduction in leaf expansion (lestres) in APSIM-Oryza could reflect the occurrence of terminal drought during the rice growing periods. This was indicated by the stress sensitive indicator of APSIM module (lstres) and the simulation reflected that upon increasing temperature by +4°C, soil moisture stress factor increased by 31% over ambient temperatures. APSIM module also indicated intermittent moisture stresses on increasing the temperatures (2-4°C), which might have exerted a cumulative drought effect on the rainfed agriculture of NEH region and thus, simulated yield reduced. With low temperature, crop maturity period increased. Analysis of historical climate data (1987-2011) revealed that delay in maturity resulted in considerable variation in rainfall distribution during crop growth seasons, more particularly from post flowering to grain filling stages reduced considerably by many folds which influenced the productivity. This was also affirmed from a very strong positive correlation (r = +0.64) between post flowering rainfall and grain yield.

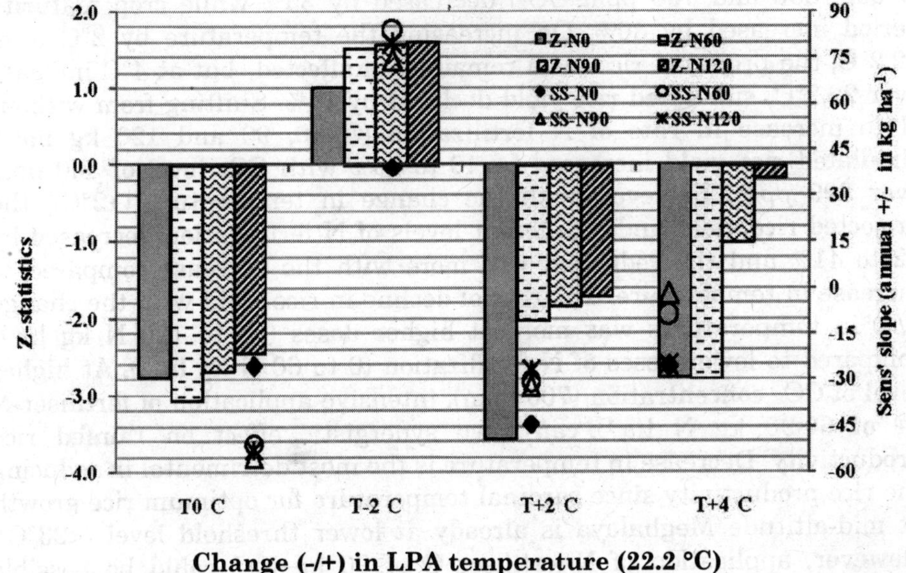

Fig. 6. Trend analysis of simulated rice yield (1987-2011) over elevated CO_2 and temperature changes. (T N0, N60, N90 and N120 are the nitrogen doses @ kg ha^{-1} & SS indicates Sens' slope)

Irrespective of fertiliser N levels, simulation studies and trend analysis of long term rice productivity using Man-Kendall non-parametric tests (Z-statistics) and Sens' slope manifested a significant (Fig. 6) declining trend in grain yield, particularly under elevated CO_2 levels (550-700 ppm) as well as temperatures (2-4°C) across all levels of fertiliser-N. However, with the decrease in temperature by 2°C from ambient 22.2°C, simulated yield reflected an increasing trend over the years (1987-2011). Since study area (Umiam, Meghalaya) is experiencing an increasing trend in average temperature annually (0.03°C/year) (Choudhury *et al.*, 2012), so, this might have exerted a beneficial effect on low temperature and thus offset the declining trend in simulated rice productivity.

Conclusion

The simulation (1987-2011) results indicated increase in rice grain yields with increase in CO_2 concentration while changes (-/+) in average temperature showed yield reduction. Elevation of CO_2 concentration from current 390 ppm to 550 ppm under farmers practice (without N), projected rice yield (average of -2, no change, +2 and +4°C temperatures) of HYV cultivar (Shahsarang) increased. On further increase in CO_2 concentration to 700 ppm, however, rice yield increased marginally (< 4%). With the decrease in average temperature by 2°C from the ambient temperature (June-November) of 22.2°C, the projected rice yield (average of 390, 550 and 700 ppm CO_2) decreased by 35% while crop maturity period increased by 30%. On increasing the temperature by 2°C from 22.2°C, the projected rice yield remained unaffected but at 4°C increase over 22.2°C, simulated rice yield declined by 14%. Shifting from without N to increase in rate of N fertilization (@ 60, 90 and 120 kg ha^{-1}), simulated rice yield increased by 10 to 19% with CO_2 level of 550 ppm over 390 ppm. However, with the change in temperature (±2°C), the projected rice yield under different levels of N fertilization decreased by 12 to 41% and the reduction was more with the decrease compared to increase in temperature. The rate of decline in rice yield with the change (-/+) in temperatures was more at higher doses (90 to 120 N kg ha^{-1}) compared to lower doses of N fertilization (0 to 60 N kg ha^{-1}). At higher level of CO_2 concentration (700 ppm), intensive application of fertiliser-N (@ of 60-90 kg N ha^{-1}) can have synergistic effect on rainfed rice productivity. Decrease in temperature is the most detrimental in reducing the rice productivity since cardinal temperature for optimum rice growth at mid-altitude Meghalaya is already at lower threshold level (<23°C). However, application of N-fertiliser @ of 90 kg ha^{-1} could be possible immediate adaptation strategies under changing climate in the sub-tropical climate of mid-altitude Meghalaya.

❑❑❑

11 AquaCrop Modelling: A Water Driven Simulation Model

Monalisha Pramanik and Rajeev Ranjan

Central Soil and Water Conservation Research and Training Institute, Research Centre, Datia (MP) 475 661

AquaCrop is free software which can be downloaded from FAO site. This model was developed to intend crop growth and yield simulation under limited water condition with minimal input. *AquaCrop* needs significantly smaller number of input parameters and maintain balance between simplicity, accuracy, and robustness. The model simulates the interrelationship of soil plant water and atmosphere under different crop management practices. The model has four major components: the atmosphere, the crop, the soil and management. Root zone water content is simulated by keeping track of inflow and outflow water fluxes. *AquaCrop* uses canopy ground cover instead of leaf area index. Canopy expansion, stomatal conductance, canopy senescence, and harvest index are the key physiological processes which respond to water stress. Low and high temperature stresses on pollination and harvestable yield are considered, as is cold temperature stress on biomass production. Evapotranspiration is simulated separately as crop transpiration and soil evaporation and the daily transpiration is used to calculate the biomass gain via the normalized biomass water productivity. The normalization is for atmospheric evaporative demand and carbon dioxide concentration, to make the model applicable to diverse locations and seasons, including future climate scenarios. *AquaCrop* accommodates fertility levels and water management systems, including rainfed, supplemental, deficit, and full irrigation. Simulations are routinely in thermal time, but can be carried out in calendar time. Future versions will incorporate salt balance and capillary raise. *AquaCrop* is aimed at users in extension services, consulting firms, governmental agencies, NGOs, farmers associations and irrigation districts, as well as economists and policy analysts in need of crop models for planning and assessing water needs and use of projects and regions.

Model Descriptions

Conceptually *AquaCrop* is the refinement of the eqn. 1 given by FAO *Irrigation & Drainage Paper* 33 (Doorenbos and Kassani, 1979) for predicting the yield response to water of annual and perennial crops. The equation is given below

$$\left(\frac{Y_x - Y_a}{Y_x}\right) = k_y \left(\frac{ET_x - ET_z}{ET}\right) \quad (1)$$

Where Y_x and Y_a are the maximum and actual yields, ET_x and ET_a are the maximum and actual evapotranspiration, and k is the proportionality factor between relative yield loss and relative reduction in evapotranspiration.

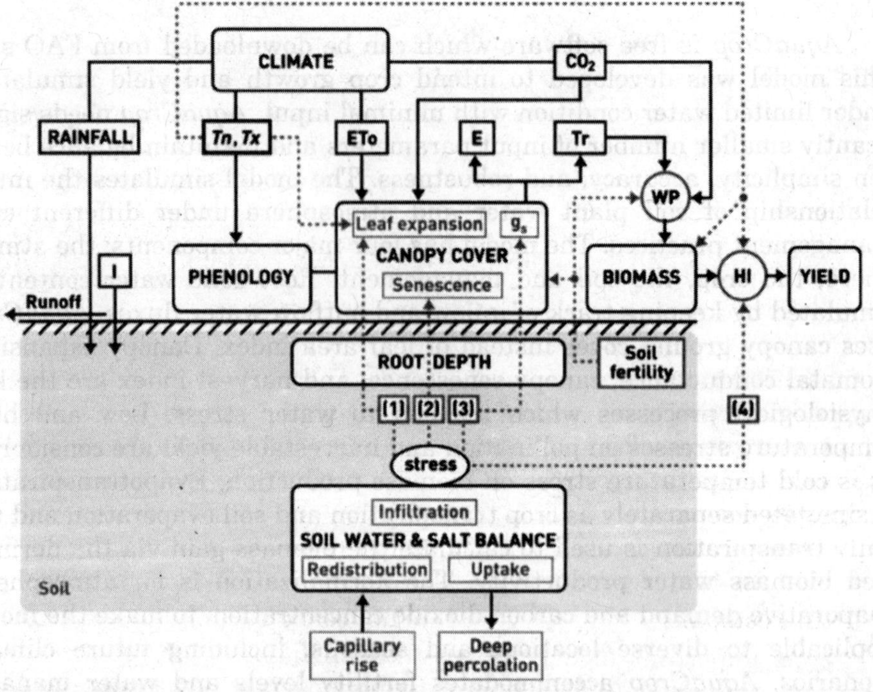

Fig.1. Chart of *AquaCrop* indicating the main components of the soil-plant-atmosphere continuum and the parameters driving phenology, canopy cover, transpiration, biomass production and final yield (I-Irrigation; T_n - Min air temperature; T_x - Max air temperature; ET_o - Reference evapotranspiration; E – Soil evaporation; T_r - Canopy transpiration; g_s - Stomatal conductance; WP- Water productivity; HI – Harvest Index; CO_2 - Atmospheric carbon dioxide concentration; (1), (2), (3), (4) - different water stress response functions). Continuous lines indicate direct links between variables and processes

In *AquaCrop* model, the crop evapotranspiration (ET) is separated into soil evaporation (E) and crop transpiration (Tr) to avoid the confounding effect of the non-productive consumptive use of water (E). This is particularly important when canopy cover of the ground is incomplete and soil E may be the major component of ET. The harvestable yield (Y) is expressed as a function of biomass (B) and harvest index (HI) to distinguish between environmental stress effects on B from those on HI. The separation of these two kinds of effects, which differ fundamentally, makes it possible to introduce functional links based on underlying physiological processes. The changes described led to the following equations at the core of the AquaCrop growth engine:

$$B = WP \times \sum T_r \tag{2}$$

and,

$$Y = B \times HI \tag{3}$$

Where WP is the water productivity parameter in units of kg (biomass) m^2 (land area) mm^{-1} (water transpired). Stepping from Eq. (1) to Eq. (2) makes the model more robust and more applicable, due to the conservative behavior of WP when normalized for climatic conditions (Steduto *et al.*, 2007), although both equations are expressions of a water-driven growth-engine in terms of crop model design (Steduto, 2003).

AquaCrop introduces daily time steps to account for dynamic changes in water supply, soil evaporation, crop transpiration and air temperature, in contrast to the use of Eq. (1) to compute production over long periods (weeks to months).

AquaCrop has a configuration that overarches the soil-plant-atmosphere continuum. It comprises the soil, with its water balance; the plant, with its development, growth and yield processes; and the atmosphere, with its thermal regime, rainfall, evaporative demand and carbon dioxide concentration. In terms of management, the model incorporates irrigation, bunding, mulching and also considers soil fertility, especially nitrogen as they affect the soil water balance, crop development and growth. Pests, diseases, and weeds are not considered.

It is important to understand the benefit of *AquaCrop* model over other existing crop models. Other existing models which require large and complex data input mainly used by the researcher, academicians and students. These models are rarely used by the target user such as extension personnel, farm managers, consulting engineers, and economist. The simplicity, robustness and accuracy makes *AquaCrop* model more users friendly.

Particular features that distinguish *AquaCrop* from other crop models are:

- Its focus on water

- The use of canopy cover instead of leaf area index makes it possible to cover large area using remote sensing data
- The normalization of water productivity, atmospheric evaporative demand and CO_2 concentration that confer the model an extended extrapolation capacity to diverse locations, seasons, and climate, including future climate scenarios
- Requires less input data
- Requires only explicit and mostly intuitive input parameters and variables;
- A well-developed user interface
- Its significant balance between accuracy, simplicity, and robustness
- Its applicability to be used in diverse agricultural systems that exists worldwide.

3. Input data requirement

AquaCrop needs relatively less number of input data for simulation. The main menu (Fig. 2) consists of mainly three panel such as environmental panel, simulation panel and project and field data panel. In the environmental panel (Fig. 3) the user has to specify (select or create) climate data (Rainfall, ET_0, maximum –minimum air temperature and CO_2 concentration), crop data (conservative and cultivar specific crop parameters), soil data (soil physical properties) and management file (Field and irrigation management). While in the simulation panel the user has to specify the simulation period, initial condition of simulation run and off-season conditions when the simulation period exceeds the growing period. In the project and field data panel, the user can select, create or update project and field data file. It is very important to understand the environment in which the crop growth happens. A brief description is given below of different components of environmental panel.

Climate data

Five weather input variables: daily maximum and minimum air temperatures, daily rainfall, daily evaporative demand of the atmosphere, expressed as ET_o and the mean annual carbon dioxide concentration in the bulk atmosphere are needed (Fig. 2). While the first 4 can be obtained or derived from data of agro meteorological stations, the CO_2 concentration uses the Mauna Loa Observatory records in Hawaii. ET_o should be calculated from daily solar radiation, temperature, humidity, and wind data following the procedures described in the FAO Paper 56 (Allen *et al.*, 1998). When not all the required input variables for ET_o are available, Paper 56 describes the methods for their estimation. *AquaCrop* does not include the routines for calculating ET_o but a separate software program (named ETo Calculator) based on Paper 56 is provided for such purposes. When daily weather data are not available, 10-day or monthly records are processed by *AquaCrop* into daily values using the downscaling procedure of Gommes (1983). This flexibility allows the use of *AquaCrop* in areas of limited weather records. The model also has a sub-routine to estimate effective rainfall from 10-day or monthly data through two options, the USDA Soil Conservation Service method (SCS, 1993), or by setting effective rainfall as a percentage of total rainfall. The temperature plays a role in influencing crop development (phenology, biomass production and pollination); the rainfall and ET_o are inputs for the water balance of the soil root zone; and the CO_2 concentration of the atmosphere influences the canopy growth rate and the water productivity (WP).

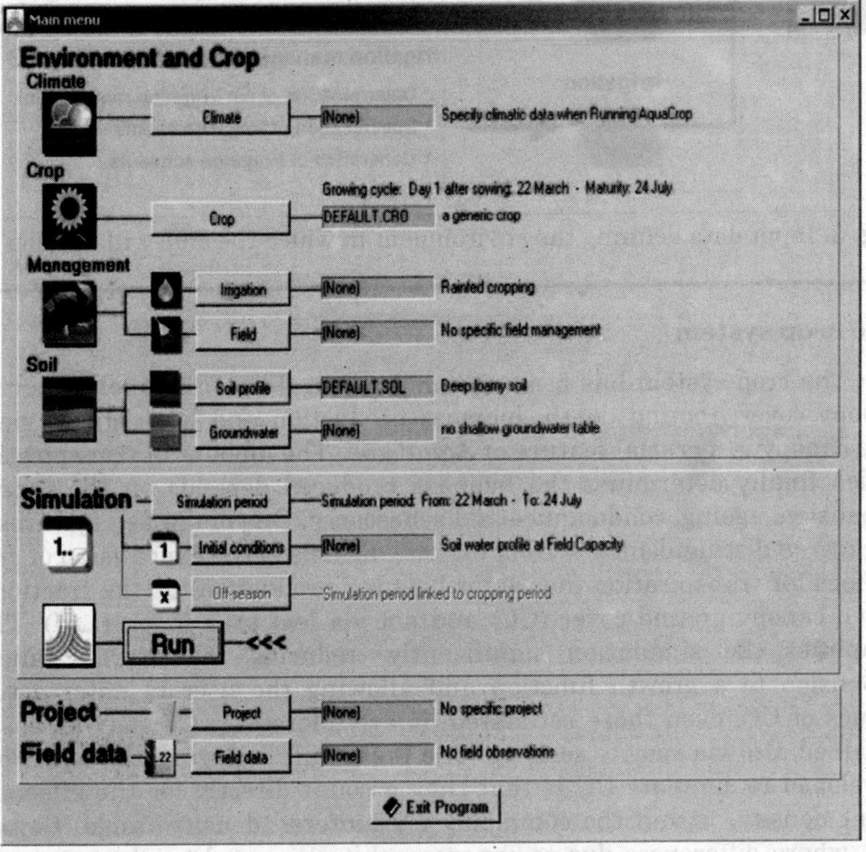

Fig. 2. The main menu of AquaCrop

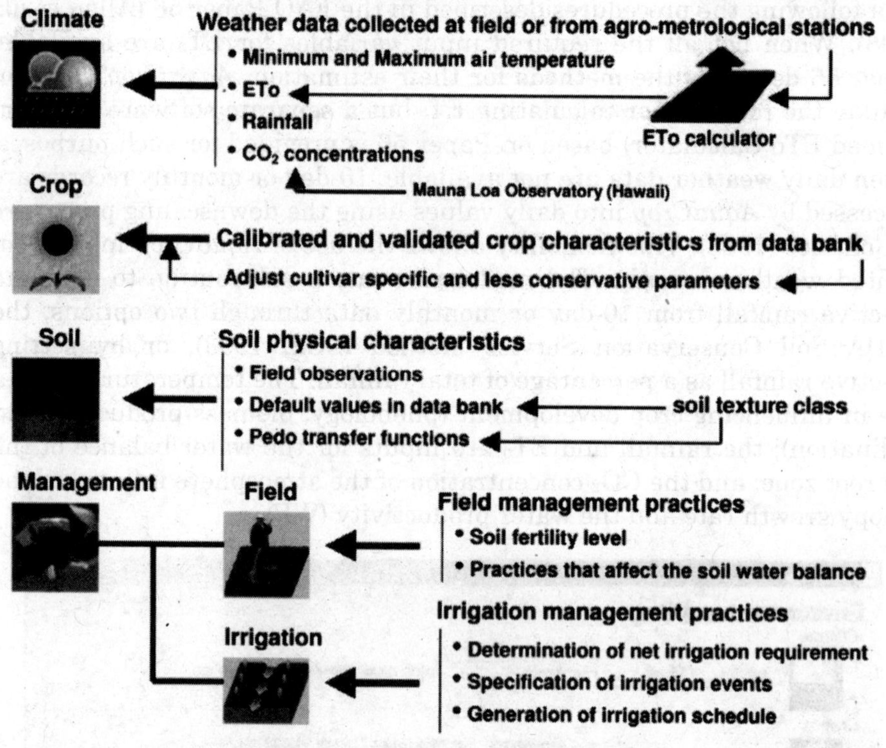

Fig. 3. Input data defining the environment in which the crop will develop

The crop system

The crop system has 5 major components (Fig. 1): phenology, green canopy cover, rooting depth, biomass production and harvestable yield. The canopy is a crucial feature of *AquaCrop*. The amount of transpiration which finally determines the biomass produced depends on the canopy expansion, ageing, conductance and senescence. One of the key features of *AquaCrop* distinguishing it from existing models is the expre-ssion of crop surface for transpiration (hence for biomass production) as the fractional green canopy ground cover (CC) and not via leaf area index (LAI). This simplifies the simulation significantly, reducing the overall canopy expansion to a growth function and allowing the user to enter actual values of CC, even those estimated by eye. Moreover, CC may be easily obtained also via remote sensing. One important feature of the equations developed to simulate CC is that they account directly for the effects of plant density, within the commonly encountered density range. Beyond CC, where differences due to canopy architecture and height may exert influence, the effects are implicitly incorporated when parameterizing the crop coefficient for transpiration of each crop species. In non-stress conditions, the canopy expansion from emergence to full canopy development follows a sigmoid-type curve (Fig. 4) constructed with an exponential function for the first half of the development, and an exponential decay function for the second half, according to the equations:

$$CC = CC_0.e^{-CGC.t} \quad (4)$$

$$CC=CC_x- (CC_x- CC_0)\ e^{-CGC.t} \quad (5)$$

Where CC is the canopy cover at time tin growing degree day (GDD) or calendar day, CC_0 is the initial canopy cover (CC at t=0), CGC is the canopy growth coefficient in fraction per GDD (or per day), and CC is the maximum canopy cover, or canopy cover at t= ∞. CC_0 represents the initial canopy cover once the seedlings are established and is equal to the plant density times the mean initial canopy cover per seedling. The latter is a crop-specific parameter provided in *AquaCrop* for each of the major crops.

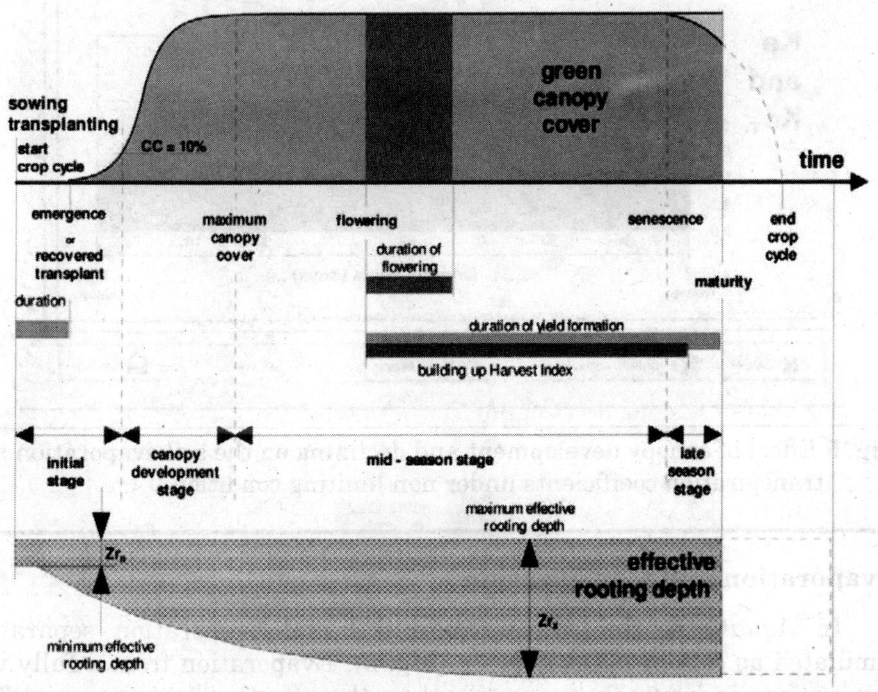

Fig. 4. Schematic representation of canopy development

After its full development, the canopy can have a variable duration before entering the senescence phase. For the fully developed canopy and before senescence, an ageing effect is applied to account for the slight reduction in the overall canopy transpiration capacity over time (e.g., 0.3% reduction per day). Once senescence starts in the late season, CC enters a declining phase following the equation:

$$CC = CC_x \, [1 - 0.05(\exp^{CDCt/CC_x} - 1)] \qquad (6)$$

Where t is time since the start of canopy senescence, and CDC is a canopy decline coefficient that reflects the speed of decline up to maturity.

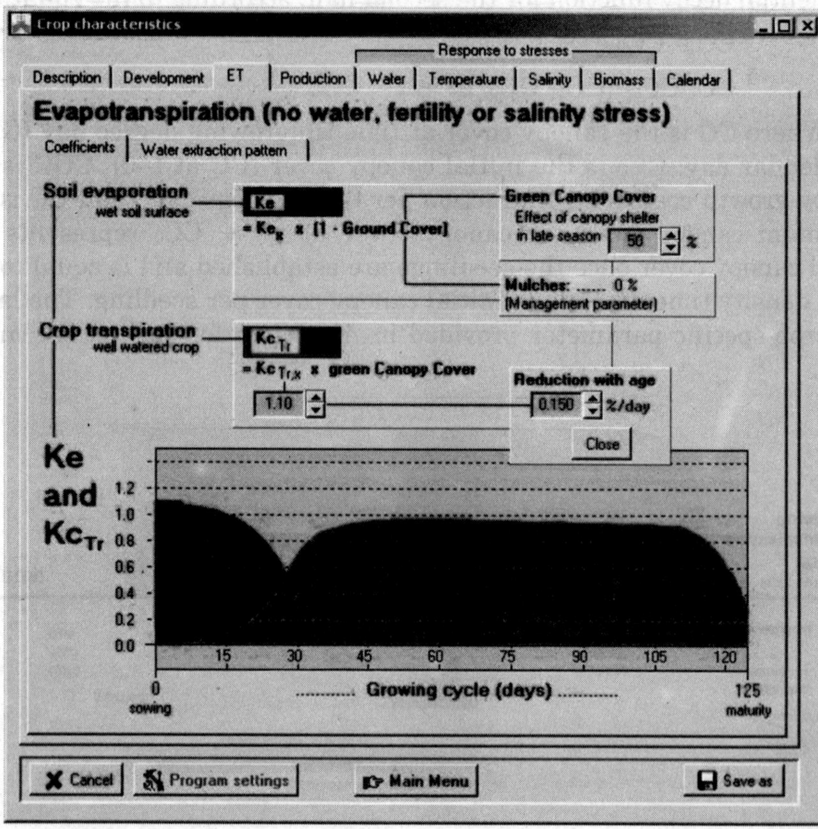

Fig. 5 Effect of canopy development and declining on the soil evaporation and transpiration coefficients under non limiting condition

Evaporation and transpiration

In *AquaCrop*, the transpiration (T_r) and evaporation separately simulated as canopy cover develop (Fig. 5). Evaporation from a fully wet soil surface is inversely proportional to the effective canopy cover. The shelter effect is also considered in the late crop growth season. The effect of canopy shelter is parameterized based on whether the senescening canopy retains more or less of its dead leaves. Transpiration is directly proportional to canopy cover which finally coverts in biomass of the product (Eq. 2). The model tries to estimate accurately T_r by accounting for the extent of CC, effects of inter-row micro-advection, and effects on stomata. Transpiration is calculated as a function of a specific crop coefficient Kc_{tr}, which is the T_r/ET_o ratio of the full canopy of the crop under optimal conditions (Kc_{trx}), adjusted for effects of stresses on stomata and for canopy aging and senescence, as follows:

$$T_r = Ks_{sto} \times Kc_{tr} \times CC_{adj} \times ET_o \quad (7)$$

Where CC_{adj} is CC adjusted to account for the interactions between T_r and E, which are particularly relevant under partial canopy cover (Ritchie 1983) and enhances Tr, and Ks_{sto} is the water stress coefficient for stomatal closure. After CCx is reached, Kc_{trx} is adjusted slightly downward per day as the canopy ages until the onset of senescence. When senescence starts and CC declines following the trend indicated by Eq. (6), KC_{trx} is adjusted further downward. If water stress affects T_r by reducing stomatal opening, then Ks_{sto} becomes smaller than 1.

Effect of water stress

Since the *AquaCrop* is a water driven simulation model, the efforts were made to include all the responses underpinning the effects of water stress on crop yield. Other models have given more emphasis on simulating stress effects on photosynthesis than on canopy expansion or senescence (van Ittersum *et al.*, 2003; Jones and Kiniry, 1986). The canopy response to water deficits (Fig. 1) into three components, namely, reduction in expansion rate, i.e., reducing the canopy growth coefficient (CGC), reduction in stomatal conductance (g_a), and acceleration of senescence, i.e., triggering the early start of senescence and adjusting the canopy decline coefficient (CDC) to the level of water stress. Water deficits are quantified using a water stress coefficient (K_s) for each response that varies from 1(no stress) to zero (frill stress). Stress occurs when the depletion in the relative soil water content of the root zone reaches an upper threshold value, p, varying between 0 and 1. When that threshold value for a specific response is reached, *Ks* for that response is computed through the following equation:

$$Ks = 1 - \frac{(e^{D_{rel} \cdot f_{shape}} - 1)}{(e^{f_{shape}} - 1)}$$

Where, the f_{shape} influences the shape of the function *Ks* and D_{rel} is the relative depletion between the upper and lower threshold. A sample response functions of Ks is presented in Fig. 6.

Shape of Ks curve: Between the upper and lower thresholds the shape of the Ks curve determines the magnitude of the effect of soil water stress on the process. The shape can be linear or convex (Fig. 6). Tests so far suggest that the thresholds and shapes of these curves may be conservative, at least to a fair degree. The shape factor can range from +6 (strongly convex) to 0 (linear).

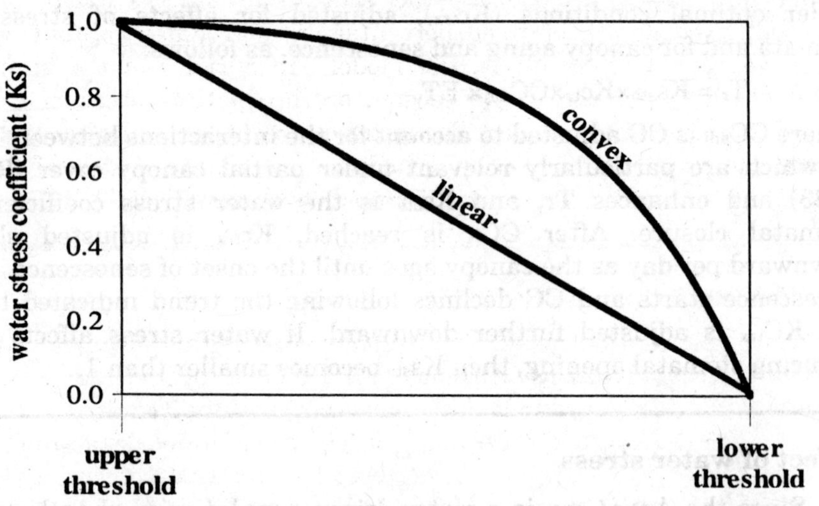

Fig. 6. Convex and linear shapes of the Ks curve

Soil components

The soil component comprises of soil physical properties such as bulk density, hydraulic conductivity at saturation (Ksat), field capacity (FC), and at permanent wilting point (PWP) and the soil water content at saturation (sat). The user can select the indicative values provided by Aqua Crop for various soil texture classes, or import locally determined or derived data from soil texture with the help of pedo-transfer functions. If a layer blocks the root zone expansion, its depth in the soil profile has to be specified as well.

Management

In management component, two management practices namely field management and irrigation management are considered. Under field management there are options to select or define the fertility level or regime of the soil, to select and define field-surface practices of mulching to reduce soil E, or the use of soil bunds to control run-off and infiltration, and to define the time of cuttings of forage crops. Under irrigation management, the user can define the crop grown under rainfed condition or irrigated. Under irrigation management the user chooses whether the crop is rainfed or irrigated. If irrigated, the user can select the application method (sprinkler, drip, or surface), the fraction of surface wetted, and specify for each irrigation event, the irrigation water quality, the timing and the applied irrigation amount. The user can define his/her own schedule on the basis of applied water depth, or timing criteria, or let the model automatically generate the scheduling on the basis of fixed interval, fixed depth, or fixed percentage of soil water content criteria. With weather data as input, the model can either run the simulation automatically over the whole season, or run in time steps of I day or longer as defined by the user. The latter option is particularly suitable for trying out different irrigation schedules by adding water of selected amounts at selected times over the crop development cycle.

Fine tuning

Every simulation model requires fine tune of the parameters for the prevailing local condition. Although FAO has calibrated many crops for different climate conditions but some parameters require adjustment according to the existing condition. Three type of input data are required namely conservative type, cultivar specific and less conservative type. The conservative parameters are crop specific but do not change materially with time, management practices, or geographical location. They were calibrated with data of the crop grown under favorable and non-limiting conditions and remain applicable for stress conditions via their modulation by stress response functions. As such the conservative parameters require no adjustment to the local conditions and can be used as such in the simulations. The cultivar specific crop parameters might require an adjustment when selecting a cultivar different from the one considered for crop calibration. Less-conservative crop parameters are affected by field management, conditions in the soil profile, or the weather (especially when simulating in calendar day mode). These parameters might require an adjustment or fine tuning after downloading to account for the local variety and or local environmental conditions. The types of parameters of different component are listed below (Table 1).

Table 1. Different input parameters and their type

1. Crop Phenology		
Symbol	Description	Type (1), (2), (3), (4)
1.1 Threshold air temperatures for growing degree days		
Tbase	Base temperature (°C)	
Tupper	Upper temperature (°C)	Conservative (1)
1.2 Development of green canopy cover		
cco	Canopy size of the average seedling at 90% emergence, or canopy size of the transplanted seedling (cm^2)	Conservative (2) Management (3)
	Number of plants per hectare	

	Time from sowing to emergence (days or GD days) or recovery time (for transplanted seedlings)	Management [3]
CGC	Canopy growth coefficient (fraction per day or per growing degree day)	Conservative [1]
CCx	Maximum canopy cover (fraction soil cover)	Management [3]
	Time from sowing to start senescence (days or GD days)	Cultivar [4]
CDC	Canopy decline coefficient (fraction per day or per growing degree day)	Conservative [1]
	Time from sowing to maturity, i.e. length of crop cycle (days or GD days)	Cultivar [4]
1.3 Flowering or start of yield formation		
	Time from sowing to flowering or to the start of yield formation (days or GD days)	Cultivar [4]
	Length of the flowering stage (days or GD days)	Cultivar [4]
	Crop determinacy linked/unlinked with flowering	Conservative [1]
1.3 Flowering or start of yield formation		
	Time from sowing to flowering or to the start of yield formation (days or GD days)	Cultivar [4]
	Length of the flowering stage (days or GD days)	Cultivar [4]
	Crop determinacy linked/unlinked with flowering	Conservative [1]
1.4 Development of root zone		
Zn	Minimum effective rooting depth (m)	Management [3]
Zx	Maximum effective rooting depth (m)	Management [3]
	Shape factor describing root zone expansion	Conservative [1]
2. Crop transpiration		
KcTr?x	Crop coefficient when canopy is complete but prior to senescence	Conservative [1]
100 fage	Decline of crop coefficient (% of CCx per day) as a result of ageing, nitrogen deficiency, etc.	Conservative [1]
Sx, top	Maximum root water extraction (m3 m^{-3} day^{-1}) in top quarter of root zone	Conservative [1]
Sx, bot	Maximum root water extraction (m3 m^{-3} day^{-1}) in bottom quarter of root zone	Conservative [1]
	Effect of canopy cover in reducing soil evaporation in late season stage (% reduction in soil evaporation)	Conservative [1]

3. Biomass production and yield formation

3.1 Crop water productivity

WP*	Water productivity normalized for ETo and CO_2 (gram/m_2)	Conservative (1)
field	Reduction coefficient describing the effect of the products synthesized during yield formation on the normalized water productivity	Conservative (1)
	Crop performance under elevated atmospheric CO_2 concentration (%)	Management (3) Cultivar (4)

3.2 Harvest Index

HIo	Reference harvest index (%)	Cultivar (4)
	Excess of potential fruits (%)	Conservative (2)
	Possible increase (%) of HI due to water stress before flowering	Conservative (1)
	Coefficient describing positive impact of restricted vegetative growth during yield formation on HI	Conservative (1)
	Coefficient describing negative impact of stomatal closure during yield formation on HI	Conservative (1)
	Allowable maximum increase (%) of specified HI	Conservative (1)

4. Stresses

4.1 Soil water stresses

pexp, lower	Soil water depletion threshold for canopy expansion – Upper threshold	Conservative (1)
pexp, upper	Soil water depletion threshold for canopy expansion - Lower threshold	Conservative (1)
	Shape factor for water stress coefficient for canopy expansion	Conservative (1)
psto	Soil water depletion threshold for stomatal control – Upper threshold	Conservative (1)
	Shape factor for water stress coefficient for stomatal control	Conservative (1)
psen	Soil water depletion threshold for canopy senescence – Upper threshold	Conservative (1)
	Shape factor for water stress coefficient for canopy senescence	Conservative (1)
	Sum(ETo) during stress period to be exceeded before senescence is triggered	Conservative (1)
ppol	Soil water depletion threshold for failure of pollination – Upper threshold	Conservative (1)

	Parameter	Type
	Vol % at anaerobiotic point (with reference to saturation)	Cultivar (4) Environment (3)
4.2 Soil fertility/salinity stress		
	Stress at calibration (%)	(calibration)
	Shape factor for the stress coefficient for canopy expansion	Management (3)
	Shape factor for the stress coefficient for maximum canopy cover	Management (3)
	Shape factor for the stress coefficient for crop water productivity	Management (3)
	Shape factor for the response of decline of canopy cover to stress	Management (3)
	Shape factor for the stress coefficient for stomatal closure	Management (3)
4.3 Air temperature stress		
	Minimum air temperature below which pollination starts to fail (cold stress) (°C)	Conservative (1)
	Maximum air temperature above which pollination starts to fail (heat stress) (°C)	Conservative (1)
	Minimum growing degrees required for full biomass production (°C - day)	Conservative (1)
4.4 Soil salinity stress		
ECen	Electrolyte conductivity of the saturated soil-paste extract: lower threshold (at which soil salinity stress starts to occur)	Conservative (1)
ECex	Electrolyte conductivity of the saturated soil-paste extract: upper threshold (at which soil salinity stress has reached its maximum effect)	Conservative (1)
	Shape factor for Soil salinity stress coefficient	Conservative (1)

(1) Conservative generally applicable

(2) Conservative for given species but can or may be cultivar specific

(3) Dependent on environment and/or management

(4) Cultivar specific

Application: AquaCrop model is designed to use in

- irrigation strategies development under water deficit conditions
- Understanding the effect on crop yield of location, soil type, sowing date
- Understanding the effect on crop yield of various land management techniques

- Prediction of climate change impacts on crop production
- Comparing the attainable against actual yields in a field, farm, or a region, to identify the constraints limiting crop production and water productivity (benchmarking tool);
- Scenario simulations and for planning purposes for use by economists, water administrators and managers.

❑❑❑

12 Conservation Agriculture: A Strategy to Cope with Climate Change

K.M. Hati, J. Somasundaram, R. Saha and M. Mohanty

Division of Soil Physics, Indian Institute of Soil Science, Nabi Bagh, Berasia Road, Bhopal - 462 038, India

Climate change is a long-term shift in weather conditions identified by changes in temperature, precipitation, winds, and other indicators. Climate change can involve both changes in average conditions and changes in variability, including, for example, extreme events. The earth's climate is naturally variable on all time scales. However, its long-term state and average temperature are regulated by the balance between incoming and outgoing energy, which determines the earth's energy balance. Any factor that causes a sustained change to the amount of incoming energy or the amount of outgoing energy can lead to climate change. As these factors are external to the climate system, they are referred to as 'climate forcers' or `radiative forcing', invoking the idea that they force or push the climate towards a new long-term state – either warmer or cooler depending on the cause of change. This radiative forcing refers to an imbalance between incoming solar radiation and outgoing infrared radiation that causes the Earth's radiative balance to stray away from its normal state. This straying causes changes in global temperatures. When radiative forcing for the atmosphere is positive, the energy of the earth-atmosphere system will ultimately increase, leading to a warming of the system. In contrast, for a negative radiative forcing, the energy will ultimately decrease, leading to a cooling of the system. The green house gases (GHG) present in the atmosphere caused a positive forcing of the atmosphere by trapping longwave radiations released from the earth surface. Since the industrial revolution the amount of GHG in the atmosphere has increased, leading to increased radiative forcing from CO_2, methane, tropospheric ozone, CFCs and nitrous oxide. The atmospheric concentrations of the greenhouse gases like, carbon dioxide (CO_2), methane (CH_4), and nitrous oxide (N_2O) have all increased since 1750 due to anthropological activities. In 2011 the concentrations of these greenhouse gases were 391 ppm, 1803 ppb, 324 ppb and exceeded the pre-industrial levels by about 40%, 150%, and 20%, respectively (IPCC, 2013). These levels are much higher than at any time during the last 800,000 years, the period for which reliable data have been extracted from ice cores (IPCC, 2013). Fossil fuel burning has produced about three-quarters of the increase in CO_2 from human activity over the past 20 years, the rest being due to land-use change, particularly deforestation and intensive agriculture. In India out of total GHG emission from the anthropogenic activities, agriculture sector contributes 28%. The emissions are primarily due to methane emission from rice fields (23%), enteric fermentation in ruminant animals (59%) and nitrous oxides from application of manures and fertilizers (Fig. 1A and 1B) (NATCOM, 2004; Aggarwal and Pathak, 2009).

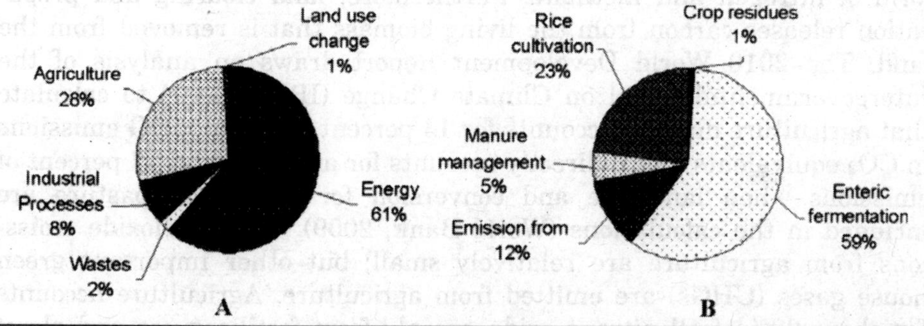

Fig. 1. Contribution of major sectors to emission of greenhouse gases in India (A), and the relative contribution of sub-sectors of agriculture to emissions in India (B).

The effects of climate change are assumed to be reaching to such an extent that an irreversible change in the functioning of the planet are feared. Some of the recent weather phenomenon occurred specially during the last decade, particularly those related to agriculture and food production like increased occurrence of storms and floods; increased incidence and severity of droughts and forest fires; steady spreading out of frost-free intervals and potential growing season; increased frequency of diseases and insect pest attacks; and vanishing habitats of plants and animals, are believed to be an outcome of the changing climatic pattern of the planet. Apparently, these modifications of previously stable climates imply an obvious warming trend and a growing climatic variability impacting the exiting ecosystem. It is thus imperative for the scientific community to use all accessible knowledge to stop or reverse this trend to the maximum extent possible to help in mitigating the problem of uncontrolled climate change in near future. There have been various attempts and viable measures in the past to bring down atmospheric greenhouse gases (GHGs) to slow down climate change (Boer *et al.*, 2000). However, if the warming trend continues at its current pace, these may soon prove inadequate. The early impacts of climate change already are being felt worldwide. Future impacts will affect abroad array of human and natural systems, with consequences for human health, food and fiber production, water supplies, and many other areas vital to economic and social well being. While certain impacts may in the nearer term prove beneficial to some, in the long-term, the effects will be largely detrimental.

Climate has obvious and direct effects on agricultural production. The effects of agriculture on GHG emissions are also large. Agriculture plays an important role in the global economy and livelihood security of a large population particularly those residing in the developing countries. However, it also uses substantial fossil fuel for farm inputs and equipment. Animal agriculture also releases substantial GHGs in the form of nitrogen and methane. Furthermore, land clearing and preparation releases carbon from the living biomass that is removed from the land. The 2010 World Development Report draws on analysis of the Intergovernmental Panel on Climate Change (IPCC, 2007) to calculate that agriculture directly accounts for 14 percent of global GHG emissions in CO_2 equivalents and indirectly accounts for an additional 17 percent of emissions when land use and conversion for crops and pasture are included in the calculations (World Bank, 2009). Carbon dioxide emissions from agriculture are relatively small; but other important green house gases (GHGs) are emitted from agriculture. Agriculture accounts for about 60% of all nitrous oxide, mainly from fertilizer use and about 50% of methane mainly from natural and cultivated wetlands and enteric fermentation. Methane and nitrous oxide emissions are projected to further increase from 35 to 60% by 2030, driven by growing nitrogen fertilizer use and increased livestock production in response to growing food demand. Besides this, the climate change is believed to have a great impact on soils. Increasing temperature would increase oxidation of the organic carbon in soil and thus its levels will go down further. Higher temperature will increase the evaporative demand causing faster drying of soil layers and water stress to crops. Incidence of runoff /wind erosion may increase due to increase in extreme events. These changes may reflect themselves in poorer soil fertility, loss of soil biota, water stress and ground water depletion (FAO, 2011). Thus to sustain agricultural productivity in future under the changed climatic scenario we need to embrace a viable adaptation strategy against the weather induced biotic and abiotic stresses. Here conservation agriculture can play a pivotal role in our effort in mitigation and adaptation to climate change.

Conservation agriculture and climate change

Conservation agriculture (CA) is as an approach to farming that seeks to increase food security, alleviate poverty, conserve biodiversity and safeguard ecosystem services Principles of CA are not confined to a single set of practices or prescription recommended to all types of crops, soil and environment. The CA encompasses a gamut of agricultural practices which will be specific to crops, areas and socio-economic condition of the people keeping in view the sustainability of the system from economic as well as from environmental perspective. Conservation agriculture promotes a series of principles to achieve conservation objectives, rather than a particular technology. As per the FAO the three pillars of conservation agriculture are (1) continuous minimum mechanical soil disturbance by reducing tillage, (2) permanent organic soil cover and (3) diversification of crop species grown in sequence or association. In the recent past conservation agriculture is gaining acceptance in many parts of the world as an alternative to both conventional agriculture and to organic agriculture. Although the practice of conservation agriculture on a large scale emerged out of US, Brazil and Argentina over the three to four decades, similar developments are occurring in many other areas of the world, notably North America in zero tillage, and Africa and Asia with technologies such as agroforestry, crop rotation, bed planting and reduced tillage. In India conservation tillage system has been practised in an area of about two million hectares in the Indo-Gangetic plains under rice-wheat system. Major practices adopted in those areas are zero-tillage, bed planting and laser levelling. Besides this, in rainfed regions different forms of conservation agricultural practices and their benefits are being tested and propagated for their adoption. Conservation agriculture is based on the principles of rebuilding the soil, optimizing crop production inputs, including labor, and optimizing profit. CA stimulates soils natural ability to recycle its resources and maintaining its health through the activities of its flora and fauna living within the ecosystem. The conservation agriculture practices through restoration of soil aggregation, improvement in soil biodiversity and sequestration of carbon in to the soil from the atmosphere, positively contribute to the mitigation of climate change. The CA practices can contribute to making agricultural systems more resilient to climate change. It has a powerful mechanism to adapt to climate change by increasing resilience to drought, increasing water-use efficiency and thermal stress resistance of crops by changing the hydrothermal regimes of the surface soil. In many cases, conservation agriculture has been proven to reduce the farming systems' greenhouse gas emissions and enhance its role as carbon sinks. The CA can increase the ability of smallholder farmers to adapt to climate change by reducing vulnerability to drought and enriching the local natural resource base on which farm productivity depends. The CA aims at increasing the annual input of fresh organic matter, controlling soil organic material losses through soil erosion, and reducing the rate of soil organic material mineralization through better aggregation.

In this chapter, climate change predictions for India are discussed. Then the potential of CA as a means to mitigate and adapt to climate change is also examined.

Climate change predictions for India

New climate projections are being developed by different modelling groups based on new representative concentration pathways (RCP) scenarios. Climate projections for the Fifth Assessment Report of the IPCC are made using the newly developed representative concentration pathways (RCPs) under the Coupled Model Inter-comparison Project 5 (CMIP5). The projections for the climate change variables for the 21st century by different scientists groups varies considerably based on the scenarios and model assemblage used. Rupakumar *et al.* (2006) project a temperature rise of 2.9°C and 4.1°C for India under the B2 and A2 scenarios of SRES respectively, in 2080s relative to 1970s while Krishankumar *et al.* (2011) projected a warming of 3.5-4.3°C over the same period for the A1B scenario. A1B scenario of SRES is generally considered close to the business as usual emission scenario. The CMIP5-based model ensemble projection of Chaturvedi *et al.* (2013) reported a warming of 2.8°C and 4.3°C under the RCP6.0 and RCP8.5 scenarios respectively, for 2080s compared to the 1970s baseline. In terms of atmospheric CO_2 concentration build-up, the A1B scenario is closer to RCP8.5 until 2050, but then diverges and is closer to RCP6.0 between 2050 and 2100.

The IPCC in their forth assessment report using the CMIP3 model output (IPCC, 2007) projected an increase of global surface temperature of the order of 0.74±0.18ºC during the 20th century while the global surface temperature is likely to rise a further 1.1 to 6.4ºC (2.0 to 11.5ºF) by the end of the 21st century. For South Asia region including India a rise in temperature of 0.5 to 1.2ºC by 2010, 0.88 to 3.16ºC by 2050 and 1.56 to 5.44ºC by 2080 has been projected by IPCC under different scenarios. The temperature increase is likely to be much higher in winter (*rabi*) season than in rainy (*kharif*) season. The most widely used set of projections available, the CMIP3 ensemble output for temperature and precipitation for the A1B emission scenario for India projected that the temperature increases are lower in the southern part of India, up to 3°C, compared to the north where changes of up to 4.5°C are projected. The agreement between models is high over almost the entire country. On the other hand, India is projected to experience increases in precipitation across most of the country. Increases of up to 20% or higher could occur in western regions with more widespread increases of 5-10% over the rest of the country. Unlike the temperature projection the ensemble agreement is moderate to low for precipitation. The frequency of occurrence of climate extremes is expected to change during the next century, with increases in the frequency of heat waves and heavy precipitation events, and decreases in the frequency of frost days, as a consequence of anthropogenically-forced climate change. A future increase in occurrence of extreme weather events including heat wave and intense precipitation events is also projected for South Asia (Lal, 2003) along with an increase in the inter-annual variability of daily precipitation in the Asian summer monsoon (May, 2004; Giorgi and Bi, 2005). For tropical cyclones, there is a projected increase of 10–20% in the intensity of storms with an increase in sea-surface temperature of 2–4°C relative to the current temperatures in South Asia. An increase in heights of storm-surges could result from the occurrence of stronger winds, increases in sea-surface temperatures and low pressures associated with tropical cyclones resulting in an enhanced risk of coastal disasters along the coastal regions of the country. A warmer climate, with its increased climate variability, will increase the risk of both floods and droughts. The areas suitable for rainfed agriculture are expected to significantly decrease affecting adversely land productivity potential of the continent (Sivakumar and Stefanski, 2011).

Conservation agriculture as an adaptive measure

It is widely predicted that due to global warming rainfall distribution pattern will change, some area will receive more rainfall while other areas will get less and occurrences of extreme weather events like draughts, flooding will increase in large part of the world particularly more in the tropical region. Conservation agriculture increases the water transmission properties like, infiltration rate and hydraulic conductivity of the soil and also improves the water holding capacity of the profile. The soil profile, under conservation agriculture retains more water which could be utilized by crops under water stress situation while it drains excess water from the soil during high rainfall events reducing the chance of water logging. Increase in temperature will also result greater loss of water through evapo-transpiration resulting in more water and heat stress to crop under conventional cultivation but in conservation agriculture mulching provided by crop residue will act as barrier between the atmosphere and soil, making the soil hydrothermal regime more congenial to crop. Loss of water through erosion under heavy rainfall event is less in CA which also makes the system more sustainable under extreme weather situation of future. Conservation agriculture is a powerful mechanism to adapt to climate change by increasing resilience to drought and flood and increasing water-use efficiency (Smith, 2005). The CA system improves the organic matter content and aggregate stability of the soil which improves the nutrient cycling and its availability to the crops resulting in better crop growth performance and improved productivity. Improved soil nutrient recycling may also help combat crop pests and diseases (Holland, 2004). Thus conservation agriculture will have better adaptability under climate change scenario.

Climate change is likely to strongly affect rice–wheat, rice–rice and maize-based cropping systems that, today, account for more than 80% of the total cereals grown on more than 100 Mha of agricultural lands in South Asia. Global warming may be beneficial in some regions, but harmful in those regions where optimal temperatures already exist; an example would be the rice–wheat mega-environments in the IGP that account for 15% of global wheat production. Agronomic and crop management practices have to aim at reducing CO_2 and other greenhouse gas emissions by reducing tillage and residue burning and improving nitrogen use efficiency. Conservation agriculture provides a viable alternative for residue management and thus checks the burning of residues, widely practised in the Indo-gangetic planes under rice-wheat system causing air pollution through aerosols and GHGs emission. In the IGP, resource-conservation technologies continue to expand in the rice–wheat cropping systems and save 50–60/l of diesel/ha^{-1} plus labour, and significantly reduce release of CO_2 to the environment (Hobbs & Gupta 2004).

Conservation agriculture to mitigate climate change

Through sequestration of carbon in the profile soil can act as an overall sink of atmospheric carbon instead of a source and thus can help in mitigating climate change impact on the ecosystem. The term "soil C sequestration" implies removal of atmospheric CO_2 by plants and storage of fixed C as soil organic matter. The strategy is to increase SOC density in the soil, improve depth distribution of SOC and stabilize SOC by encapsulating it within stable micro-aggregates so that C is protected from microbial processes or as recalcitrant C with long turnover time. In this context, managing agroecosystems is an important strategy for SOC/terrestrial sequestration. Appropriate soil and crop management practices can enhance SOC stock of the soil and contribute to the mitigation effort against the climate change. Barring the initial few years fertilizer nutrient requirement under conservation agriculture are generally low than the conventional system as more nutrients are recycled to the system from the added residues. Nitrous oxide is an important GHG which is 310 times more potent than carbon-di-oxide. A part of fertilizer N applied to the field is lost to atmosphere as N_2O through the process of denitrification. Denitrification in field occurs under anaerobic conditions in water logged areas. As aeration is normally improved under CA due to improved drainage it is envisaged that N_2O production will be less under CA. Methane emissions that have a warming potential 21 times that of CO_2 are common and significant in puddled anaerobic paddy fields and also when residues are burnt. This GHG emission can be mitigated by shifting to an aerobic, direct seeded or NT rice system. Among the different soil and crop management practices conservation tillage, crop rotation and cover crops all part of conservation agriculture package contribute to the buildup of SOC stock of the soil through their influence on soil hydrothermal regimes and biophysical properties.

(1) *Conservation tillage:* One aspect of conventional agriculture is its ability to change the landscape through clearing of native vegetation, levelling of land, repeated ploughing and keeping the loose soil bare and thus vulnerable to the eroding effect of wind and water. The destruction of the vegetative cover affects the plants, animals and micro-organisms. Some species profit from the change and turn into pests. However, most native organisms are negatively affected and either they disappear completely or their numbers are drastically reduced. With the conservation of soil cover in conservation agriculture a habitat is created for a number of species that feed on pests, which in turn attracts more insects, birds and other animals. The rotation of crops and cover crops restrains the loss of genetic biodiversity, which is favoured under mono-cropping. Presence of crop residues as mulch makes the soil congenial for a thriving and healthy bio-ecosystem though their effect on improved soil hydrothermal regime and supply of food to the soil faunal population. Conventional tillage and erosion deplete SOC pools in agricultural soils. Thus, soils can store C upon conversion from plow till to no till or conservation tillage, by reducing soil disturbance, decreasing the fallow period and incorporation of cover crops in the rotation cycle. Eliminating summer fallowing in arid and semi-arid regions and adopting no till with residue mulching improves soil structure, lowers bulk density and increases infiltration capacity (Shaver *et al.*, 2002). However, the benefits of no tillage on SOC sequestration may be soil/site specific, and the improvement in SOC may be inconsistent in fine textured and poorly drained soils (Lal, 2004). Some studies have also shown more N_2O emissions in no tillage. Smith *et al.* (1998) estimated that adoption of conservation tillage has the potential to sequester about 43 Tg C/year in the Europe including the former Soviet Union. In addition to enhancing SOC pool, up to 3.2 Tg C/year may also be saved in agricultural fossil fuel emissions. Smith *et al.* (1998) concluded that 100% conversion to no till agriculture could mitigate all fossil fuel C emission from agriculture in Europe. Conservation agriculture increases the energy efficiency of the production system through reduction in energy consumption by tillage operations and consequently reduces the amount of fossil fuel burnt for cultivation. Lal (2005) suggested that by adopting improved management practices on agricultural land like, use of NT and crop residues, food security would not only be enhanced but also offset fossil fuel emissions at the rate of 0.5 Pg/C yr^{-1}. Potentially one-third of the carbon emitted in current fossil fuel use could be offset by implementing conservation agriculture globally in the next decade.

(2) *Cover crops:* The benefits of adopting conservation agriculture for SOC sequestration are greatly enhanced by growing cover crops in the

rotation cycle. Growing leguminous cover crops enhances biodiversity, the quality of residue input and SOC pool. Legume-based cropping systems reduce C and N losses from soil. It is well established that ecosystems with high biodiversity absorb and sequester more C than those with low or reduced biodiversity. Soil biodiversity has a positive impact on the SOC pool. All other factors being equal, ecosystems with high biodiversity sequester more C in soil and biota than those with reduced biodiversity. In managed ecosystems, soil biodiversity is likely to increase with conversion to conservation tillage, replacement of toxic chemicals with viable alternatives, substitution of monoculture with mixed crop rotations and complex/ diverse systems, restoration of degraded soils and ecosystems, and conversion of crop or pasture land to a restorative land use. However, the use of cover crops as a short-term green manure may not necessarily enhance the SOC pool (Lal, 2004).

Limitations of conservation agriculture

Initial lack of knowledge about the long-term benefits and technical details of different conservation agricultural practices are the most important limitation in all areas where conservation agriculture is practised. There is no blueprint available for conservation agriculture, as all agro-ecosystems are different. Site and crop specific techniques for conservation agriculture practices utilizing locally available resources and peoples preference are to be developed for different agro-ecoregions of the country.

Some barriers that hinder the implementation of conservation agriculture in India include:

- cultural factors (farming traditions, prejudice)
- lack of technical know-how about how to implement conservation agriculture (e.g. different methods for controlling weeds and managing residues);
- competitive uses of crop residues and lack of negotiation between competing users
- limited knowledge of appropriate chemical weed management options and of integrated weed management practices without herbicides
- limited availability of effective herbicides for controlling weeds in different agro-climatic regions under different cropping system
- shortage of adequate equipment for sowing and crop management under CA
- the lack of supportive policies and adequate networks for distribution and sharing of machinery

- Higher levels of surface residue sometimes results in higher plant disease and pest infestations, if not managed properly.
- Benefits of CA accrue only in the long-term. In most of the cases, the yield level under CA practices in the first few years are either declined or just maintained causing practitioners to lose interest in continuing this new sustainable practice of cultivation.

Conclusion

Conservation agriculture could come out as a useful technology to partially mitigate the deleterious effect climate change on humanity through sequestration of carbon in to the soil and reduction of green house gases emission from agricultural activities to atmosphere. The adaptability of arable crops to climate change induced abiotic stresses like temperature and moisture stress is higher under conservation agricultural system as mulching practised in conservation agriculture improves the soil hydrothermal regime. Besides this, conservation tillage system improves soil health and its resilience to extraneous stresses. In India through adoption of conservation agriculture the long-term sustainability of agricultural productivity could be attained and energy and nutrient use efficiency could be improved. However, for acceptability of this system to larger audience, suitable implements, cropping systems and cover crops are to be assessed and also a crop specific sustainable weed management package is to be developed for different agro-ecoregions.

❑❑❑

13 Climate Change Impact Assessment using DSSAT model

Nishant K Sinha[1], M Mohanty[1], Smita Gupta[2], Pankaj K Tiwari[1], Bharat P. Meena[1], and Rohit Patidar[3]

1 Indian Institute of Soil Science, Nabibagh, Berasia Road, Bhopal-462038.
2 Govind Ballabh Pant University of Agriculture and Technology, Pantnagar-263145
3 R.A.K. College of Agriculture, Sehore- 466001.

"Climate change is the Everest of all problems, the thorniest challenge facing humankind"

— Lewis Gordon Pugh

Introduction

The earth is dynamic planet that has continually evolved since its origin. Owing to its dynamic nature, earth's climate is also changing continuously. In the last 650,000 years, earth has passed through seven ice ages and with the abrupt end of the last ice age about 7,000 years ago marking the beginning of the modern climate era- and of human civilization. Most of these climate changes are attributed to very small variations in earth's orbit that change the amount of solar radiation coming to our planet (Anonymous, 2014a). Besides these natural periodic changes in climate, it has been observed that anthropogenic activities play substantial role in recent climate change. Enzler (2013) quoted in his research article that "Svante Arrhenius (1859-1927), Swedish scientist, was the first to claim in 1896 that fossil fuel combustion may eventually result in enhanced global warming. He proposed a relation between atmospheric carbon dioxide concentrations and temperature. He found that the average surface temperature of the earth is about 15°C because of the infrared absorption capacity of water vapor and carbon dioxide. This is called the natural greenhouse effect. Arrhenius suggested a doubling of the CO_2 concentration would lead to a 5ºC temperature rise. He and Thomas Chamberlin calculated that human activities could warm the earth by adding carbon dioxide to the atmosphere". The findings of Arrhenius and Chamberlin were forgotten for very long time span. At that time it was thought that human influences were insignificant compared to natural forces, such as solar activity and ocean circulation. It was also believed that the oceans were such great carbon sinks that they would automatically cancel out our pollution. Water vapor was seen as a much more influential greenhouse gas (Maslin, 2004*)*. It was proven only after the developments of infrared spectroscopy during 1940's for measuring long-wave radiation that increasing the amount of atmospheric CO_2 resulted in more absorption of infrared radiation. A chronological events associated to understanding of greenhouse gases effect on global warming is presented in table 1.

Table 1. Chronological development in understanding of climate change effects.

S. No.	Scientist/organization name	Year	Breakthrough
1.	Jean-Baptiste Joseph Fourier	1824	Likens atmosphere to a greenhouse; introduces anthropomorphic climate change
2.	Claude Servais Mathies Pouillet	1838	Measures solar constant; identifies water vapor as a greenhouse gas
3.	John Tyndall	1861	Conducts experiments measuring thermal absorption of water vapor,CO_2, ozone and hydrocarbons.
4.	Svante Arrhenius	1896	Develops the concept of the atmospheric CO_2 energy budget.
5.	Thomas C. Chamberlain	1897 1899	Links glacial periods to atmospheric composition.
6.	Niles Ekholm	1901	Identifies CO_2 as a dominant atmospheric gas; notes differential heating of poles and equator.
7.	G. S. Callendar	1938	Studies artificial production of CO_2 and its influence on climate
8.	G. S. Callendar	1949	"Can CO_2 influence climate?"
9.	Harlow Shapley	1953	Climate Change: Evidences, Causes, and Effects
10	Gilbert N. Plass	1956	Carbon dioxide and climate
11.	Time magazine	1956	One big greenhouse

S. No.	Scientist/organization name	Year	Breakthrough
12	National academy of science of the USA	1966	Weather and climate modification: problem and perspectives
13.	World meteorological organization	1979	First world climate conference
14.	International panel on climate change (IPCC)	1990	First assessment report
15.	Kyoto protocol	1997	International treaty, that commits to reduce greenhouse gases emissions
		2005	Goes into effect

Source: Adapted from Fleming (2010)

Climate change impacts society and ecosystems in a broad variety of ways. For example change in climate can increase or decrease rainfall, influence agricultural crop yields, affect human health, cause changes to forests and other ecosystems, or even impact our energy supply. Climate change is seen as the main threat to agricultural sector over the globe; as vulnerability of this sector is high and adaptation measures are restricted by the limited resources of the developing countries (Mendelsohn and Dinar, 1999; Al-Bakri *et al*, 2010). The vulnerability of the agriculture sector to both climate change and variability is well established (Fig. 1).

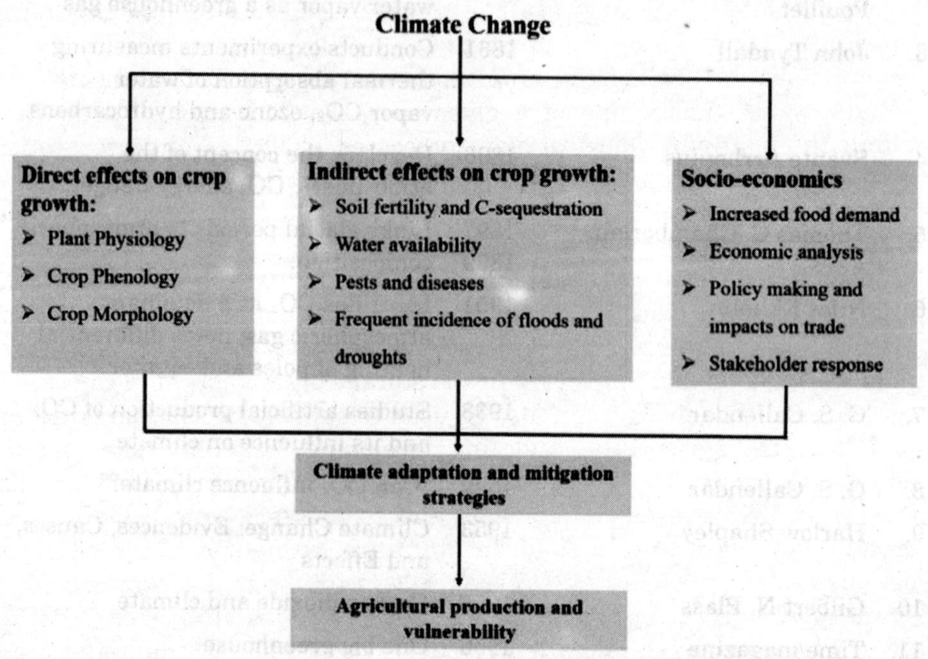

Fig. 1: Assessment of vulnerability of agriculture to climate change (Source: Anonymous, 2014b).

The general consensus is that changes in temperature and precipitation will impact plant growth and crop yield. In many developing countries, climate change is also expected to change in farming systems and will put more pressure on the rural community to cope with these changes and to build up their adaptive capacities (Reilly and Schimmelpfenning, 1999; Liwenga, 2008). Overall, climate change could result in a variety of impacts on agriculture. Some of these effects are biophysical, some are ecological, and some are economic, including:

- A shift in climate and agricultural zones towards the poles
- Changes in production patterns due to higher temperatures
- A boost in agricultural productivity due to increased carbon dioxide in the atmosphere
- Changing precipitation patterns
- Increased vulnerability of the landless and the poor

It is now well established that the net effect of climate change on world agriculture is likely to be negative and significantly effects food supply of burgeoning population over the globe. Hence, it is high time to develop adaptation strategies to regain the increasing trend in food production. There are many ways or methods by which effects of climate change on crop productivity can be assessed. Out of those methods the crop growth simulation model is the most preferred because it reduces the need for expensive and time consuming field experimentation. Crop growth simulation models are one of important tool for understanding impact of climate on agriculture and its allied sectors. The European commission (2014c) defined crop model as a group of algorithms that simulates the functioning of a given crop. Those groups of algorithms mimic the main physiological plant processes –such as light interception, respiration, carbon assimilation, grain production– through a set of assumptions and calibration parameters. Basically, the input datasets of crop models integrate meteorological data (temperature, rainfall, solar radiation, etc.), soil information (soil water capacity, soil depth) and management practices (e.g. irrigation). The outputs are usually indicators of crop development such as the biomass produced, the leaf area produced, the biomass allocated in the storage organs (grain in the case of cereals), etc. The models can also be used to support decision making and planning in agriculture and in developing different site specific adaptation strategies. Developed in the early 1960s, crop models are also used to address research problems and to increase knowledge on crop growth, development and yield. In literature, numerous crop models can be found, some of them are Decision Support System for Agrotechnology Transfer (DSSAT), Agricultural Production Systems sIMulator (APSIM), Cropping Systems Simulation Model (CropSyst), Simulateur multi disciplinaire pour les cultures standard (STICS) and many more. Out of these DSSAT is the one of most widely used crop simulation model in climate change research.

An overview of DSSAT

The DSSAT is a microcomputer software program combining crop, soil and weather databases and programs to manage them to simulate growth, development and yield as a function of the soil-plant-atmosphere dynamics (Figure 2). DSSAT was first released (v2.1) in 1989; additional releases were made in 1994 (v3.0) (Tsuji *et al.*, 1994) and 1998 (v3.5) (Hoogenboom *et al.*, 1999). Recently, DSSAT v 4.5 was released with the capability to simulate 26 crops. DSSAT have been used for many applications ranging from on-farm and precision management to regional assessments of the impact of climate variability and climate change. It has been in use for more than 20 years by researchers, educators, consultants, extension agents, growers, and policy and decision makers in over 100 countries worldwide. DSSAT integrates the effects of soil, crop phenotype, weather and management options, and allows users to ask "what if" questions by conducting virtual simulation experiments in minutes which would consume a significant part of an agronomist's career if conducted as real experiments.

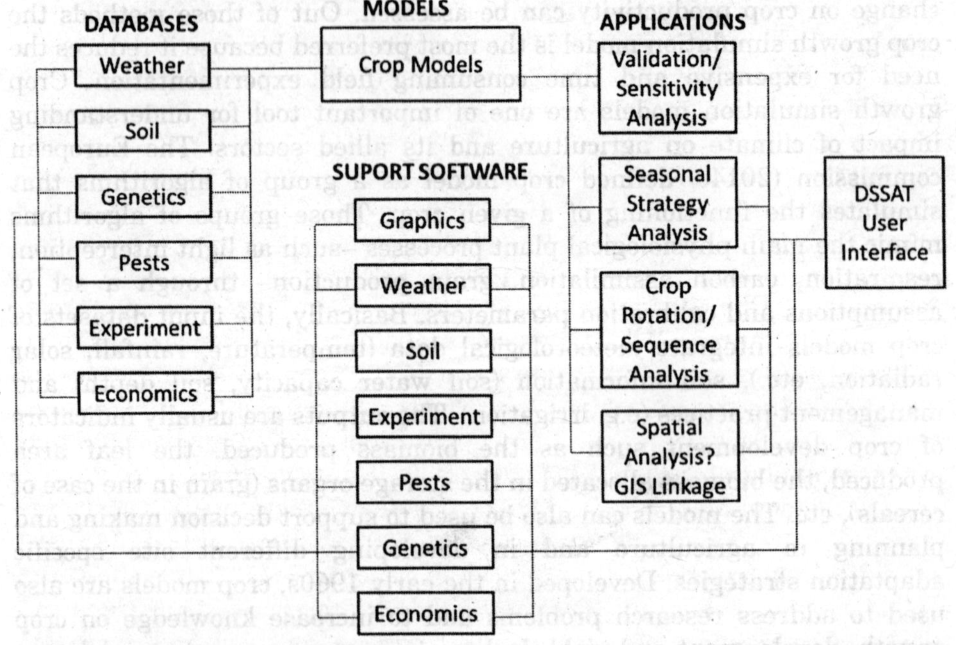

Figure 2: Description of DSSAT crop simulation model (Source: Jones *et al.*, 2003)

Climate change assessment of different parts of world using DSSAT

Presently, climate change is considered as the main threat to agricultural sector around the world especially in developing countries; as vulnerability of this sector is high and adaptation measures are restricted by the limited resources of these countries (Allen *et al.*, 1998, Mendelsohn and Dinar, 1999, Ma *et al.*, 2013). The vulnerability of the agriculture sector to both climate change and variability is well established. The general consensus is that changes in temperature and precipitation will influence plant growth and crop yield (Ma *et al.*, 2013). In many developing countries, climate change is also expected to lead to changes in farming systems and will put more pressure on the rural community to cope with these changes and to build up their adaptive capacities (Liwenga, 2008; Deng, 2010). Since development, DSSAT has been widely used over the globe for assessing the effect of climate change on agriculture and in developing adaptation strategies.

In India, soybean is the one of the major crop grown during the *kharif* season. It has seen phenomenal growth in area and production in India in the past few decades. This increasing trend of fast adaptation of the crop by the farmers in India points out that soybean is going to be the future leading commercial venture in the country (Mall *et al.*, 2004). Future climatic change is likely to have substantial impact on soybean production depending upon the magnitude of variation in CO_2 and temperature. Increased temperature significantly reduces the grain yield due to accelerated development and decreased time to accumulate grain weight (Seddigh and Joliff, 1984a, b; Baker *et al.*, 1989). In India, Mall *et al.* (2004) used DSSAT model to test performance of soybean under the changing climatic scenarios. The sites selected for this study were Coimbatore, Dharwad, Ludhiana, Hisar, Pantnagar, Delhi, Pune, Hyderabad, Ranchi, Indore, Raipur, Jabalpur and Gwalior. Climate change scenarios for the selected regions of the Indian subcontinent were developed using three widely known GCMs namely, Goddard Institute of Space Studies Model (GISS-2; Russell and Rind, 1999), Geophysical Fluid Dynamics Laboratory Model (GFDL-R30; Knutson *et al.*, 1999) and United Kingdom Meteorological Office, Hadley Climate Prediction Centre Model (UKMO, HadCM3, Mitchell *et al.*, 1998). There was a decrease (ranging between about 10 and 20%) in soybean yield in all the three future scenarios when the effect of rise in surface air temperature at the time of the doubling of CO_2 concentration was considered. The results obtained on the mitigatory option for reducing the negative impacts of temperature increases indicate that delaying the sowing dates would be favorable for increased soybean yields at all the locations in India. Sowing in the second season would also be able to mitigate the detrimental effects of future increases in surface temperature due to global warming at some locations. Similarly, Mishra *et al.* (2013) used DSSAT model to simulate the effects of climate variability and climate change on rice and wheat yields in Indo-Gangatic plain of India. The emission scenario used for this study was A1B (The A1 scenario family develops into three groups that describe alternative directions of technological change in the energy system. The three A1 groups are distinguished by their technological emphasis: fossil intensive (A1FI), non-fossil energy sources (A1T), or a balance across all sources (A1B), where balance is defined as not relying too heavily on one particular energy source, on the assumption that similar improvement rates apply to all energy supply and end-use technologies) for the period 2011 to 2040 and projected climate data for this scenario were developed from two RCMs, namely REMO and HadRM3. Both the RCMs used initial and boundary conditions of the ECHAM5 Global Climate Model to develop the long term future plausible climate scenarios at a 0.22° (~25 km) grid scale over the entire IGP. These projections have been found to simulate the important climatic processes fairly well for the Ganga river basin and have been used to assess possible climatic changes and implications for adaptation responses for this region (Moors *et al.*, 2011; Mathison *et al.*, 2012). Results indicated a significant gap between actual and potential yield for upper, middle and lower Indo-Gangetic basin (IGB) stations. The analysis based on RCM projections shows that during 2011–2040, the largest reduction in rice and wheat yields will occur in the upper IGB (reduction of potential rice and wheat yield respectively by 43.2% and 20.9% by REMO, and 24.8% and 17.2% by HadRM3). In the lower IGB, however, contrasting results are obtained, with HadRM3 based projections showing an increase in the potential rice and wheat yields, whereas, REMO based projections show decreased potential yields (Fig 3a & b). From figure 3, under HadRM3 RCM projected scenario, it is marked that there is an increase in rice and wheat yields from base period yield in the lower IGB as compared to the middle and upper IGB. This decrease in yield of rice and wheat in middle and higher IGB attributed to the shortening of rice and wheat maturity period (harvest date), was most pronounced at upper IGB (6–7 days for rice and 3–6 days for wheat) followed by Middle IGB (2–7 days for rice and 2–3 days for wheat). This shortening in the base period of crops has led to a reduction in the yield in both the regions. However, in the lower IGB, the results show an increased rice maturity period with the REMO projections and an almost 2 to 4 days decrease in the maturity period for HadRM3 projections. The variations in the specific period between anthesis and maturity explain the changes in crop yield, i.e., when the period is simulated to increase from the base period, there is an increase in crop yield while a decrease results in reduced crop yield (Mishra *et al.*, 2013). An important factor, Leaf Area Index (LAI) also shows a marked decrease for both rice and wheat at upper IGB and middle IGB stations as compared to lower IGB. An increase in heat stress is also indicated by a greater increase in maximum temperature values across the three stations. Temperatures greater than 34°C have been found to decrease wheat yields by up to 50% due to increased leaf senescence (Asseng *et al.*, 2011). Maximum and minimum temperatures in upper IGB and middle IGB demonstrate a greater increase as compared to lower IGB and therefore, may be one of the important causes of better crop yield response in lower IGB.

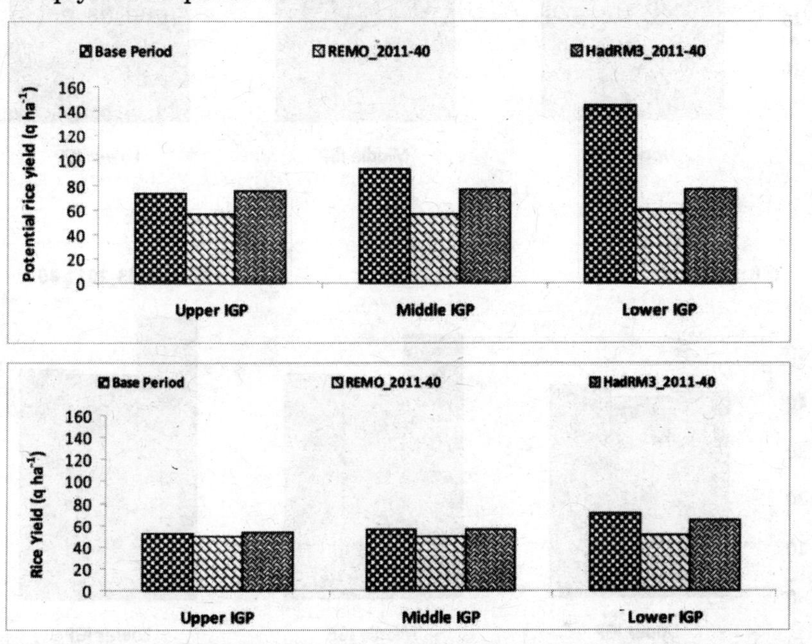

Figure 3a. Effect of climate change in the future (2011–2040) on rice yields in the upper, middle and lower IGB.

Agriculture in semi-humid and semi-arid region has been characterized as highly to climate change. The North China Plain (NCP), located in the semi-humid region, is the main food supply area and contributes approximately 41% of the total wheat yield and more than 30% of the total maize yield in China and most susceptible to climate change. In future, evapotranspiration and water use efficiency of crop will alter with climate change (Thomas, 2008; Mo *et al.*, 2007). To adapt crop systems to the changing climate, it is important to know how climate change affects agricultural production and water use efficiency. Hence, it is important to know how climate change affects agricultural production and water use efficiency to develop the adaptation strategies in this region. To explore the responses of wheat and maize yields and water use efficiencies in North China Plain to climate change, Guo *et al.* (2010), conducted a study with the DSSAT CERES model. The Climate change projections used in this was A2A, B2A and A1B from HadCM3 model

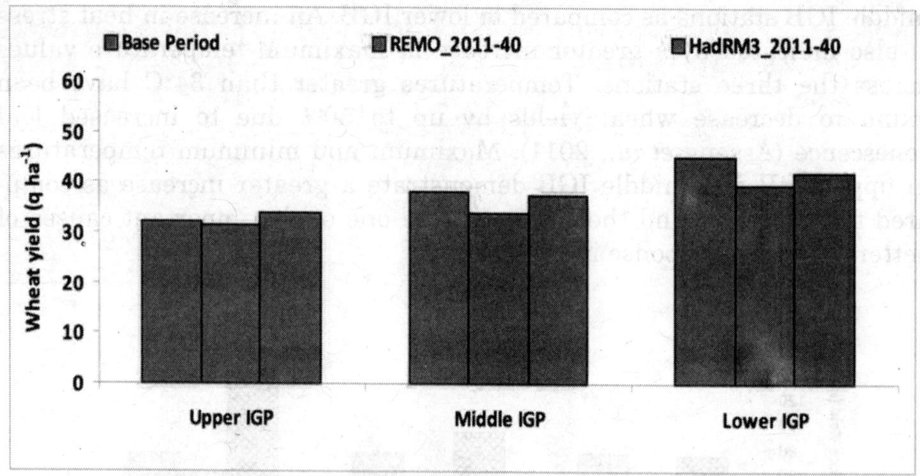

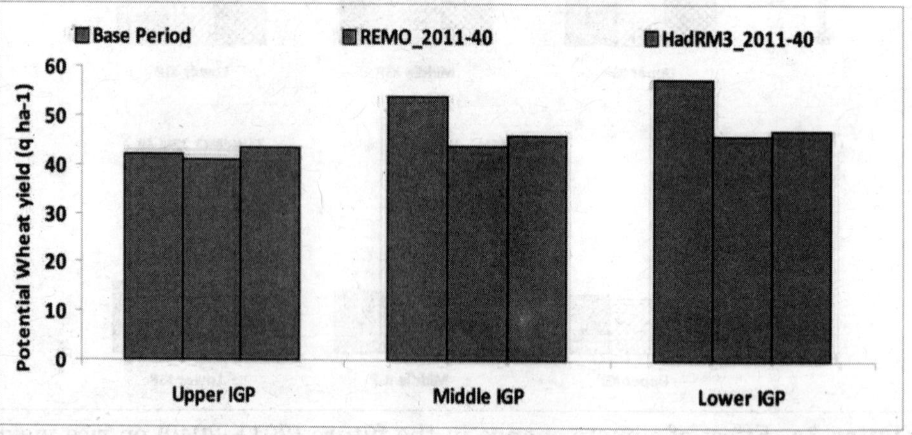

Figure 3b. Effect of climate change in the future (2011–2040) on wheat yields in the upper, middle and lower IGB.

(Table 2) and the climate generator (CLIGEN) was used to create daily weather series with monthly projection data. The results obtained by Guo *et al.* (2010) indicated that higher temperature significantly shortens the growing period which leads to early crop maturity by accelerating crop growth and development process (Figure 4). It has been also observed that changes of growing period in different stations impacted by temperature are inconsistent and show obvious spatial differences. For wheat, the maximum difference among seven stations of shortened days reaches 15 days. The least difference appeared under B2A scenario in 2090s, being 9 days. For maize, the maximum difference happens under A2A in 2090s, being 11 days. Under B2A, the difference is still the least in 2060s (Guo *et al.*, 2010). Further, Guo *et al.* (2010) concluded that the

Table 2. CO_2 concentrations under future climate projections predicted by HadCM3 model.

	A2A	B2A	A1B
2030s	488	445	479
2040s	632	521	588
2050s	775	597	697

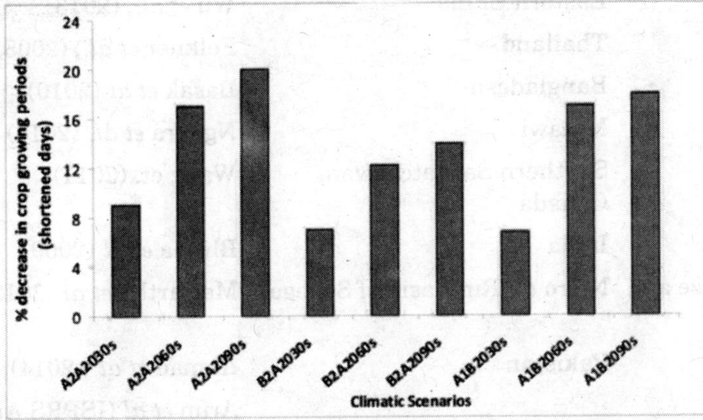

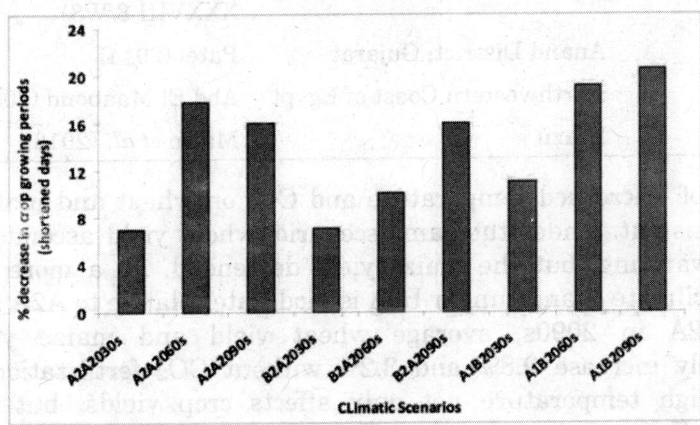

Figure 4. Changes of crop growing period under future climate scenarios relative to baseline (a: wheat, b: maize).

Table 3. Climate change effects on different crops in different region using DSSAT.

Crops	Region	References
Rice	Tamil Nadu (India)	Ramraj *et al.*, (2009)
Groundnut	Tamil Nadu (India)	Ramraj *et al.*, (2009)
Rice–wheat system	Indo-Gangetic Plains of India	Subhash and Ram Mohan (2012)

Crops	Region	References
Wheat	Punjab (India)	Timsina *et al.*, (2008)
Rice	Cauvery basin of Tamil Nadu	Geethalakshmi *et al.*, (2011)
Winter durum wheat and tomato	Southern Italy	Ventrella *et al* (2012)
Pearl Millet	West Africa	Hussaini and Halilu (2013)
Wheat	Eastern China	Wu *et al.*, (2013)
Rice	Thailand	Felkner *et al.*, (2008)
Boro rice	Bangladesh	Basak *et al* (2010)
Maize	Malawi	Ngwira *et al.* (2014)
Wheat	Southern Saskatchewan, Canada	Wang et. (2011)
Groundnut	India	Bhatia *et al* (2009)
Millet, Maize and Peanut	Nioro du Rip Basin of Senegal	MaCarthy *et al* (2013)
Rice	Pakistan	Ahmad *et al* (2014)
Tomato	-	Arun *et al* (ISPRS Archives XXXVIII-8/W3)
Potato	Anand District, Gujarat	Patel (2014)
Wheat	Northwestern Coast of Egypt	Abd El Maaboud (2004)
Sugarcane	Brazil	Marin *et al.* (2013)

impacts of increased temperature and CO_2 on wheat and maize yields are inconsistent. Under the same scenario, wheat yield ascended due to climatic warming, but the maize yield descended. As a more probable scenario, climate change under B2A is moderate relative to A2A and A1B. Under B2A in 2090s, average wheat yield and maize yield will respectively increase 9.8% and 3.2% without CO_2 fertilization in this region. High temperature not only affects crop yields, but also has positive effect on water use efficiencies, mainly ascribing to the evapotranspiration intensification. There is a positive effect of CO_2 enrichment on yield and water use efficiency. If atmospheric CO_2 concentration reaches nearly 600 ppm, wheat and maize yields will increase 38% and 12% and water use efficiencies will improve 40% and 25% respectively, in comparison to those without CO_2 fertilization. However, the uncertainty of crop yield is considerable under future climate change scenarios and whether the CO_2 fertilization may be realized is still needed further research. In Jordan Al-Bakri *et al.* (2010) used DSSAT to assess the impact of different climate change scenarios on rainfed wheat and barley. Results from DSSAT model for years showed that model was able to capture the trend of yield over the years realistically well. The model predicted an average yield of wheat of 1176 kg ha^{-1}, compare to observed average of 1173 kg ha^{-1} and an average predicted yield of barley was 927 kg ha^{-1} as compared to observed 922 kg ha^{-1}, with RMSE for barley (476 kg ha^{-1}) than for wheat (319 kg ha^{-1}). Results for predicting future yield of both crops showed that the responses of wheat and barley were different under different climate change scenarios. The reduction of rainfall by 10–20% reduced the expected yield by 4–8% for barley and 10–20% for wheat, respectively. The increase in rainfall by 10–20% increased the expected yield by 3–5% for barley and 9-18% for wheat, respectively. The increase of air temperature by 1, 2, 3 and 4°C resulted in deviation from expected yield by -14%, -28%, -38% and -46% for barley and -17%, +4%, +43% and +113% for wheat, respectively. Over the past few decades, several authors have used DSSAT crop models to simulate crop performance under future climatic scenarios and climatic variability, some of them are presented in table 3. Apart from studying impact of climate change research on crop productivity, DSSAT can be used for developing precision agriculture decision support system (Thorp *et al.*, 2008), assessment of fertilizer rates and split applications (Jeong *et al.*, 2014), modelling soil and plant phosphorus (Dzotsi *et al.*, 2010), simulating water content, crop yield and nitrate-N loss (Liu *et al.*, 2011) and many more.

Table 4. Applications of the DSSAT crop models in different regions of the world

Region	Application Type	References
Africa	Crop management	Mbabaliye and Wojtkowski (1994), Wafula (1995)
	Fertilizer management	Jagtap *et al.* (1999), Thornton *et al.* (1995),
	Irrigation management	Kamel *et al.* (1995), MacRobert and Savage (1998)
	Precision management	Booltink *et al.* (2001)
	Food security	Pisani (1987), Thornton *et al.* (1997)
Asia	Crop management	Alagarswamy *et al.* (2000), Salam *et al.* (2001)
	Fertilizer management	Godwin *et al.* (1994)
	Irrigation management	Hundal and Prabhjyot-Kaur (1997)
	Pest management	Luo *et al.* (1997), Pinnschmidt *et al.* (1995)
	Yield forecasting	Kaur and Hundal (1999), Singh *et al.* (1999)
Europe	Crop management	Ruiz-Nogueria *et al.* (2001), , Zalud *et al.* (2000)
	Fertilizer management	Gabrielle *et al.* (1998), Zalud *et al.* (2001)
	Irrigation management	Ben Nouna *et al.* (2000)

Region	Application Type	References
	Tillage management	Castrignano *et al.* (1997)
	Variety evaluation	Brisson *et al.* (1989), Colson *et al.* (1995)
	Precision farming	Booltink and Verhagen (1997), Booltink *et al.* (2001)
	Environmental pollution	Kovacs and Nemeth (1995)

(Adapted from Jones *et al.*, 2003)

Mitigation and adaptation

Intergovernmental Panel on Climate Change (IPCC) defines adaptation as 'adjustment in natural or human systems to a new or changing environment. Adaptation to climate change refers to adjustment in natural or human systems in response to actual or expected climatic stimuli or their effects, which moderates harm or exploits beneficial opportunities', whereas mitigation involves reducing the magnitude of climate change itself and, as we will see in the final two lessons, can be subdivided into two alternative strategies: *emissions reductions* - dealing with the problem at its very source, and *geoengineering* - somehow offsetting the effects of greenhouse gas emissions. Relations between mitigation and adaptation strategies can be better understood by Figure 5. A crop growth model is very much useful in evaluating and developing different adaptation strategies to reduce severity of climate change.

Ventrella *et al.*, 2012 evaluated different agronomic adaptation strategies using DSSAT model in southern Italy to reduce climate change vulnerability. The agronomic adaptation options evaluated by Ventrella *et al.*, 2012 were sowing and transplanting time for wheat and tomato, respectively. The optimal sowing or transplanting dates were considered as November 28th (330 day of year, DOY) and April 30th (120 DOY) for wheat and tomato, respectively. In order to better identify the response curve of the two crops to sowing/transplanting time, other four sowing dates were also considered for each crop. In particular, for wheat they considered 2 spaced sowings of 15 days, while in the case of tomato they considered a 10-day interval under changing climatic scenarios. Results indicated that no negative yield effects of climate change were observed for winter durum wheat with delayed sowing (from 330 to 345 DOY) increasing the average dry matter grain yield under forecasted scenarios. Instead, the warmer temperatures were primarily seen to accelerate the phenology, resulting in decreased yield for tomato under the + 5°C future climate scenario. In general, under global temperature increase by 5°C, early transplanting times could minimize the negative impact of climate change on crop productivity but the intensity of this effect was not sufficient to restore the current production levels of tomato cultivated in southern Italy. In an another study conducted by Bhuvaneswari *et al.*, 2014 in western zone of Tamil Nadu, resulted in reduction of rice yield by 4 -56% with increase in temperature from 1 to 5°C respectively from the current climate under different dates of planting from 1st June to 15th July. The different sowing windows tested as adaptation strategy to climate change indicated that the change in yield from current condition under early, normal, late planting during kharif season was -21.2, -15.0 and -16.3% respectively for 3°C increase in temperature and 650 ppm CO_2 enrichment, during winter season, it was -9.2, +10.2 and +11.0% respectively (Table 5).

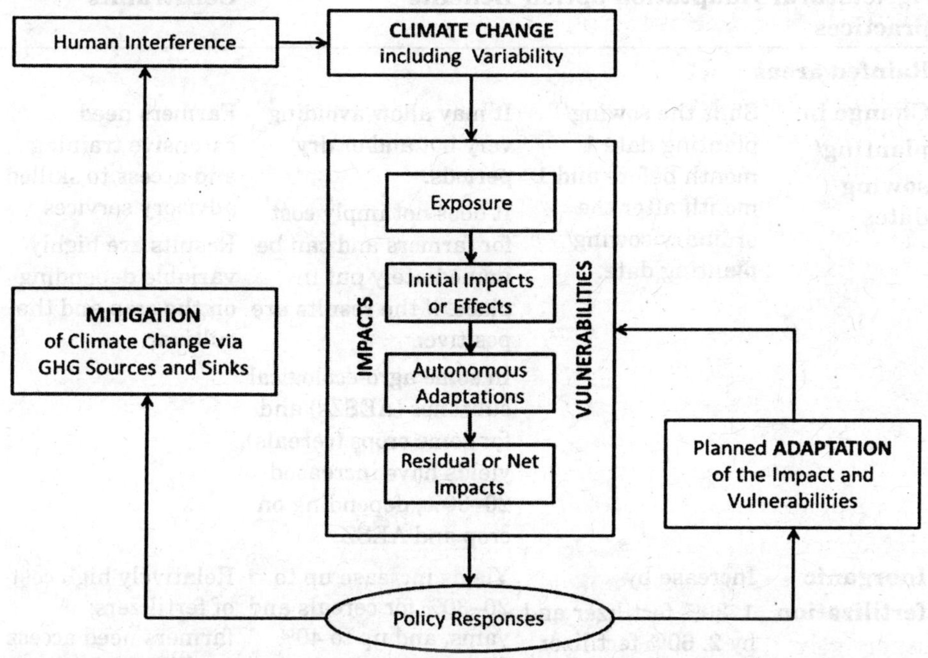

Figure 5. Place of adaptation and mitigation in response to climate change. (IPCC TAR 2001 WG2 after Smith *et al.*, 1999)

Table 5. Impact of elevated temperature on rice productivity (t ha^{-1}) under different dates of planting

Date of planting	Present climate	Changes in temperature				
		1°C	2°C	3°C	4°C	5°C
1st June	6.6	- 4	- 15	- 22	- 37	- 53
15th June	5.8	- 4	- 13	- 23	- 39	- 53
1st July	5.2	- 4	- 12	- 24	- 40	- 54
15th July	4.6	- 6	- 15	- 25	- 40	- 56

Source: Bhuvaneswari *et al.* (2014)

In Nigeria, under a World Bank project on "Towards Climate-Resilient Development", several practices were analyzed in terms of their potential to offset-across space (different AEZs); time (2020 and 2050); and crops-the negative impacts of climate change on yields. For rain-fed and irrigated areas different adaptation options were considered with focused on yield improvements (Table 6).

Table 6. Differed adaptation option with their benefits and constraints: A Nigerian study (Adapted from Cervigni *et al.*, 2013).

Agricultural practices	Adaptation option	Benefits	Constraints
Rainfed areas			
Change in planting/ sowing dates	Shift the sowing/ planting date 1 month before and 1 month after the ordinary sowing/ planting date.	It may allow avoiding very hot and/or dry periods. It does not imply cost for farmers and can be immediately put in place, if the results are positive. In some agro-ecological subzones (AESZs) and for some crops (cereals), yields have increased 20–30%, depending on crop and AESZ.	Farmers need extensive training and access to skilled advisory services. Results are highly variable depending on the crop and the cultivar.
Inorganic fertilization	Increase by 1. 30% fertilizer and by 2. 60% fertilizer over the ordinary fertilization amount.	Yields increase up to 20–30% for cereals and yams, and up to 40% for cassava.	Relatively high cost of fertilizers; farmers need access to skilled advisory services. 2. There may be an impact on the environment.
Conservation agriculture	Application of manure or residues from crop production (residue) to comple-ment baseline nutrient manage-ment; Complete substitut-ion of inorganic ferti-lization with manure.	Yields increase up to 25% for sorghum and millet, up to 35% for rice, and up to 50% for maize and cassava.	Farmers need extensive training and access to skilled advisory services. There may be a relatively high up-front cost for the purchase or application of manure and residues.

Agricultural practices	Adaptation option	Benefits	Constraints
Irrigated areas			
Combining shift in growing period and irrigation	Shift the sowing/ planting date 1 month before and 1 month after the traditional date, in addition to irrigation practice	Yields increase for cassava and yams, and there is a positive synergy between irrigation and the shift in growing period.	Farmers need extensive training and access to skilled advisory services.

Conclusions:

According to IPCC, climate change will have broad range of impacts on agricultural crops and livestock which consequently affects the food security around the globe. This is mainly because, food production is critically dependent on local temperature and precipitation conditions. Any changes in climate compel farmers to adapt these changes, and this adaptation requires resources that could be used for other purposes. For a few, the adaptations might be beneficial, but for many, major challenge is to increase productivity and more difficult is to manage the risk associated with climate change. The agricultural system as a whole will have difficulty in supplying adequate quantities of food to maintain constant real prices. The DSSAT crop simulation model, developed by an international network of scientists, has been in use for the more than 25 years by researchers worldwide for assessing impact of climate change and devising adaptation strategies to maintain and/or enhance crop productivity. Presently, this system incorporates models of 28 different crops with software that facilitates the evaluation and application of the crop models for different purposes. Various crop adaptation strategies such as choice of crop (use of drought, heat, pest resistant varieties, use of early/late maturing varieties altering mix of crops), tillage and time of operations (change planting date, terracing, ridging, land levelling, reduced tillage, deep ploughing, change fallow, mulching practices, alter cultivations and switch seasons for cropping), crop husbandry (alter row and plant spacing intercropping), irrigation and water harvesting, and input of agrochemicals (change amounts of fertilizer application and alter time of application) can be evaluated by DSSAT and best evolved/ suited strategies may be recommended to farmers to cope up with future climate change.

14 Integrating Remote Sensing Data in Crop Process Models

Nirmal Kumar[1], *N.K. Sinha*[2] *and M. Mohanty*[2]

[1]National Bureau of Soil Survey & Land Use Planning, Amravati Road, Nagpur -440 033, India
E-mail: urwithnirmal@gmail.com
[2]Indian Institute of Soil Science, Nabi Bagh, Berasia Road, Bhopal - 462 038, India

Introduction

Reliable and timely prediction of crop yield is important for agricultural land management and policy making. Country-level crop yield estimation using remote sensing data has been carried out operationally for more than two decades (MacDonald and Hall 1980) like Large Area Crop Inventory Experiment (LACIE) in the USA, which aimed at estimating wheat productions over the world using Landsat MSS information (Erickson 1984), to the Monitoring Agriculture with Remote Sensing (MARS) programme of the European Union, which issues real-time estimations of crop areas and productions at European level using high spatial resolution satellite data (Meyer- Roux 1990). In India, well-planned programmes for crop acreage and production forecasts like Crop Acreage and Production Estimation (CAPE) and Forecasting Agricultural Output using Space, Agro-meteorology and Land-based Observations (FASAL) culminated in the establishment of a dedicated Mahalanobis National Crop Forecasting Centre (MNCFC) by Ministry of Agriculture (GoI), for crop inventorying and drought assessment. Crops covered are rice, wheat, cotton, mustard, sugarcane, etc.

In operational monitoring, satellite data are used only for crop acreage estimations, that is an important part of the agricultural production assessment at a national or continental level (Sharman *et al.*, 1992). Crop growth simulation models have a high potential in crop growth monitoring and yield prediction. However, the need of a large number of input parameters, concerning genetic characteristics of the crop, soil properties, weather data and crop management information, hampers the application of such models at large spatial scale and limits the confidence in their output. Numerous field experiments and theoretical studies have shown that remotely sensed measurements in the different wavelengths can provide information on crop activity (Moulin *et al.*, 1998). Remote sensing is able to provide spatial and temporal distributed information on vegetation cover, as well as on state variables of the models, such as biomass and leaf area index (LAI). Under these circumstances, an integrative use of remote sensing and crop growth models can be beneficial to improve model predictions.

Here we discuss an overview of how satellite observations are used to feed or tune crop models and improve their capability to predict crop yields in a region. Relations between crop characteristics which correspond to models state variables and satellite observations are briefly analysed, together with the various types of crop models commonly used.

Remote sensing inputs for crop models

The development of remote sensing methods has led to a better understanding of as to how leaf reflectance changes in response to leaf thick-ness, canopy characteristics, leaf age and water status. Leaf chlorophyll absorption at various wavelengths provides the basis for measuring reflectance with either typical broad-band radiometers, current satellite platforms or hyper-spectral sensors that measure reflectance in narrow bands (Ahmed *et al.*, 2011).

Remote sensing tools can provide information about vegetation in various wavebands: solar spectrum, active and passive microwave (Preavot *et al.* 1993) and in thermal range (Moran *et al.*, 1994, Seguin *et al.* 1994). Remotely sensed measurements can be related to instantaneous values of various canopy state variables. The NDVI has proven to be a proxy for the status of the aboveground biomass at the landscape level due to the high correlation with green-leaf density, net primary production and CO_2 fluxes (Wylie *et al.*, 2003, Li, 2012). Canopy state variables can be connected to remotely sensed information by using physical radiative transfer models or empirical relationships. Canopy variables appear in physiological functions which determine final yield when integrated in time. Crop simulation models (Whistler *et al.* 1986) provide formal descriptions of state variables and time integration of biophysical processes. In the microwave range, information about soil moisture and plant water content is obtained (Ines *et al.*, 2013); whereas information on plant water status can be derived from thermal infra-red (TIR) range (Jones *et al.*, 2009). Solar spectral data are related to leaf area index and the photosynthetically absorbed radiation (Moriondo *et al.*, 2007; Ines *et al.*, 2013). Radar, which gives information on plant structure, is sometimes used to supplement optical data sets (Bouman 1992; Ines *et al.*, 2013). Airborne and satellite radar data have been successfully coupled with crop models (Prévot *et al.*, 2003; Ines *et al.*, 2013).

Detection of crop phenology using remote sensing

Monitoring seasonal changes in vegetation activity and crop phenology over wide areas is essential for deciding time boundary conditions for crop yield modelling (Bauman *et al.*, 2001).

Normalized difference vegetation index (NDVI)

In particular, NDVI-based time series are fundamental to the remote sensing of vegetation phenology and extraction of numerical observations related to vegetation dynamics. The vegetation phenology characterization can be addressed by calculating "NDVI metrics" such as the date of greenup (sudden increase of NDVI or NDVI surpassing a threshold value), date and magnitude of maximum NDVI, temporal integration of NDVI, length of growing season, and rates of NDVI change (Hall-Beyer, 2003). All these features can be obtained from NDVI time series. Various methods using daily NDVI data have been developed for monitoring crops and natural vegetation (Xiao *et al.*, 2002). Jakubauskas *et al.* (2002) stated that the application of harmonic analysis to time series data is useful for detecting changes of land use and land cover, such as at planting and harvest.

Net primary productivity (NPP), measured as carbon absorption by plants through the process of photosynthesis per unit time and space, is usually adopted as a straightforward indicator when vegetation vigour is referred to as biological productivity. NDVI has a strong relationship with the fraction of photosynthetically active radiation (PAR) absorbed by the plants, which sets the upper limit for NPP. Furthermore, it has been well documented that NPP is correlated with NDVI across a range of geographical areas and ecosystems, which leads to a common use of NDVI as a surrogate estimator of NPP.

Enhanced Vegetation Index (EVI)

The EVI can be linearly correlated with the leaf area index, and has a higher sensitivity than that NDVI in high biomass areas (Huete *et al.*, 2002). Furthermore, EVI is more practical than NDVI when humidity is high. Sakamoto *et al.*, 2005 developed a new method for remotely determining phenological stages of rice. The method consists of three procedures: (i) prescription of multitemporal MODIS/Terra data; (ii) filtering time-series EVI data by time-frequency analysis; and (iii) specifying the phenological stages by detecting the maximum point, minimal point and inflection point from the smoothed EVI time profile. They proposed this new method using the wavelet transform; Wavelet based Filter for determining Crop Phenology (WFCP). Since the biomass of rice fields is low compared with forests, EVI retains linearity in rice fields. Therefore, EVI can be used to evaluate the process of rice growth adequately with a wide dynamic range. Since the rice growing stage falls in the rainy season when humidity is high, EVI is a more practical than NDVI for monitoring the rice fields in Japan. Zhang *et al.* (2003) identified phonological dates by finding points on the time profile curve of the Vegetation Index (VI) where the rate of curvature is maximum.

Leaf Area Index (LAI) and Fraction of Absorbed Photosynthetically Active Radiation (fAPAR)

The ratio of leaf surface area to unit ground surface area, called leaf area index (LAI), describes the potential surface area available for leaf gas exchange between the atmosphere and the terrestrial biosphere (Cowling and Field, 2003). Therefore, it is an important parameter controlling many biological and physical processes of the vegetation, including the interception of light and water (rainfall and fog), attenuation of light through the canopy, transpiration, photosynthesis, autotrophic respiration, and carbon and nutrient (e.g. nitrogen, phosphorus, etc.) cycles. LAI obtained across a range of spatial scales, from individual plants to entire regions or continents (Sellers *et al.*, 1986) has been used extensively in interactive models of land surface processes (Pielke *et al.*, 1998). Another simplified approach to regional crop yield assessment from satellite imagery is to use the fraction of absorbed photosynthetically active radiation (fAPAR) (Doraiswamy *et al.*, 2005). Lobell *et al.* (2003) incorporated fAPAR into a model based on crop light-use efficiency to predict biomass and crop yields.

Through LAI and APAR, optical remote sensing can provide information on the actual status of agricultural crops during the growing season, thus offering the possibility of calibrating the growth modelling. Clevers, 1997 developed a simple approach based on the fraction APAR, estimated from optical remote sensing measurements, for crop growth monitoring and yield prediction in practice. Parameters of this linear relationship were not estimated empirically, but a complex physiological crop growth model was used to derive these parameters. Results were similar to the accuracy of yield predictions obtained from directly calibrating a complex crop growth model using a time-series of optical reflectance data. Kang *et al.* (2003) used 30-year mean air temperature and MODIS/LAI data to detect the onset of vegetation greenness in Korea. Their results suggested that the phenological stages can be detected more precisely through a combination of climatic data and MODIS data. Doraiswamy *et al.*, 2005 evaluated the applicability of the MODIS 8-day, 250 m resolution composite imagery for operational assessment of crop condition and yields.

Satellite data products mostly used in crop simulation models

Moderate Resolution Imaging Spectroradiometer (MODIS) and the Advanced Very High Resolution Radiometer (AVHRR) sensor products are the most popularly used data as inputs for crop simulation modelling. AVHRR sensor provides the longest time series of satellite images to study vegetation dynamics. Most frequently, the well-known maximum value composites of daily NDVI data are used to reduce the influence of atmospheric effects on images. The MODIS instrument includes seven spectral bands that are designed exclusively for monitoring earth's land surfaces. When combined, the Terra and Aqua MODIS provide sub-daily global coverage at 250- and 500-m spatial resolutions and offer enhanced spectral, spatial, radiometric, and geometric quality for improved mapping and monitoring of vegetation activity. The LAI/fPAR products from AVHRR, the MODIS sensor and the Système Pour l'Observation de la Terre (SPOT) sensor have a large earth science community user base and the ease of access, provision of pixel quality and validation information have greatly aided the use of these products (Ganguly *et al.*, 2014).

Multi-decadal, validated, consistent global and regional data sets of LAI/FPAR from the AVHRR, MODIS, and the SPOT-VGT sensors are now available at resolutions of 1 km to 1° in service of several national and international initiatives (Ganguly *et al.*, 2008). The Landsat series of sensors also provide a unique opportunity to characterize terrestrial ecosystem processes at a spatial scale at which most natural resources management decisions are made. Although regional- to continental-scale multitemporal mosaics of Landsat data have been constructed for pilot studies of national land use change monitoring and disturbance mapping (Chander *et al.* 2009), the Landsat archive has not yet been exploited to derive long-term biophysical products (Ganguly *et al.*, 2014).

Data availability and sources

MODIS is an extensive program using sensors on two satellites that each provides complete daily coverage of the earth. The data have a variety of resolutions; spectral, spatial and temporal. The MODIS web site, http://modis.gsfc.nasa.gov/index.php, is a good place to begin learning about this important program. Because the MODIS sensor is carried on both the Terra and Aqua satellites, it is generally possible to obtain images in the morning (Terra) and the afternoon (Aqua) for any particular location. Night time data are also available in the thermal range of the spectrum. While daily products are available for any specific date and location, many consolidated products have been developed with MODIS data. These include 8-day or 16-day composite images, a variety of indices, and a range of other global products with different time scales. The standard MODIS collection of images, NDVI, BRDF, GPP, LAI/ FPAR products are available via http://earthexplorer.usgs.gov at a spatial resolution of 250 m, 500 m and 1 km depending on the product. The long-term multi-year (1981 till present) monthly AVHRR LAI/FPAR dataset based on a scaled version of the MODIS algorithm is available upon request at the Climate and Vegetation Research Group at http://cliveg.bu.edu/modismisr/index.html. The GEOV2 LAI/ FPAR products can be downloaded freely from the GEOLAND2 web portal located at http://www.geoland2.eu/portal/. They are available at the decadal time step with 0.05° and 0.0089° spatial sampling interval in lat-lon geographic projection system. NASA Earth Exchange (NEX) in collaboration with USGS EROS is also currently making Landsat derived LAI available to research community on demand basis (Ganguly *et al.*, 2014).

Integration of remote sensing data in to crop models

Some basic ways to integrate the radiometric observations with a crop model were described by Maas (1988), Delecolle *et al.* (1992) and Moulin *et al.* (1998), who described four main integration methods: (i) the direct use of driving variables derived from remote sensing data; (ii) the updating of state variables within the model (forcing); (iii) the reinitialization of the model; and (iv) the re-calibration of the model. Re-parametrization is similar to reinitialization except that the model parameters are adjusted (Plummer 2000). All four methods, however, show strong limitations, especially when applied on a regional scale Moriondo *et al.*, 2007. In particular, the direct use of a driving variable is generally difficult because the temporal resolution of remote sensing data does not match the time step requirement of crop models. The availability of estimated input parameters describing crop canopy characteristics is the key liming factor in the application of the other methods in regional scale studies.

Pre-processing

In general, time-series NDVI data obtained by satellite include various noise components such as aerosols and bidirectional reflectance distribution factors. Thus, noise reduction or fitting a model to observed data is necessary before phenological stages can be determined. Tools for such pre-processing include smoothers such as polynomial and median filters (VanDijk *et al.*, 1987), a moving window to select the local maximum VI (Viovy *et al.*, 1992, Temporal Window Operation (TWO; Park *et al.*, 1999), and logistic curve fitting (Zhang *et al.*, 2003). Several frequency analysis studies have used the Fourier transform for separating the high-frequency components of noise and the low-frequency components of seasonal changes of VI (Azzali & Menenti, 2000). Sawada *et al.* (1999) developed the Local Maximum Fitting method by applying the Fourier transform to 10-day composite data of NOAA/ AVHRR NDVI to reconstruct a cloud-free seasonal profile of NDVI. Hara *et al.* (2003) modified that method and applied it to 10-day composite data of SPOT-VEGETATION NDVI. The wavelet transform also can be used to remove noise (Sakamoto *et al.*, 2005). Its strength is the feasibility of identifying the timing of events such as localized objective signals in the presence of noise. Sakamoto *et al*, 2005 compared the Fourier and wavelet transforms as filtering methods in the identification of rice phenological stages from daily VI changes.

Direct use of empirical models

Different methods have been developed to estimate crop yields by means of satellite data, and one commonly applied is the development of empirical relationships between NDVI and crop yield. Rossini and Benedetti (1993) found a strong relationship between wheat yield and NOAA-AVHR NDVI data over an Italian region. Hamar *et al.* (1996) established a linear regression model to estimate corn and wheat yield at a regional scale based on vegetation indices computed with Landsat MSS data. Maselli and Rembold (2001) demonstrated the potential of using multi-year NOAA-AVHRR NDVI data to estimate wheat yield in North African countries. Similar relationships were obtained for various crops in other areas (for example: Rasmussen (1992) for millet yield, Manjunath and Potdar (2002) for wheat yield). This approach is fundamentally simple, but the relationships found have only a local value and cannot be easily extended to different areas.

Studies have shown that the cumulative seasonal normalized difference vegetation index (NDVI) values were significantly correlated with reported crop yields. Doraiswamy and Cook (1995) demonstrated that cumulative NDVI values for spring wheat during the grain fill period improved estimates of crop yields in North Dakota. However, the regression based approach is applicable only for a given region and the same range of weather conditions where it was developed.

Houlès *et al.* (2007) used simple statistical relationships to derive biomass and N uptake from LAI, and to calculate the deficit of N absorption that has to be filled by N fertilization. A similar approach has been incorporated into the algorithm adopted in the Farmstar operational service (Blondlot *et al.*, 2005) for nitrogen fertilization advice, based on LAI maps generated from SPOT satellite imagery. In the latter case, specific relationships between LAI and biomass for different wheat development stages and varieties have been obtained and calibrated using a large number of experimental trials, mostly carried out in the north of France (Blondlot *et al.*, 2005). Empirical relationships are suitable in a specific operational context, although, given the complexity of the processes that link crop variables; they have a limited range of validity.

Methods of coupling remote sensing data in to models

Remote sensing had been incorporated into crop simulation models either as a forcing function or simulation steering (Bouman *et al.*, 1997). Forcing function is applied to replace simulated state variable with the remote sensing observation while simulation steering is used to re-initialize (e.g., sowing date, planting density) or re-parameterize (e.g., canopy and growth parameters) the crop model in a way that minimizes the difference between simulated and measured data. Coupling a radiation transfer model to a crop model makes it possible to simulate reflectance for those times in crop growth for which remote sensing data are available. The inversion of the combined model on these data then makes it possible to estimate new values for certain sensitive parameters of the crop model. Of the several algorithms (e.g., particle filter, Kalman filter) capable of performing data assimilation to update sequentially model states and parameters, the Monte Carlo-based Ensemble Kalman Filter (EnKF) is the one that is widely used (Evensen, 2003). EnKF received a lot of attention in the geosciences because of its ease of implementation, computational efficiency and optimum performance. Many studies have implemented EnKF to assimilate RS data in meteorological and hydrological models with considerable success (Evensen, 2003). Scattering by Arbitrarily Inclined Leaves (SAIL) is another popular RT model used for assimilating remote sensing data to crop model. Weiss *et al.* (2001) successfully assimilated the radiative transfer model SAIL and a crop growth model (STICS); simulated and measured reflectances agreed very well. Combining the process model (PROMET-V) with SAIL, Bach *et al.* (2001) obtained very good results for LAI, dry biomass and canopy height. Fang *et al.*, 2011 coupled Cropping System Model (CSM)–Crop Environment Resource Synthesis (CERES)–Maize and the RT model Markov Chain canopy Reflectance Model (MCRM) (Kuusk 1998) for forecasting yield.

Forcing

The forcing strategy consists of updating at least one state variable in the model using remote sensing data. The crop model requires a value for this state variable at each time step. On the other hand, remote sensing estimations are available only at acquisition dates, generally less frequent than the model step. Gaps between dates must therefore be filled by some interpolation procedure, usually by fitting an empirical curve of time evolution of the state variable to the `observed' data. Thorp *et al.* (2010) assimilated measured LAI in the DSSAT-CSM-Wheat model using forcing and updating mechanisms. Delecolle *et al.* (1992) used high spatial resolution satellite data (SPOT/HRV) over wheat fields to improve the predictive ability of a complex deterministic model (ARCWHEAT, Weir *et al.* 1984). Since the temporal frequency of SPOT data is limited, the derived LAI values are interpolated to a daily time step using a simple model of LAI time course (Baret, 1989). The resulting daily LAI values are then used as input variables to ARCWHEAT. This coupling leads to a reduction of the mean error on yield estimation. Casa *et al.*, 2012 evaluated the performances of approaches based on empirical relationships and forcing LAI within the STICS model (Brisson *et al.*, 2009) with emphasis on the effect of the accuracy and frequency of LAI data used. Both actual and simulated case studies on wheat for northern France conditions were investigated under several levels of knowledge of the model input parameters and initial conditions. The results highlight the interest of using model based approaches for the estimation of agronomic variables. Forcing LAI data into the crop model allows compensation for the lack of detailed knowledge on management practices or soil characteristics.

Re-initialization/ re-parameterization

This consists of minimizing the difference between a derived state variable or the radiometric signal and its simulation by the re-parameterization and/or reinitialization of the crop production model. Examples of this approach include the works of Bouman (1992), Olioso *et al.* (2005), Fang *et al.* (2011) and Thorp *et al.* (2012) who linked radiative transfer models with crop models. Ines *et al.* (2006) used remotely sensed evapotranspiration to reparameterize soil properties, crop and water management parameters of a pseudo-regional Soil–Water–Atmosphere – Plant (SWAP) model. Vazifedoust *et al.* (2009) conducted a simple sequential data assimilation using a constant gain Kalman filter to assimilate LAI and the ratio of actual evapotranspiration (ET) to potential ET (ET/ETp) in SWAP-WOFOST and found significant improvements in simulated total dry matter but only one among three of their observation fields showed significant improvements in simulated yield. It should be noted that timing of the use and frequency of LAI data assimilation in crop model is critical as LAI (or NDVI) is more directly related to yield at silking and grain filling (Ozalkan *et al.*, 2010) stages. Yuping *et al.*, 2008 re-initialized the emergence date and the biomass at turn-green stage using MODIS data based on coupling the SAIL model to WOFOST model through the state variable LAI.

Conclusion

Both of crop growth simulation models and remote sensing method have a high potential in crop growth monitoring and yield prediction. However, crop models have limitations in regional application and remote sensing in describing the growth process. Therefore, many researchers try to combine those two approaches for estimating the regional crop yields. The techniques based on multi-temporal, multi-spectral, satellite-sensor-acquired data at different spatial resolution serve to detect, identify, map and monitor crop activities and changes. At very fine temporal resolution, the vegetation changes result from its phenology (intra-annual or seasonal changes), and a lower temporal resolution, they are mainly known as land-cover changes (inter-annual changes). These observations give help to extend crop models to a regional scale.

❑❑❑

15 Decision Support System for Managing Soil Fertility and Productivity in Agriculture

A.K. Vishwakarma, Brajendra, K.M. Bujarbaruah, S.V. Ngachan and K.A. Pathak*

ICAR Research Complex for NEH Region, Umiam-793103 Meghalaya.

Introduction

Soil is one of the most important natural resource for agriculture, finite in nature and subject to deterioration. It requires effective and efficient utilization for maintenance and conservation for future generations. Crop harvests tend to exhaust soils of their nutrient pool, resulting in nutrient mining which if not adequately replenished results in decline of agricultural productivity over a period of time. The long term sustainability of production in agricultural systems depends upon proper management of soil fertility. It is often reported that a wide range of variation is observed between the potential and actual yield of the crops in farmers field. There are several factors like soil factors, climatic factors, plant factors and biological factors and interaction of these factors ultimately determines the final outcome in the form of economic yield. Among these factors nutrient management holds the key in terms of soil fertility and sustainable productivity.

Nutrient management is one of the most important and critical factors that determines the final output of the crops in the form of grain yield or economic yield. Nutrient recommendations to the farmers are usually made on the basis of soil test values and the crop, which the farmer would like to grow. Under such type of blanket recommendations the important factors like the target yield which farmer would like to obtain from the crop, the genetic potential of the crop variety selected, soil efficiency and the added fertilizer nutrient use efficiencies which play an important role in the uptake of nutrients absorbed by the crop plants to the total quantity of the nutrients applied in the soil, are not properly taken into consideration. This resulted in huge differences in the yield, non-judicious application of nutrients either insufficient or excessive, which resulted in either poor crop yields or toxic effects of excessive fertilization. These ultimately resulted in poor economics of cultivation; deterioration of soil health which affects the long term sustainability of the production system and farmers suffer huge losses.

The Decision support system (DSS) is of immense importance as it can provide immediate information to solve the problem of the user and to provide assistance in making decisions related to crop nutrient requirement and their management. The farmers/land users and researchers urgently need information about the crop nutrient requirements, nutrient uptake, and the quantity of various fertilizers and manures to be applied to the crop for obtaining desired production levels under a certain soil fertility level. The architecture of DSS should be in such a manner that wide range of users including farmers/agriculture extension officers/ subject matter specialists/researchers with different demands to crop and soil related data and methods can utilize it and obtain informations. It may also be capable of giving warnings and alerts to the user related to target yield and the quantities of different fertilizers to be applied to the soil wherever required.

Basics of the DSS developed

Several models have been developed to work out the crop nutrient requirements, but most of them take into consideration only single plant nutrient or a specific crop under specific management condition. This software system takes into account most of the parameters (nutrient uptake by the crop, soil nutrient content, soil efficiency, which influences crop nutrient requirement and their respective fertilizer needs. The system takes into consideration soil test crop response equations as the key basic concept to work out the nutrient requirement of the crop and then takes into account different fertilizers and their fertilizer use efficiencies to work out the quantities of various fertilizers required to be applied

The various components of these equations are furnished either as customized equations for different crops to generate results based on users input or the user can also submit his values as per the requirement of the software to obtain results. The basic equations utilized in the software are as under.

$$\text{Nutrient requirement (NR)} = T \times NU - SNC \times SE$$

Where,

T = Target yield (q/ha)

NU = Nutrient uptake (kg/q)

SNC = Soil nutrient content (kg/ha)

SE = Soil efficiency (%)

This equation when used in conjunction with fertilizer use efficiency will provide near estimations of quantities of the fertilizers, which are required to obtain a certain desired yield level, which forms the basis of this software.

$$\text{Fertilizer requirement (FR)} = \frac{NR \times 100}{FNC} \times \frac{100}{FUE}$$

Where,

NR = Nutrient requirement to obtain a target yield (kg/ha)

FNC = Nutrient content in fertilizer (%)

FUE = Fertilizer use efficiency (%)

This software can determine the crop nutrient requirements for different crops and the quantities of different fertilizers and manures to be applied to the crop under different types of nutrient management practices such as inorganic, organic, integrated and intercropping situations for obtaining a realistic desired target yield along with efficiency calculations.

The architecture of this software is designed in such a way that a wide range of users with different requirements can easily use the software and generate desired informations by furnishing certain soil and crop related parameters. Each module has been supplied with options to determine nutrient and fertilizer requirements depending upon the choice of the user and the amount of information he/she can furnish:

1. Crop nutrient and fertilizer requirements based on blanket recommendations can be obtained in the absence of any soil and crop related data and as such can be used for places where no information is available.
2. Crop nutrient and fertilizer requirements based on built in readymade nutrient uptake equations that can be obtained by furnishing the soil test values and the target yield for a particular crop and the options for availability of different fertilizers. The user can select any of the available fertilizer material for supplementation of particular nutrient and the required quantities of different fertilizers can be obtained at a set fertilizer use efficiency of 50%.
3. Crop nutrient and fertilizer requirements based on actual soil test values crop nutrient uptake, target yield and soil efficiency can also be obtained by furnishing the desired information for a particular crop. After selecting the nitrogen, phosphorus and potassium fertilizer sources the required quantities of different

fertilizers can be obtained at a site specific fertilizer use efficiency.

Operation of the software:

The software can be easily installed on any PC or can be run directly from the compact disk. To operate the software one needs to double click on it. The first screen of the software will appear displaying the informations about the author and the space for furnishing the software code the password and after furnishing the desired software code and password and selecting a language one can enter the main menu of the software the module from the choice of nutrient management on pressing the enter button the desired module can be accessed (Fig. 1 & 2).

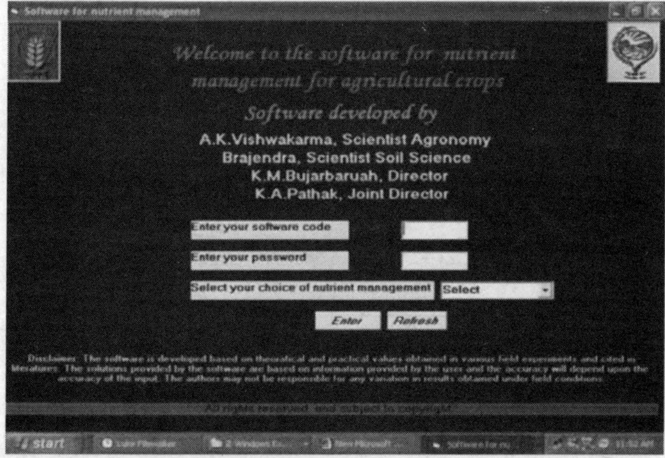

Fig. 1. Start up menu of the software

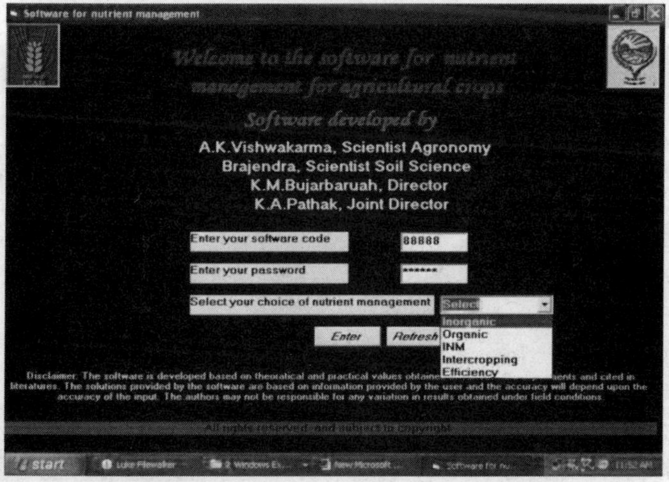

Fig. 2. Initiation of the programme

The different modules of the software and their operations are as under

1. Inorganic nutrient management

Selecting the inorganic nutrient management option from the choice of nutrient management forms the main menu of the software one can access this module of the software. The screen will appear having options for furnishing the informations required to obtain results depending upon the quantity of data available with the user (Fig. 3). This module provides three options to the user for getting information.

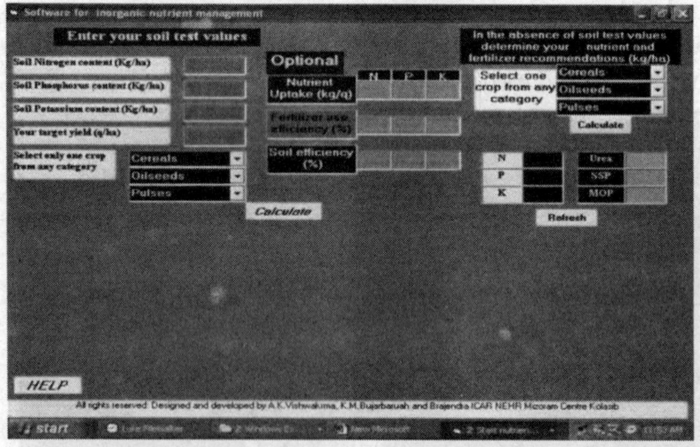

Fig. 3. Menu for inorganic nutrient management

STEP I- One can generate information on nutrient requirement, respective fertilizer requirement and the expected yield that can be obtained by application of that quantity of nutrients to the crop in the absence of availability of any soil and crop related data based on blanket recommendations (Fig. 4 & 5)

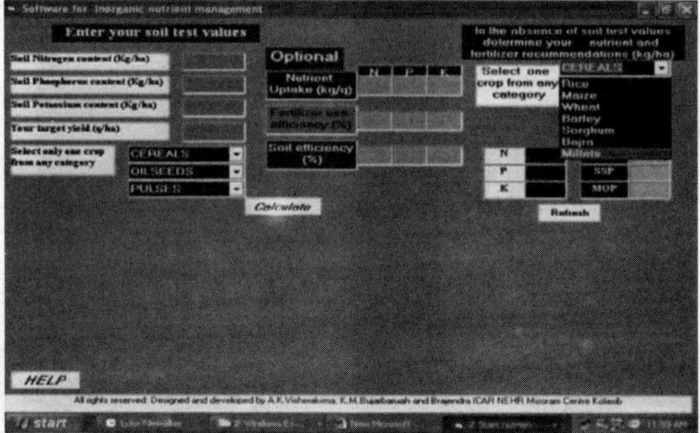

Fig. 4. Procedure of crop selection for blanket recommendations.

Fig. 5. Blanket nutrient recommendation based on crop selection.

STEP II- The user can furnish soil test values, set a target yield and select a crop from the options and pressing the calculate button to determine the quantity of nutrients required to obtain the set target yield based on readymade nutrient uptake equations provided with the software and from the options for selection of different fertilizer for nutrient supplementation the quantities of different fertilizers required can be obtained at a set fertilizer use efficiency of 50% (Fig. 6-8).

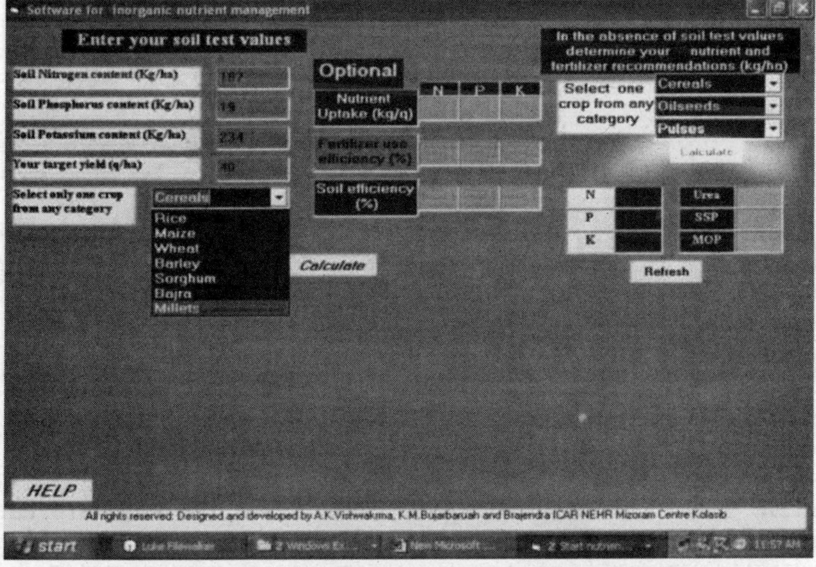

Fig. 6. Soil test and target yield based crop nutrient recommendation

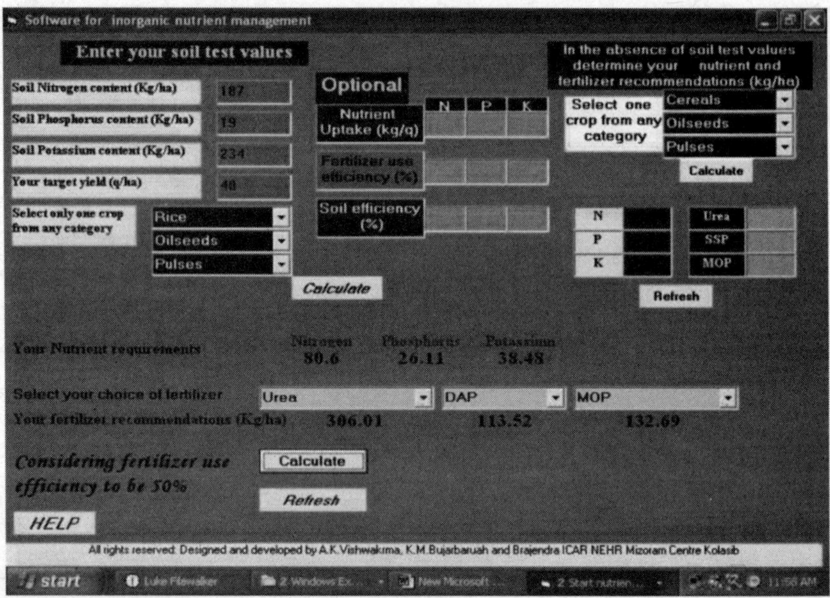

Fig. 7. Nutrient and fertilizer recommendation based on soil test and target yield

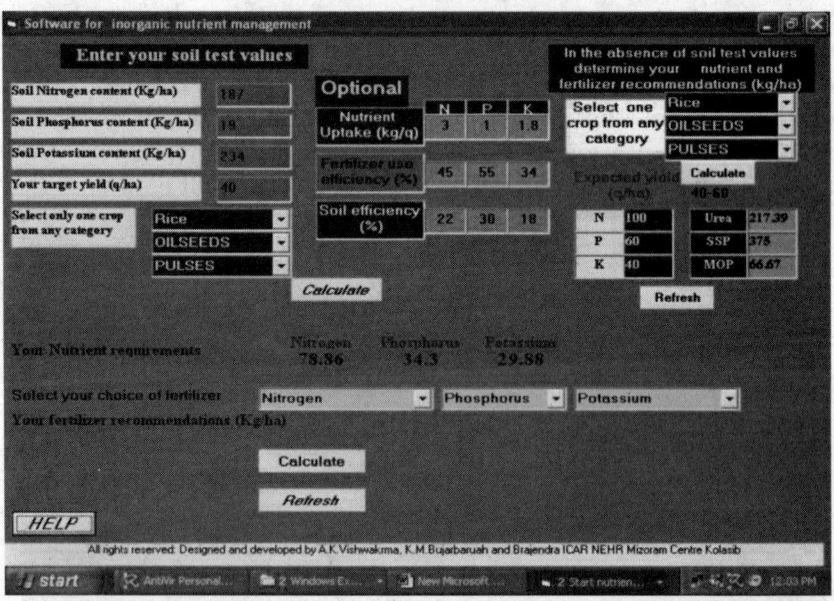

Fig. 8. Nutrient recommendation based on soil test, target yield and nutrient uptake.

STEP III - The optional space provided in the software can be utilized in the presence of desired data of crop nutrient uptake, soil efficiencies and fertilizer use efficiencies. To obtain information from this module one needs to provide the

- Soil test values
- Set a target yield
- Select a crop
- Nutrient uptake values for different nutrients
- Soil efficiency
- Fertilizer use efficiency

after furnishing all the informations in the optional column and then press the calculate button on the screen. Based on these information accurate site specific recommendations can be made available to the user and accurate estimates of the quantities of the nutrients required to obtain the set target yield can be determined. Then by selecting the appropriate fertilizers for nutrient supplementation one can determine the required quantities of the different fertilizer materials based on the fertilizer use efficiency furnished by the user (Fig. 9).

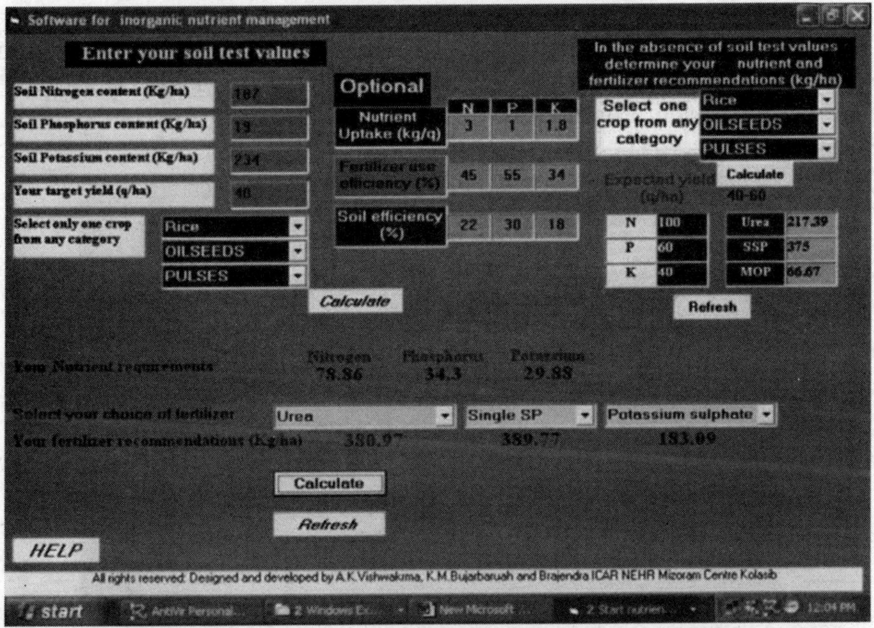

Fig. 9. Nutrient and fertilizer recommendation based on nutrient uptake, soil test and target yield.

2. Organic nutrient management

Selecting the organic nutrient management option from the choice of nutrient management from first screen of the software one can access this module of the software. The screen will appear having options for furnishing the informations required to obtain results depending upon the quantity of data available with the user (Fig. 10). This module provides two options to the user for getting information.

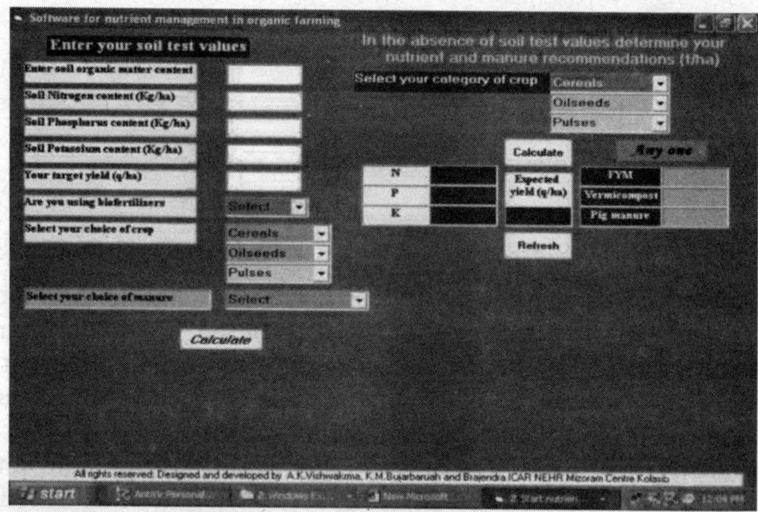

Fig. 10. Menu for organic nutrient management.

STEP I- One can generate information on nutrient requirement, respective manure requirement and the expected yield that can be obtained with the application of that quantity of nutrients through organic manures to the crop (Fig. 11).

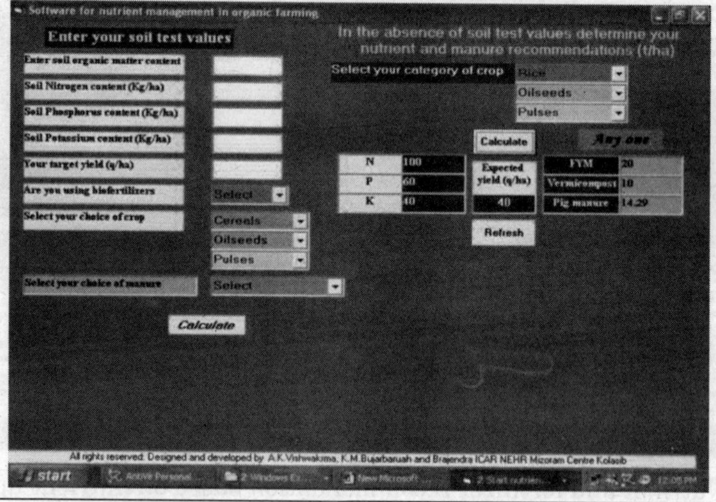

Fig. 11. Crop based blanket manure recommendation.

STEP II- To obtain recommendations with this module the user can furnish

- Soil test values
- Set a target yield
- Use of biofertilizers
- Select a crop from the options

- choice of manure

after furnishing these information's and pressing the calculate button one can determine the quantity of manures along with the quantity of rock phosphate required to obtain the set target yield based on readymade nutrient uptake equations provided with the software. The recommendations on organic nutrient management are based on five-year basis to obtain consistently the set target yield of the particular crop with nutrient supplementation from a selected source of organic nutrient. Since the quantities of the phosphorus required by the crop could not be met through the application of manures alone, therefore the recommendations for application of rock phosphate will also be made by the software, which is considered to be a organic source of phosphorus (Fig. 12 & 13).

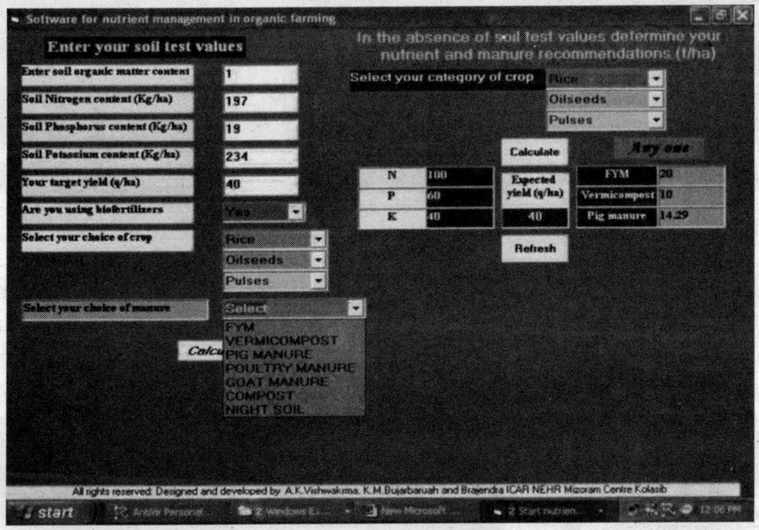

Fig. 12. Soil test based organic manure recommendation.

3. Integrated nutrient management

Integrated nutrient management is considered to be the best approach for crop nutrient management, which takes into consideration both the organic as well as inorganic sources of nutrient supplementation but the estimation of different nutrients and calculations of their fertilizer and manure equivalents is a complex process. Selecting the integrated nutrient management option from the choice of nutrient management from the first screen of the software one can access this module of the software. The screen will appear having options for furnishing the information's required to obtain results depending upon the quantity of data available with the user. This module provides three options to the user for getting information.

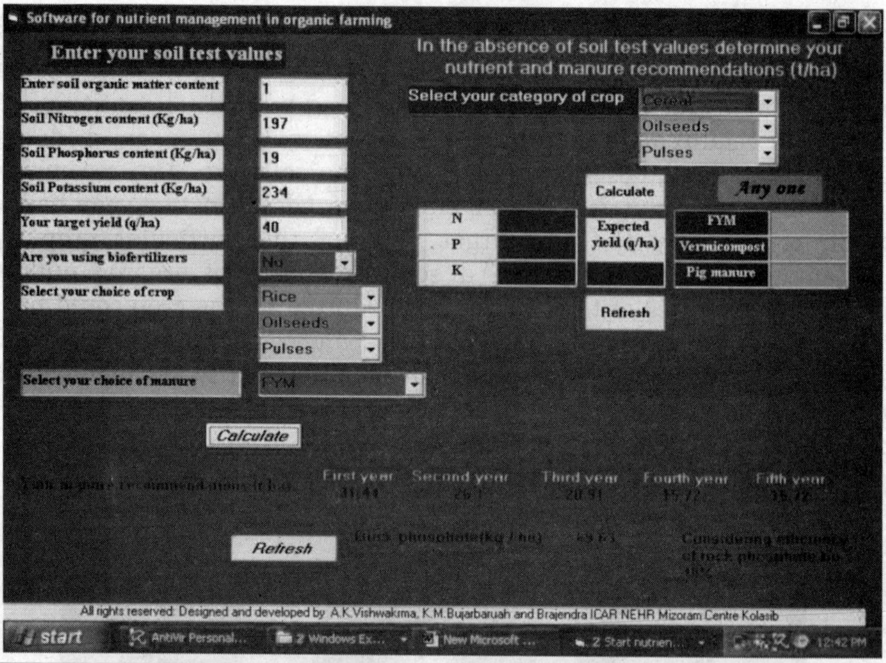

Fig. 13. Manure recommendation under organic nutrient management.

STEP I- One can generate information on nutrient requirement, respective fertilizer and manure requirement and the expected yield that can be obtained by application of that quantity of nutrients to the crop in the absence of availability of any soil and crop related data (Fig. 14 & 15).

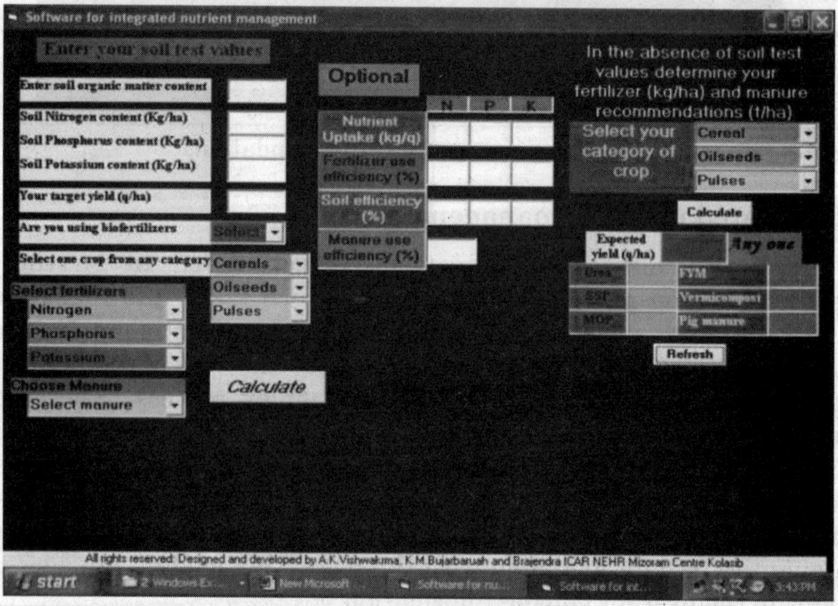

Fig. 14. Menu for INM

Fig. 15. INM under blanket recommendation.

STEP II- The user can furnish soil test values, set a target yield, select a crop from the options, application of bio fertilizers, different fertilizers available for use, the selection of manures to be applied to the crop can be obtained by pressing the calculate button one can determine the quantity of nutrients required to obtain the set target yield based on in built readymade equations provided with the software and the quantities of different fertilizers and manures required can be obtained at a set fertilizer use efficiency of 50% (Fig. 16).

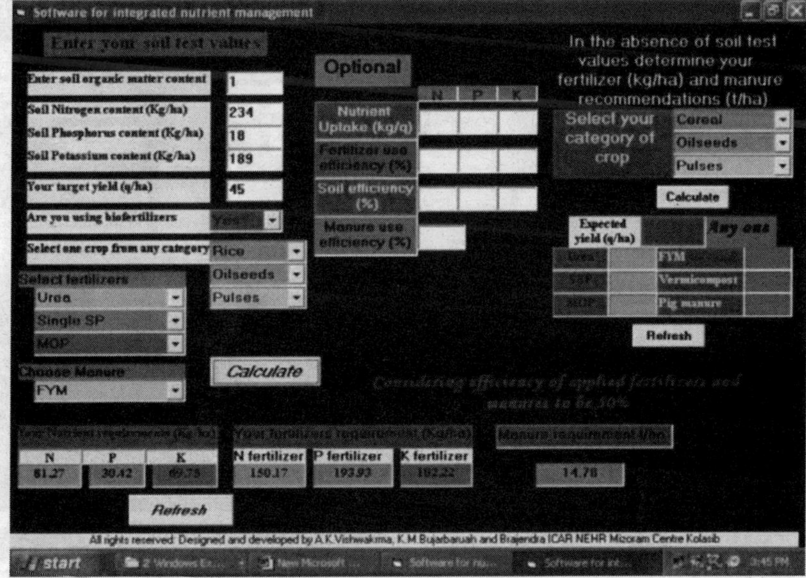

Fig. 16. INM based on soil test and target yield.

STEP III - The optional space provided in the software can be utilized in the presence of desired data and accurate site specific recommendations can be made available to the user. To obtain information from this module one needs to provide the following information.

- Soil test values
- Set a target yield
- Select a crop
- Application of bio fertilizers
- Different fertilizers available for use
- The choice of manures to be applied to the crop
- Nutrient uptake values for different nutrients
- Soil efficiency
- Fertilizer use efficiency
- Manure use efficiency

after furnishing all the information's in the optional column press the calculate button on the screen. This will provide the user with accurate estimates of the quantities of the nutrients required to obtain the set target yield and the required quantities of the different fertilizer and manures to be applied based on the fertilizer use efficiency and manure use efficiency furnished by the user (Fig. 17).

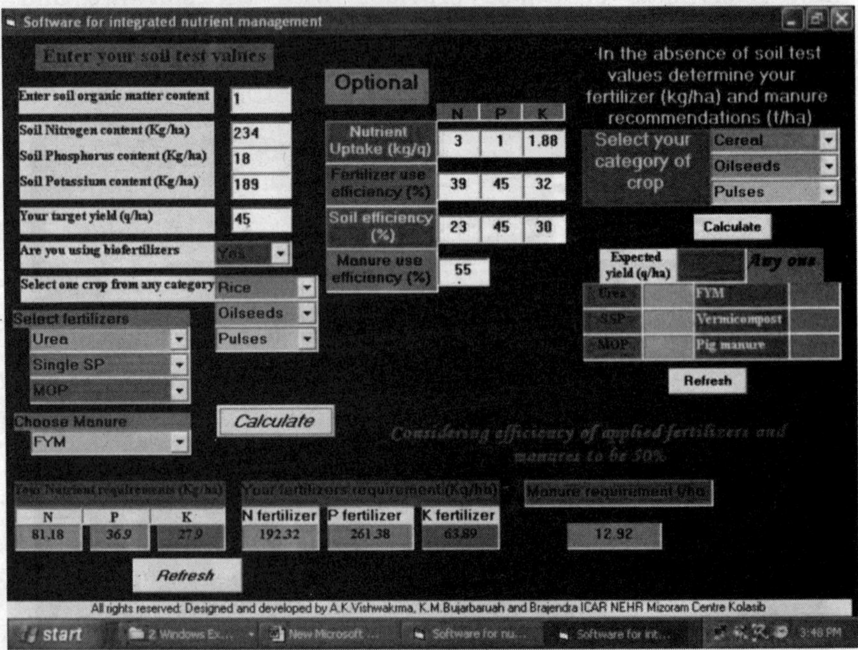

Fig. 17. INM recommendation based on soil test, target yield and nutrient uptake.

4. Nutrient management under intercropping conditions

Nutrient management under intercropping conditions is a very complex process and it is very difficult to workout the quantities of nutrients to be applied to the crop under intercropping, because of different nutrient requirement of the crops, their planting proportions and the expected yield from individual crop component. This resulted in either over fertilization to one crop and under fertilization to another crop. Thus poor crop yields were obtained, to solve the problem of fertilization under intercropping conditions this module has been designed taking into consideration the yield component of the crops and their nutrient uptake pattern (Fig. 18).

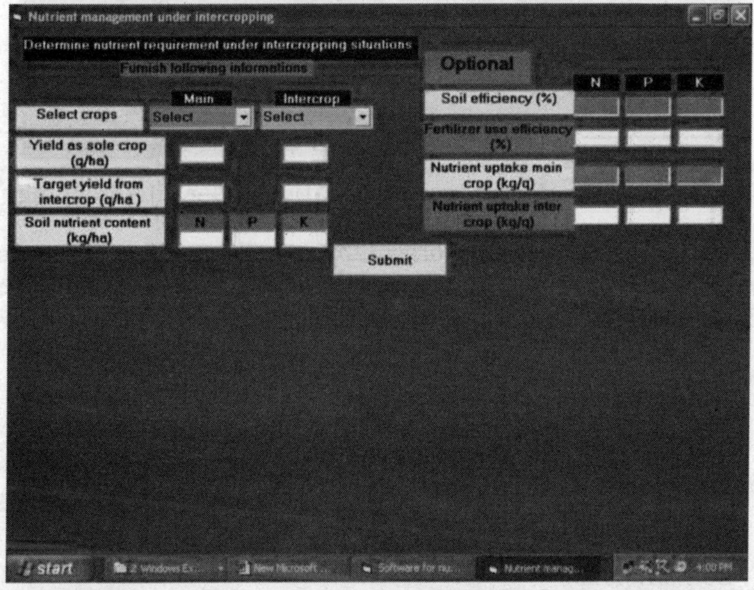

Fig. 18. Menu for nutrient management under intercropping.

This module can provide solutions in two ways

STEP I- When the soil test values are available but the efficiency and nutrient uptake values are not available. The user can select the main crop and intercrop from the list of crops available in the select column then yield as sole crop and intercrop needs to be furnished the soil test values.

On pressing the submit button, it will calculate the different nutrients requirement for sole and intercrops to obtain the desired yield levels based on readymade uptake values for different crops. Then different fertilizers can be selected from the list of fertilizers available for nutrient supplementation and on pressing the calculate button the quantities of different fertilizers can be determined based on the set fertilizer use efficiency (Fig. 19).

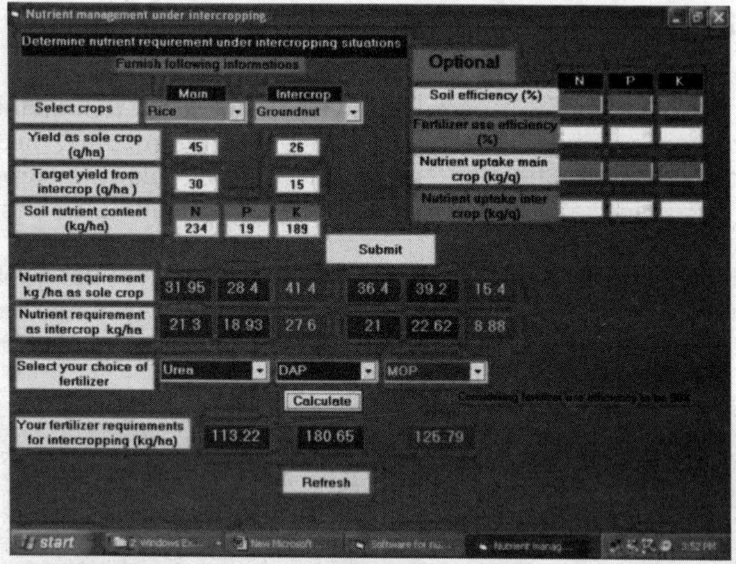

Fig. 19. Nutrient and fertilizer recommendation under intercropping.

STEP II- The user with more information's can obtain nutrient and fertilizer recommendations based on site specific conditions by furnishing the desired informations on main crop and intercrop from the list of crops available in the select column

- Yield as sole crop and intercrop
- Soil test values
- Nutrient uptake by main crop
- Nutrient uptake by intercrop
- Soil efficiency
- Fertilizer use efficiency

by furnishing the desired informations in the optional columns, on pressing the submit button the informations related to nutrient requirement can be obtained, and by selecting the fertilizers from the menus and pressing the calculate button the quantities of different fertilizers required can be determined based on the fertilizer use efficiency (Fig. 20).

5. Efficiency calculations

Efficiency calculations are complex, time consuming, exhaustive and are subject to miscalculations often, which result in driving incorrect results, and all the calculations and results obtained will be faulty. Therefore, this module has been provided with the software to enable the users to obtain soil and fertilizer use efficiencies by furnishing the desired data on

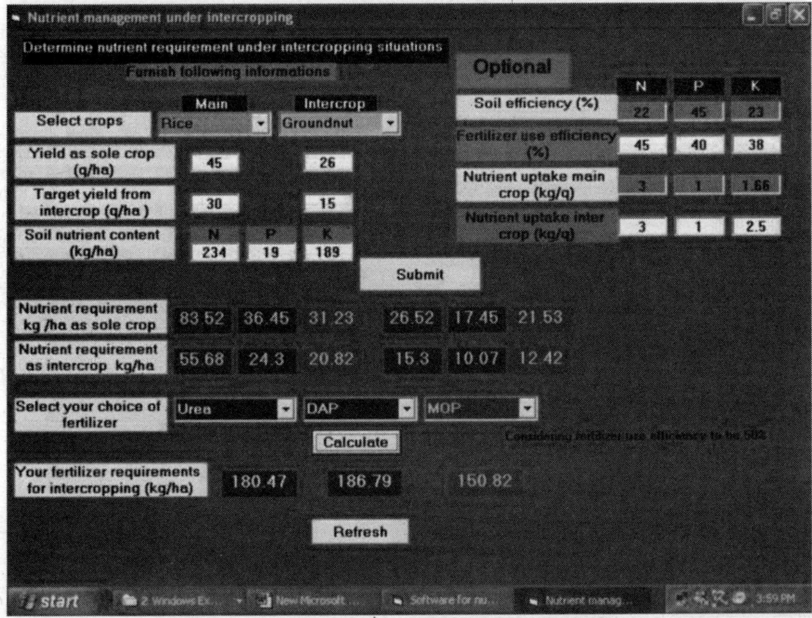

Fig. 20. Nutrient and fertilizer recommendation for site specific condition.

- Select a crop
- Soil nutrient status
- Doses of fertilizer application
- Crop yield under the fertilized plots.
- Crop yield under the control plots

after furnishing all the required information's on pressing the calculate button the information's related with nutrient uptake in the control plot and the fertilized plot will be generated along with apparent nutrient recovery, agronomic efficiency and production efficiency for nitrogen, phosphorus and potassium. This information's can be generated for the nutrients supplied through soil, nutrients derived by the crop from the fertilizer application and the total nutrient recovery and efficiencies from the soil and applied fertilizer based on the readymade nutrient uptake values for the selected crop (Fig. 21).

Secondly for the users having nutrient uptake values for the crop under control and fertilized conditions the software is provided with options to furnish the values for nutrient uptake in the grain and the straw under fertilized and control plots along with all other information's and then pressing the calculate all the total nutrient uptake values and the different efficiencies can be obtained. This, in turn can be fitted in any of the previous modules to obtain the desired nutrient and fertilizer recommendations to the crop (Fig. 22 & 23).

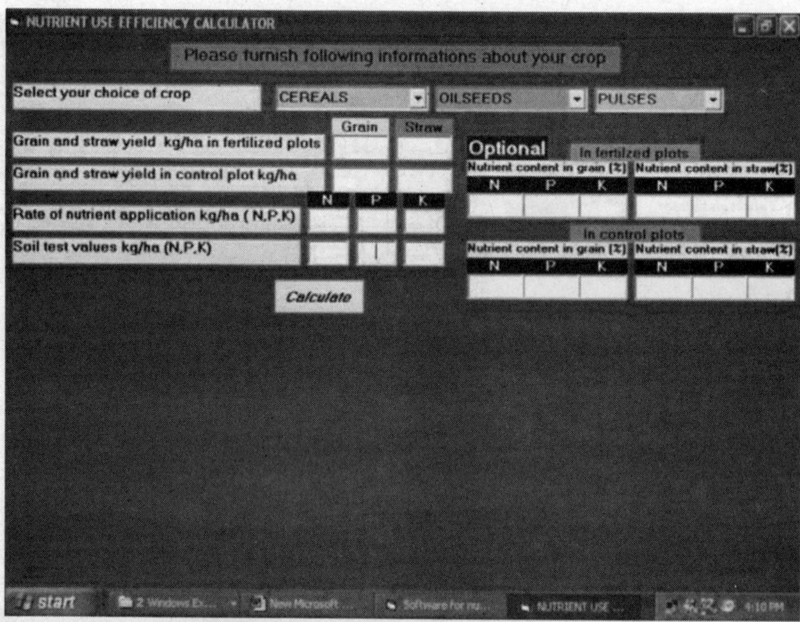

Fig. 21. Menu for efficiency estimation.

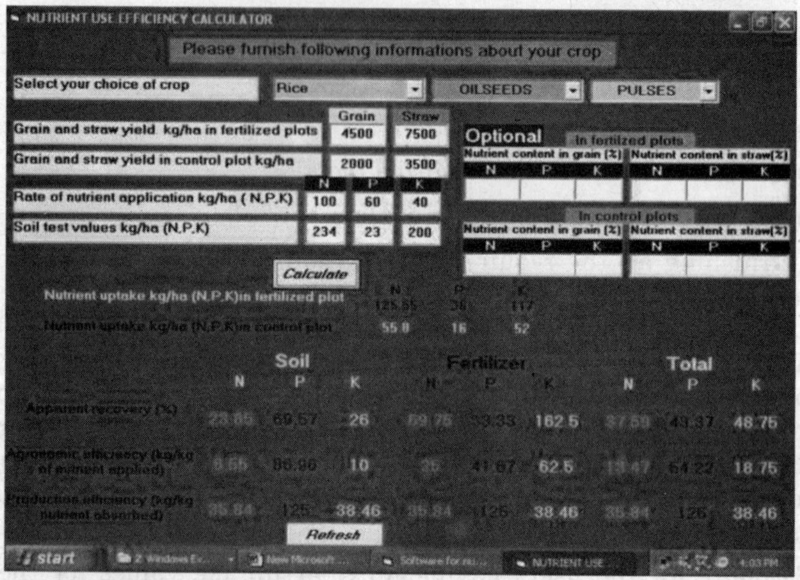

Fig. 22. Soil and fertilizer use efficiency estimation.

Limitations of the software

1. The results obtained may subject to change under varying agro-climatic conditions influencing crop growth and development.
2. The target yield may deviate with reference to various biotic and abiotic stresses.

3. The results may differ, if the target yield is set at a higher yield level then the potential yield of the particular variety. Though the software is capable enough to provide warnings on various critical levels.
4. Furnishing of incomplete and incorrect values may lead to vague and faulty results.

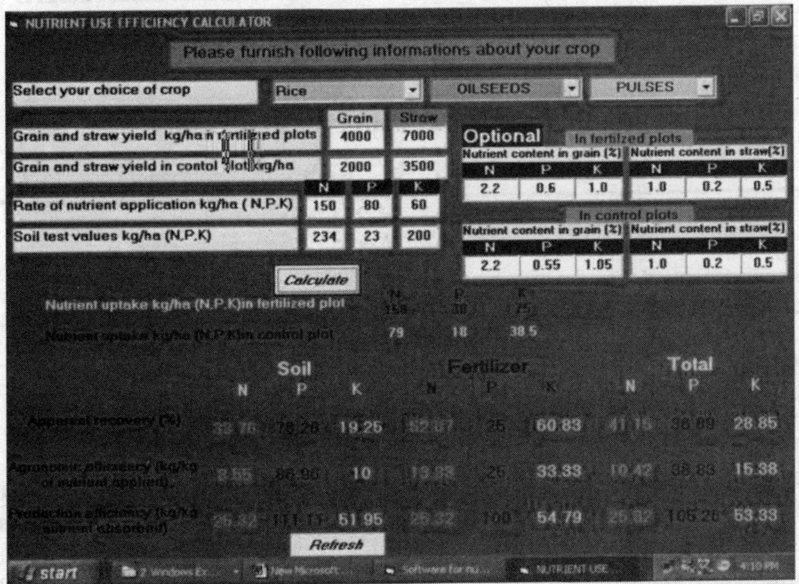

Fig. 23. Soil and fertilizer nutrient use efficiency based on nutrient uptake.

Conclusion

The software is of immense importance as it can provide immediate information to solve the problem of the user and to provide assistance in making decisions related to crop nutrient requirement and their management. The farmers/land users and researchers urgently need information about the crop nutrient requirements, nutrient uptake, and the quantity of various fertilizers and manures to be applied to the crop for obtaining desired production levels under a certain soil fertility level. The architecture of this software is organised in such a manner that wide range of users including farmers/agriculture extension officers/subject matter specialists/researchers with different demands to crop and soil related data and mehtods can utilize it and obtain desired. informations.

❑❑❑

16 De-Nitrification De-Composition Model-An Introduction for SOC Simulations

Pramod Jha

Division of Soil Chemistry & Fertility,
Indian Institute of Soil Science, Bhopal-462038

Introduction

The C stored in soils is mainly in the form of organic matter. SOC content is highly and dynamically affected by ecological drivers (e.g., climate, vegetation, and anthropogenic activity), soil environmental factors (e.g., temperature, moisture, pH, redox potential, and substrate concentration gradients), and biochemical or geochemical reactions (e.g., decomposition, assimilation, leaching etc.) (Li, 2000, 2001; Li *et al.*, 2003). Any small change in SOC may greatly impact atmospheric CO_2 concentration. Consequently there is much scientific interest in soil carbon stocks and their potential feedbacks to climate change (Davidson and Janssens, 2006). In order to predict the SOC change in response to climate change, many models (like Century, DNDC (DeNitrification DeComposition), Daisy and RothC) were developed and most of these models are based on several conceptual carbon pools (with different turn-over rates) within the soil profile (Coleman and Jenkinson, 1999; Jensen *et al.*, 1997; Li *et al.*, 1997; Parton *et al.*, 1995). SOC pools are generally large relative to their annual changes, so long-term experiments (measur-ement and/or modelling) are required to detect significant changes in SOC.

Model descriptions

DNDC simulates carbon and nitrogen cycling in agro-ecosystems at a daily or subdaily time step. It consists of four interacting submodels: soil climate (including water flow and leaching), plant growth, decomposition, and denitrification (Figure 1). The soil climate submodel simulates soil temperature and moisture profiles based on soil physical properties, daily weather, and plant water use. The plant growth submodel calculates daily water and nitrogen uptake by vegetation, root respiration, and plant growth and partitioning of biomass into grain, stalks and roots. Optimal crop yield and biomass allocation are input parameters; actual yield simulated by the model will generally be suboptimal due to limitations in water and/or nitrogen availability. The decomposition sub-model simulates daily decomposition, nitrification, ammonia volatilization, and carbon dioxide production by soil microbes. The denitrification submodel operates at an hourly time step to simulate denitrification and the production of nitric oxide, nitrous oxide, and dinitrogen.

Data input

To simulate soil organic matter dynamics in agricultural land, DNDC requires as input: (1) climate data (daily air temperature and precipitation); (2) soil properties (bulk density, texture or clay content, pH, and initial organic carbon content in the surface soil); and (3) cropping practices (crop rotation, tillage, fertilization, manure application, and irrigation). The model simulates turnover of organic matter in the soil, soil respiration, crop growth, and partitioning of crop biomass into roots, stems, and grain. When the crop is harvested, all of the root biomass and a specified fraction of the stem biomass are added to the soil litter pool of carbon.

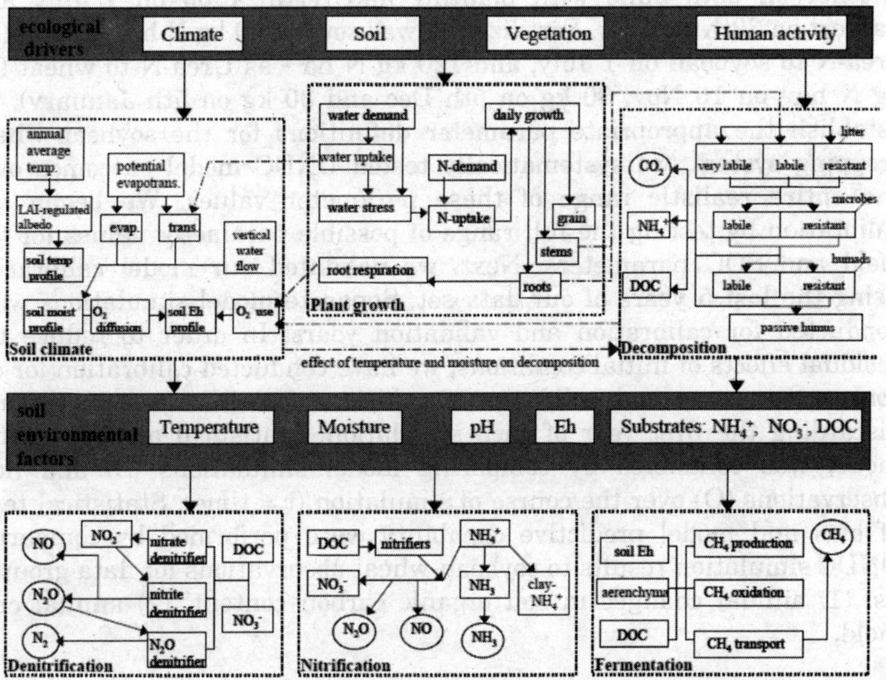

Figure 1. Structure of the DNDC model

Model calibration and validation

The DNDC model has been validated throughout the world by using long-term and short-term experimental data to test its behavior on the modelling of the carbon biogeochemical process in agricultural soils (Li 2000). The sensitivity studies are carried out to ascertain DNDC behavior in simulating soil carbon response to changing climate, soils, and agricultural practices. The input parameters for running the model can easily be collected. Caution to be exercised, in defining the model carbon pools as the final output of SOC is sensitive to carbon allocation in different pools of soil organic matter. The model must be calibrated with respect to soil organic carbon before using it for simulation purpose. For calibration of the model, datasets of long term experiments should be used. We parameterized DNDC for different cropping systems. Daily precipitation and temperature files were created to allow multi-year runs of the model. The clay soils of Jabalpur were used, with a clay fraction of 57%, porosity of 0.482, saturated conductivity of 0.0046 m per hour, and field capacity and wilting points of 0.75 and 0.45 (as water-filled porosity), respectively. DNDC was calibrated and validated using the 10-year data set described previously of crop yields and soil organic carbon (Fig 2). The management schedule used was one typical for soybean and wheat production in central India. For soybean, one tillage was on 25 April and another on 28th June with planting and fertilization on 1 July and harvest on 30th October. Fertilization was sowing 20 kg N ha^{-1} applied as urea-N to soybean on 1 July, and 120 kg N ha^{-1} as Urea-N to wheat (60 kg N ha^{-1} on 15 Nov, 30 kg on 5th Dec and 30 kg on 5th January). To establish the appropriate parameter definitions for the soybean-wheat cropping system, we systematically tested DNDC model outcomes over the entire realistic range of these parameter values. We began our calibration by testing the full range of possible parameter values for the yield and SOC parameters. Next, we validated our model calibration using the last 5 years of our data set. Separate model simulations were conducted for calibration and validation years. In order to reduce the residual effects of initial conditions, we have conducted calibration for six consecutive years and validation simulations for six consecutive years, discarding the first year of each simulation. Statistical analysis of the model was conducted by comparing model simulations (S) and field observations (O) over the course of simulation (t = time). Statistical tests of bias and model predictive capability were performed by comparing DNDC simulation results to soybean-wheat observations for data grouped as: (1) annual changes in soil organic carbon content, (2) annual crop yield.

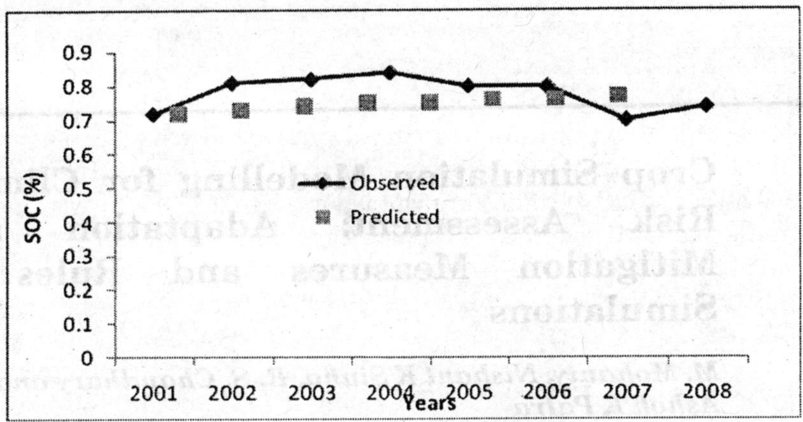

Figure 2. Temporal changes in SOC as predicted by the model.

Implications related to model calibration

Simulation accuracy of global bio-geochemical carbon model depends on the initial carbon content of soil and the relative distribution of soil carbon pools. No definite method of quantification of soil carbon pools differentiation has been proposed by the model developer. Basso *et al.* (2011) developed iterative procedure for computation of soil carbon pools for initialization of DSSAT-Century model. Predicted changes in SOC should also be compared with measured data that represent the spatial and temporal range of model inferences to assess uncertainty and bias (Falloon and Smith, 2003; Ogle *et al.*, 2007). The first step in calibration of DNDC model for simulating changes in soil carbon content is the yield calibration. The yield calibration is vital step because the belowground input to soil is directly related to above ground biomass production. The model is very sensitive with respect to amount of carbon in different pools. As such the model allocates carbon in different pools by default. However, some tuning is required for calibration of yield with respect to carbon allocation in different pools. Furthermore, the decay constant of different pools has to be adjusted for yield and soil organic carbon calibration. The lower the C/N ratio, the more of the residue will be partitioned into very labile or labile pool. Each of the SOM pools has a specific decomposition rate subject to temperature, moisture and N availability. The organic matter in the litter pools will be broken down by the soil microbes. When the microbes die, their biomass will turn into humads pool. Humads can be further utilized by the soil microbes and turned into passive humus. During the sequential decomposition processes, a part of the organic C becomes CO_2, and a part of the organic N becomes ammonium (Adapted from Zhang *et al.*, 2006).

❑❑❑

17 Crop Simulation Modelling for Climate Risk Assessment: Adaptation and Mitigation Measures and Rules of Simulations

M. Mohanty, Nishant K Sinha, R. S. Chaudhary and Ashok K Patra

Indian Institute of Soil Science, Nabibagh, Berasia Road, Bhopal-462038.

Introduction

Climate change is changing the face of agriculture - rising temperatures, changing rainfall patterns and the evolution of pests and diseases all pose threats to food production. Modelling has been used extensively to predict the likely impacts of climate change on crop yields and ranges, but different studies give different answers, often as a consequence of the variety of scenarios tested and analytical approaches used.

There have been many more projections of the impact of climate change on crop production since 2007. Report of the Intergovernmental Panel on Climate Change (IPCC), used an update of the 2007 data set with over twice as many studies, and set out to answer three related questions: What are the likely impacts of differing degrees of climate change on yields, by crop and by region? How far can adaptation measures offset the impacts of changes in temperature and rainfall? Will yields increase or decrease in the remaining decades of this century, and by how much?

There would be a clear benefit for wheat and rice in adopting adaptation measures, such as irrigation, fertilizers, or changes in cultivated varieties. The benefits would be less apparent for maize. Overall, crop-level adaptation measures could help to avoid yield declines in wheat, and could even increase yields by 7–15% in some situations, relative to yields of non-adapted crops.

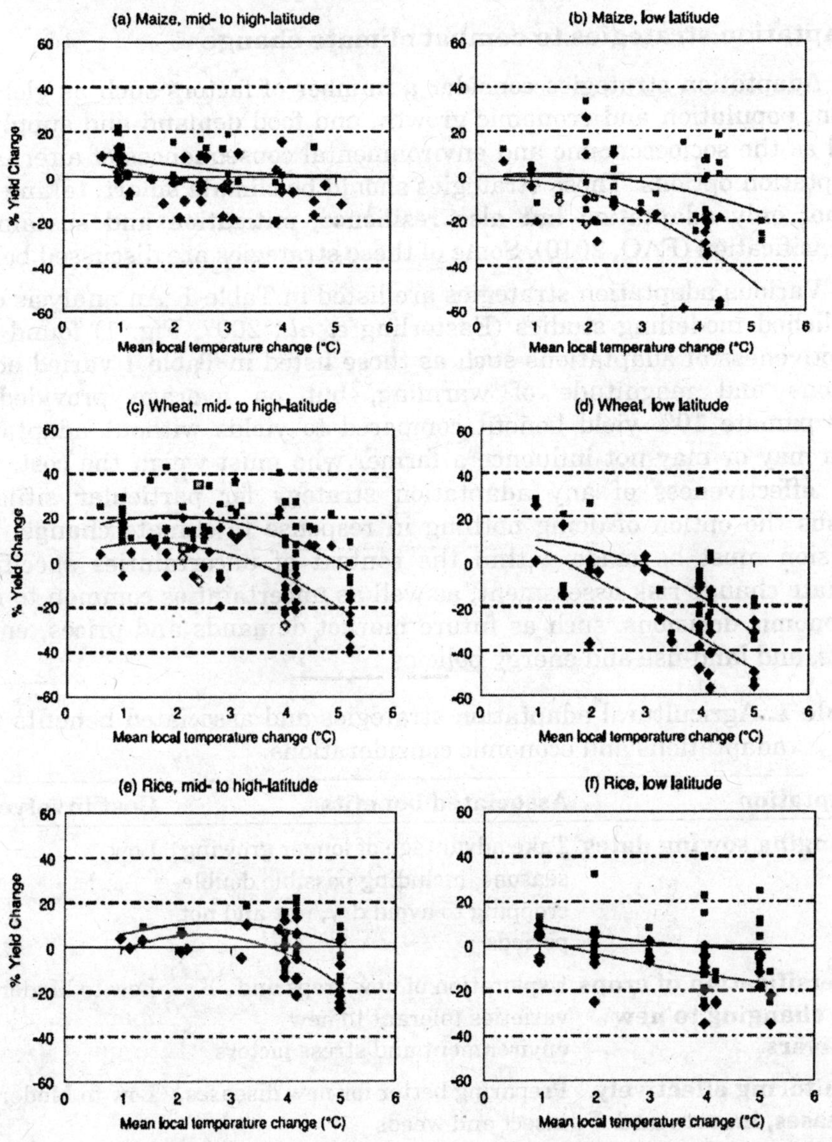

Figure 1. Sensitivity of cereal yield to climate change for maize, wheat and rice, as derived from the results of 69 published studies at multiple simulation sites, against mean local temperature change used as a proxy to indicate magnitude of climate change in each study. Responses include cases without adaptation (red dots) and with adaptation (dark green dots). Adaptations + represented in these studies include changes in planting, changes in cultivar, and shifts from rain-fed to irrigated conditions. Lines are best-fit polynomials and are used here as a way to summarize results across studies rather than as a predictive tool (Adopted from Easterling *et al.*, 2007)

Adaptation strategies to combat climate change

Adaptation strategies consider a number of factors such as globalization, population and economic growth, and food demand and supply, as well as the socioeconomic and environmental consequences of alternative adaptation options. These strategies should be climate smart, taking care of not only adaptation but also resilience, mitigation and sustainable intensification (FAO, 2010). Some of these strategies are discussed below.

Various adaptation strategies are listed in Table 1. An analysis of 69 published modelling studies (Easterling *et al.*, 2007; Fig. 1) found that effectiveness of adaptations such as those listed in Table 1 varied across regions and magnitude of warming, but on average provided an approximate 10% yield benefit compared to yields without adaptation. This may or may not influence a farmer who must weigh the cost, risk, and effectiveness of any adaptation strategy for particular situation versus the option of doing nothing in response to climate change. This decision must be made within the context of uncertainties specific to climate change risk assessment, as well as uncertainties common to most agronomic decisions, such as future market demands and prices, energy costs, and land-use and energy policies.

Table 1. Agricultural adaptation strategies and associated benefits from adaptations and economic considerations.

Adaptation	Associated benefits	Cost involved
Changing sowing dates	Take advantage of longer growing seasons, including possible double-cropping to avoid dry, wet and hot periods	Low
Diversification of crops and changing to new cultivars	Exploration of new crops and varieties tolerant to new environment and stress factors	Low to Moderate
Monitoring effectively diseases, insects and weed dynamics	Preparing better for new diseases, insect and weeds	Low to Moderate
Irrigation systems	Maintain yield and quality during dry periods	Moderate to High
Water logging	Maintain yield and quality during wet periods	Low to High

Bridging the yield gaps

The gap between the poetical and actual yield of many crops are big in many Asian and south Asian countries. In the eastern parts of South

Asia, gaps are especially large, and this region can be a future source of food security for the sub-continent. For meeting food demands from the huge population in future and to overcome the physical constraints (due to temperature and rainfall, CO_2) poised by climatic changes, there is an urgent need to raise yield potential for regions and bridge yield gaps. Availability of seed or planting materials, poor technology dissemination in farm sectors, and lack of adequate capital and information are the principal reasons for yield gaps. Reducing even some of the yield gap could strengthen the food security in the region and reduce the vulnerability of the populations to climate change.

Change in sowing dates

Among various adaptation measures, changing sowing date can be effective to take advantage of longer growing season (Wolfe *et al.*, 2008), including possibly sowing two crops (Wolfe *et al.*, 2011). Sowing date shifts can also be used to avoid crop exposure to drought. When uncertainty is high, diversification in cropping systems by changing to different sowing dates in the same season may be an effective strategy. Effectiveness of such planning always depends on region and the extent of change in climate and climate variability. In some parts of Asia and Africa, farmers routinely modify sowing dates in relation to the onset of the monsoon season (Tadross *et al.*, 2005), although this is likely to become more challenging with climate change (Naylor *et al.*, 2007). In much of the moisture-limited tropics, climate change is going to reduce the length of growing period and thus, sowing date change may not always work for such situations. In a maize crop model simulation study, Burke and Lobell (2010) found no shift in optimal planting date with future climate scenarios for southeastern Kenya, where sowing dates are based on arrival of the long and short rains.

The things may be different in high-elevation or high-latitude regions of the world where cool season crops are grown, changing sowing dates may be considered to avoid negative effects from high temperature stress. A potato crop model simulation study (Rosenzweig *et al.*, 1996) found that delaying planting date by two weeks so that tuber development would occur at more favorable cooler temperatures was effective at mitigating yield reductions associated with 1.5 and 2.5°C of warming in some regions. But this did not work to overcome potato yield reductions with a 5°C warming scenario regardless of location.

Altering the time at which fields are sown or planted can also help farmers regulate the length of the growing season to better suit the changed environment. But in the Indo-Gangetic Plains, this option will be limited by low winter temperatures. Sowing date adjustment to minimize the effect of temperature increase-induced spikelet sterility in wheat can be effective against yield reduction and instability, by avoiding flowering period coinciding with the hottest period. Reliable seasonal weather forecasts should also be seen as one supportive measure to optimize sowing, planting and applying irrigation (Gadgil *et al.*, 2002).

Changing crop varieties developing climate-smart crops

Plant breeding program should be intensified with the objectives of developing more climate smart green crop cultivars/varieties which are tolerant not only to heat or water stress but also to water logging. Future crop breeding efforts need to address other stresses like pest and disease loads to crops imposed by changing climate. The developed varieties need to show several adaptive traits in a suitable agronomic background. Conservation of germplasms of wild relatives, landraces, sensitivity of wild and cultivated species, and genetic material currently in the gene banks which are tolerant/resistant to multiple biotic and abiotic stresses. Exploitation of genetic engineering and molecular breeding becomes important in ongoing breeding (Varshney *et al.*, 2011).

Changing varieties and crops is an off-cited adaptation strategy to take advantage of opportunities and reduce yield losses. It often may go hand-in-hand with changing sowing dates, such as switching sowing date and crop or crop cultivar in response to a longer growing season in high-latitude regions. Varieties with improved tolerance to heat or drought will be available for some crop species. New molecular-assisted crop breeding strategies may provide new genetic types more tolerant of environmental stress and pests and pathogens (Reynolds *et al.*, 2010). Genotypic variation in plant response to elevated atmospheric carbon dioxide (CO_2) exists (Wolfe *et al.*, 1998), but concerted efforts to take advantage of this for yield gains have been few. An evaluation of new wheat varieties released during the twentieth century suggests that responsiveness to CO_2 has not increased but declined with traditional breeding programs (Ziska *et al.*, 2004).

Changing crop cultivars and diversifying the cropping systems might be an expensive or ineffective strategy in case of perennial fruit and nut crops, as new crops take several years to reach maximum productivity. Even for annual crops, changing cultivars is not always a low-cost option because: new stress-tolerant cultivars are sometimes expensive and may not be available ; new varieties often require investments in many areas such as new sowing/planting equipment, and also require new cultural/ management practices. In some cases, it may not be possible to identify an alternative variety that is adapted to the new climate, as well as to local soils and farming practices, and quality features for the market also.

In south Asia, farmers will need crop varieties with greater tolerance to stresses such as drought and heat, excess soil moisture, as well as photo- and thermal-insensitive varieties (Agrawal *et al.*, 2010). Adapted seeds/planting materials may no longer be available in the existing local seed certifying agency/corporations and, therefore, interventions must involve long-distance seed exchange programme to be facilitated by genebanks. Modern GIS-based approaches can be very useful in identifying materials that have been successful around the world in regions with similar climatic profiles. There is still a big need for scientific tools to assist in regional land use planning that can help farmers and societies to adapt to climate change, meet food demands and environmental goals leading to higher income in farming communities.

Suitable water management strategies

Water availability will be in short supply as projected for many agricultural regions of the world due to reductions in rainfall, and gradual loss of glacial mass supplying agricultural regions. Even in many temperate humid regions where growing season rainfall is not projected to decline, crop water demand (i.e., potential evapotranspiration) will increase due to increase in summer temperatures and longer growing seasons, and thus, increases in the demand for supplemental irrigation (Hayhoe *et al.*, 2007;Wolfe *et al.*, 2011). In tropical and subtropical regions of world will also suffer supply of irrigation water due to high temperature resulting in high evapo-transpirations. Irrigation systems are a relatively expensive option, and a challenge for farmers will be determining when the yield losses due to summer water deficits have or will become frequent enough to warrant such a capital investment. For some regions, where water supplies are low, the farmers may opt for low water-use systems such as drip irrigation, sprinkler irrigation in some cases to minimize water use by optimizing irrigation scheduling. For poor regions of the world like some Asian and African nations where water supplies for agriculture are constrained, increasing soil water-holding capacity by increasing application of crop residues, manures of soil and other organic amendments, conserving soil water by maintaining surface residue, or mulch and water "harvesting and recycling" are other low-cost options that can reduce the impact of short-term water deficits on field crops.

Many areas in south Asia and other flood plains areas of the world will be subjectd to increased frequency of severe flooding due to sea level rise and high intensity rainfall. A more widespread problems will be agricultural crop lands that are subjected to continuous flodding due to increase in frequency of high rainfall events (e.g., more than 5 cm in 48 hours) as a result of climate change. In addition to causing damage to field crops in arable lands, other cultural events like delayed sowing and reduction in the growing season; lack of access to the field during critical periods for farm operations; soil compaction because of tractor use on wet soils; increase in diseases and pests, root disease; increased soil erosion; and increased runoff of fertilizers particularly nitrogenous or organic manures applied on the surface lost into waterways or crop-growing areas, with negative implications for human health (Hatfield *et al.*, 2011; Wolfe *et al.*, 2011).

A low-cost option that can buffer against minor or short-term flooding problems is to maintain or improve soil drainage by increasing soil organic matter. Also, maintaining vegetative cover year-round with winter cover crops can minimize soil erosion losses during heavy rainfall events. Changing planting date to avoid wet periods (if they can be predicted) or switching to more flood-tolerant crops or crop varieties will be other relatively low-cost options, when and where they are feasible.

Changes in land-use management

Land use changes are considered to be an important source of adaptation to climatic stresses. Intercropping is such an example where farmers switch from monocropping to cope with a drought period. Strategic land use changes are, however, guided not only by the need to cope with climatic stresses but also to adapt to opportunities provided by technology, institutions and markets. A typical example is the shifting apple cultivation to higher altitude in Himachal Pradesh due to rise in temperature (Agrawal *et al.*, 2010) and this led to cultivation of vegetable crop in the lower altitude area. Such a land use change helps the farmers to adapt to changing climate and also to increase their farm incomes (Ranbir *et al.*, 2010).

Adjustment in cropping sequence/pattern, including changing the cultural operations such as timing of sowing, planting, spraying, fertilizing and harvesting, to take advantage of the change in duration of crop growing seasons and associated heat and water stress is another option for increasing adaptation to climate change.

Resource conservation technologies and use efficiency

There is an increasing competition for resource use (such as land, water, capital and labor) in agriculture due to continuously increasing populations and income. Climate change will further add fuel to fire this competition. Efforts need to be made now to conserve natural resources or increase their use efficiencies. Agriculture in South Asia uses more than 70% available water. This fraction needs to be reduced in future. The efficiencies of water-use, therefore, need to be improved at the farm, community and regional scales. It has been demonstrated that proper leveling of farms could improve water application efficiencies by over

20%. Greater realism in water and energy pricing could also promote efficient use of natural resources (Shah, 2009). Use of modern irrigation methods such as microirrigation could also be promoted to enhance water-use efficiency. Small changes in climatic parameters can often be better managed by altering dates of planting, spacing and input management. Early planting of wheat, facilitated after rice harvest by surface seeding or zero-tillage may offset most of the losses associated with increased temperatures in South Asia. This reduces costs of production and the use of natural resources such as fuel, and improves the efficiency of water and fertilizers. Other resource-conserving technologies such as systems of rice intensification, bed planting, and alternate wetting and drying also need to be carefully assessed.

On a regional scale, Managed Aquifer Recharge has large potential to enhance natural recharge rates of groundwater (Shah, 2009). The use of efficient water utilization methods such as micro-irrigation coupled with groundwater use may lead to reduction in depletion of groundwater. Conjunctive management of surface and groundwater in Punjab offers large opportunities for improving water productivity as well as for saving energy (Shah, 2009). Such resource-conserving technologies in rice-wheat system also have pronounced effects on mitigation of greenhouse gas emission and adaptation to climate change (Aggarwal and Pathak, 2009).

Harnessing the local technical knowledge of farmers

Farmers are the best source of indigenous technical knowledge (ITK) which they gather from their father and fore fathers. Combination of modern technologies and ITK from farmers can work wonder in adapting crops to climate change. Several regions have earlier minimized their exposure to climatic stresses by resorting to mixed cropping, changing varieties and planting times, by diversifying sources of income for farmers, and maintaining buffer stocks of food for managing periods of scarcity. These management strategies would also help in the future climate change scenarios but may not be enough in view of the increasing intensity of climatic risks and pressure on land to produce more food and also with much higher efficiency. In some parts of the world such as in Bangladesh and eastern parts of India, recurring cyclones and floods have almost become a rule. In such instances, traditional risk minimization approaches might no longer apply. Instead, a focus on adaptation action is needed and would require considerable investment to increase resilience of the communities. It must also be pointed out that although local knowledge can often help individual subsistence farmer to survive periods of climatic stress, but in today's world, farmers need to produce food not only for themselves but also for 50% of South Asia's population living in urban areas. It remains to be evaluated if local knowledge can assist the farmers in ensuring food security for all.

Mitigation

Climate Change Mitigation refers to efforts to reduce or prevent emission of greenhouse gases. Mitigation can mean using new technologies and renewable energies, making older equipment more energy efficient, or changing management practices or consumer behavior (UNEP).

Mitigation management options for agriculture are many, beginning with reducing deforestation and crop residue burning (a major issue in many developing regions using slash-and burn-practices), and improving energy use efficiency and reducing inputs that are energy-intensive to produce, such as synthetic nitrogen (N) fertilizers (a major issue in many intensive operations in developed regions). Other key solutions include reducing tillage, improving fertilizer N use efficiency and manure management, improving water management, winter cover-cropping and, including legumes and perennials in rotation schemes. Table 2 summarizes key mitigation options associated with the three primary greenhouse gases. Smith *et al.* (2007) and others have provided detailed reviews of this topic. Many of the management options in Table 2 are cost neutral or could potentially increase farm profits, while benefiting the environment and increasing climate change resilience (Scherr and Sthapit, 2009). Many of them involve increasing soil C sequestration, which not only plays an important role in climate change mitigation (Lal, 2004), but also improves soil health, crop productivity, and resilience to climate change (Hobbs and Govaerts, 2010). Corn, wheat and rice production systems are of particular interest because collectively they account for about 50% of all N fertilizer produced worldwide, and there are substantial opportunities for improving N use efficiency (Cassman *et al.*, 2002). Rice production systems are also an important case because of the high methane emissions from flooded paddies. There are opportunities for reducing methane emissions by reducing the flooded period by midseason drainage, or various forms of dryland production with irrigation (Ortiz-Monasterio *et al.*, 2010).

The agricultural sector in South Asia contributes to about 20% of total GHG emissions. The emissions from agriculture are primarily due to enteric fermentation in ruminant animals, paddies, and nitrous oxides from application of manures and fertilizers to agricultural soils. In future, emissions from soils are likely to increase significantly because of increased intensification of input use required to meet increased food demand. Increasing resource-use efficiency also has huge co-benefits and implications for GHG mitigation (Aggarwal, 2008).

Table 2. Agronomic mitigation strategies for three primary greenhouse gases.

Greenhouse Gases(GHG)	Agricultural sources of emissions	Mitigation strategies to be employed
N_2O	— Excessive or non-synchronizing N fertilizer applications — Water logged soils and manures	— Split applications of N fertilizers, optimize timing and amount applied by soil tests & crop demand — Appropriate application of nitrate-N in upland conditions and ammoniacal-N in wetland soils can help decrease N_2O emissions — Use nitrification inhibitors, such as nitrapyrin and dicyandiamide (DCD), along with N fertilizers to curtail the process of nitrification is also important — Organic products such as neem oil, neem cake and karanja seed extract which can also act as nitrification inhibitors. — Improved manure management practices improve soil drainage — Use of legumes (biological N fixation) in crop rotations — Use of green manures or green leaf manures and biofertilizers
CH_4	— Flooded rice soils — Enteric fermentation by cattle	— Altering water management in paddy, particularly promoting mid-season drainage and improving organic matter management by promoting aerobic degradation have the potential for reducing methane emissions. However, drainage of rice fields may lead to increased emission of nitrous oxides. Making a balance between two processes is very important. — Employ various dryland production options with irrigation where feasible — Incorporating organic materials during dry period of rice production

		— Develop new feeding strategies and feed amendments to reduce methane emissions from livestock — Use covered or tank storage of manures and store at low temperature
CO_2	— Deforestation and residue burning — Soil organic matter decomposition — Fossil fuel and electricity use for transportation — Fossil fuel use in manufacture of energy-intensive farm inputs such as synthetic N fertilizers, pesticides, and herbicides	— Improving the efficiency of energy use in agriculture by designing machinery, increasing fuel efficiency in agricultural machinery — Reduce or minimize slash-and-burn agricultural practices — No-tillage or reduce soil tillage (slows organic matter decomposition and reduces fuel use for tractors) — Retain and incorporate crop residues — Increase crop residues by increasing yields and biomass and rotation crops — Use C-rich sources of fertilizers, manures like FYM, composts, or biochar — Minimize use of energy-intensive farm inputs (e.g., use organic fertilizers rather than synthetic) — Use alternative energy sources, such as biofuel crops, anaerobic digestion or pyrolysis of farm manures and wastes, solar or wind power.

Resource-conservation technologies restrict release of soil carbon, thus mitigating increase of CO_2 in the atmosphere (Grace *et al.*, 2003). According to an estimate, no-tillage saves about 30 L/ha of diesel compared to conventional tillage, thereby leading to a reduction of CO_2 production by 80 kg/ha/year. If such savings could be translated even partially to larger arable lands, substantial reduction in CO_2 emission to the atmosphere could be achieved. Lal (2004) estimates that such practices can lead to carbon sequestration of 25–50 Tg C/year in soils of South Asia for several decades, including 11–22 Tg C/year on croplands. Changing land use by increasing the area under biofuels or agro-forestry could also mitigate GHG emissions (Smith *et al.*, 2007a). This, however, may have trade-offs with the goal of increasing food production. South Asian countries that are scarce of land need to examine the consequences of diverting land to biofuels on their food security (Aggarwal *et al.*, 2007).

Models and the climate adaptation studies

Most studies of climate change impacts on crop yields apply either statistical models (Lobell and Burke 2010) or process-based crop simulation models (Rotter *et al.* 2011b; White *et al.* 2011; Osborne *et al.* 2013). Most process-based models are also capable of simulating, effects of enhanced CO_2 concentration and management practices on biomass, seed yields and water use of crops (Rosenzweig and Parry 1994; Nelson *et al.* 2009; Ewert *et al.* 2011; Angulo *et al.* 2013). However, even the more complex process based crop simulation models cannot take all important interactions between the environment and management (E x M) into account, such as effects of heavy rainfall on harvested yield. Neither do they include all interactions between genotype and environment (G x E) such as yield reduction due to weather-induced pest and/or disease occurrence. On the other hand, crop growth simulation is the only meaningful practical way for analysing the interactions between the many options of combining different crop cultivars with diverse management practices under a wide range of possible new environmental conditions (Semenov and Halford 2009; Rotter *et al.* 2011a,b; Rosenzweig *et al.* 2013). Usually, crop-climate models do not cover all important crops and soils in a region. For this reason, agroclimatic indicator approaches are sometimes applied to provide a more comprehensive picture of the agroclimate for larger areas and its shifts under climate change (Harrison and Butterfield 1996; Trnka *et al.* 2011). Knowledge of the broad-scale agroclimate can also provide a useful basis for upscaling site specific crop simulation results, offering a strong argument for combining the two approaches. Such a combination can provide information on shifts in the suitability and potential for crop production in a region or country under climate change (Carter and Saarikko 1996; Challinor 2011).

According to the present state of climate science, the increase in greenhouse gas concentrations will result in changes of the climate parameters; particularly, an increase of the global temperature and changes in precipitation are expected. These changes in climate, which are likely to occur during future decades, may have significant consequences (positive or negative) on the development, growth, and yields of crops important to a particular region or country. DSSAT model has been used to investigate physiological responses of the winter wheat and maize crop to changes in climate (Cuculeanu *et al.*, 1999). They have investigated different potential adaptation measures to analyze by alternative simulations of agricultural management systems, such as changes in crop variety, sowing date, crop density, as well as changes in the level of fertilization, and application of irrigation. It has also been reported that winter wheat could benefit from the interaction of a double CO_2 concentration with higher temperatures, while maize appears to be vulnerable to climate change, especially in the case of a warm and dry scenario. The negative effect of the temperature increase that causes a shorter growth period would be counterbalanced by the positive effect of the doubling CO_2 concentration.

APSIM has been verified and used widely around the world, particularly in Australia, to study crop production, nutrient cycling, carbon sequestration and impacts of climate change and variability on farming systems (e.g. Probert *et al.*, 1998, 2005;Wang *et al.*, 2011; Yang *et al.*, 2013). For the prediction of SOC dynamics under different cropping and management systems, several investigators demonstrated that the performance of APSIM simulations is at a high level of accuracy (Luo *et al.*, 2010, 2011; Probert *et al.*, 2005).

If one is confident that the estimates of climate change impacts are reliable, they can be used to create information relevant to the adaptation actions taken by stakeholders. Challinor (2009) discusses this topic at length, and assesses the potential for adaptation to climate for a crop in India. Genotypic variation might be the one that can be used to adapt to climate change. The requisite crop genetic properties determined from the simulations are compared with those of existing germplasm. Interestingly, a separate study showed that under doubled CO_2 in India, the uncertainty in the simulation of adapted crops may be greater than that of non-adapted crops (Challinor *et al.*, 2008a).

When considering adaptation, it is important to consider how weather and crop yield forecasts will be used, and what spatial and temporal scales will be the most appropriate for the users. Useful weather/climate forecasts can range from a few days ahead for some crop management decisions, to decades in the future for infrastructure and strategic planning. For example, ensemble climate modelling can be used with crop models in order to predict crop yield a season ahead of the harvest (Challinor *et al.*, 2005c). Information should also be provided in relevant formats (Stone and Meinke, 2005). Whether the information best suited to users is based on computer-intensive systems, or on less high-tech systems such as observational networks and capacity building, depends to a large extent on the particular users considered (see, for example, Patt *et al.*, 2005). In Africa, for example, a prudent way to address the threat of climate change may be to focus on strategies for coping with climate variability, rather than longer term climate change (Washington *et al.*, 2006). This may mean a greater focus on *in situ* and remotely sensed observations as well as consideration of the multiple stresses that act on food security (Gregory *et al.*, 2005; Haile, 2005; Verdin *et al.*, 2005).

Approaches for simulation study i.e. Assessment of crop yield or rules of simulations

Most of the crop simulation models include crop modules like cereals, legumes, oil seeds, sugar crops etc. The crop models are composed of parameterization of important physiological processes responsible for plant growth and development, evapotranspirations, partitioning of photosynthates to produce economic yield. The simplified functions enable prediction of growth of crops as influenced by major factors that affect yield i.e. genetics, climate daily solar radiation, maximum and minimum temperature, rainfall etc), soils and management practices. The models include soil water sub-models so that they can be used to predict crop yield both under irrigated and *rainfed* conditions. The models also simulate effects of nitrogen on crop growth. The following steps to be sued before one could use crop simulation models for climate studies.

- Defining the representative crop management (i.e. variety, fertilizer input, rainfed or irrigated production systems, number crop rotations, soils etc)
- Definition of base line climate data. More the years betters the predictions. Usually current climate to past thirty years weather data are considered for this scenario.
- Validation of crop models under current climate from the experimental data from field trials to the extent possible.
- Simulation of crop response to modified climate scenarios i.e. change in temperature, rainfall and CO_2 concentration.
- Testing for farm level adaptations i.e. shifting of planting dates, additional application of irrigation water and N to the crops under stress and changing of crop varieties to adapt to different climate change scenarios.
- Crop simulation results for main crops are aggregated to agro-climatic regions by weighting their representative contribution to current regional production.
- Statistical analyses are used to derive regional yield response functions.
- The yield functions are used to the spatial climate change data (scenario changes in temperature, CO_2 concentration and rainfall) to derive scenario yield change estimates for individual crops

Methodologies for simulating impacts of climate change on crop production

Crop simulation models are widely used to predict potential impacts of climate change on future crop production and to examine viable options for adaptation by farmers and other stake holders. However, methodologies followed in such assessments vary to such an extent that they increase the potential for bias in projected impacts. It is important that suitable measures should be taken before using simulation models for climate assessment studies. A few guidelines adopted from White *et al.* (2011) are as follows. These are target crops and regions; the crop model(s) used and their characteristics; sources and application of data on [CO_2] and climate; impact parameters evaluated; assessment of variability or risk; and adaptation strategies.

Recommended procedures to improve assessments of climate change in agro-ecosystems

- Wheat was the crop most often assessed, followed by maize, rice and soybean. Based on FAO data (FAO, 2008), the number of crops assessed showed rough agreement with their importance based on respective areas under cultivation, although rice and soybean were underrepresented relative to wheat and maize. This may reflect difficulties in accessing rice studies, notably papers from China and India, and underrepresentation of rice and soybean producing countries in climate change research. Most of studies conducted on climate change research on crop production follows the order: the USA > European countries. Other countries played nonsignificant role compared to USA and EU countries. There are also evidence of not representating the crops properly for the study conducted i.e. justification and also the importance of the geographical area considered for the study. Sometimes it is important that the whole geographical regions may be considered for the study rather than a point location.
- Justify the selection of the simulation model(s) and describe the model(s) with sufficient detail to allow a reader to understand how key processes are represented. Specifically, describe whether transpiration is affected by [CO_2], whether canopy temperature is estimated, and whether specific heat stress effects (e.g., reducing pollen fertility, increasing embryo or grain abortion or accelerating leaf senescence) are modeled. Clearly specify the assumed baseline [CO_2] and corresponding time period, ensuring that baselines assumed for the simulations and for climate data are consistent with each other. In the last decades CERES group of models were used mostly for climate

change studies around the world. The logical way to understand how process completeness and detail affect model accuracy and usability is through comparisons of models using common datasets, and calls for improved analysis and comparisons of models are hardly original. Both the IPCC assessment reports have cited the need for continued improvement of process-based models.

- Ensure that the end date of the baseline climate data is as close to the date of paper submission as possible.
- When appropriate, use current IPCC greenhouse gas scenarios and identify these with the recognized abbreviations in addition to a text description.
- State which weather variables are modified and if applicable, how outputs from GCMs and/or RCMs are downscaled. The most commonly used GCMs were HadCM2 and HadCM3, which together represented about 35% of GCM usage. The three most widely used regional climate models were NCAR RCM, PRECIS and UKTR. We note that there was ambiguity in nomenclature for climate models per se, versions and climate simulation experiments. For example, UKCIP is the UK Climate Impacts Programme, and their associated climate datasets have been based on results from HadCM1, HadCM2 and HadCM3 (UKCIP, 2010).
- Assess impacts of soil variability, which might include season to season differences in initial conditions and local spatial variation.
- Ensure that adaptation strategies represent as likely a set of alternatives as possible, preferably by consulting with producers and other stakeholders familiar with the target production environments. Simulation models can readily test crop management options (Tsuji *et al.*, 1998) and thus examine potential for technological adaptations to climate change. Such options include improved varieties, shifts in recommended planting dates and rates, novel cropping sequences, change in the number of fallow years required for soil-water recharge in rainfed systems, and introduction of alternative or new crops. Planting date was the most frequently changed option. The foremost problem is in prioritizing among potentially adaptive changes in management, especially for interacting practices such as planting dates and irrigation management. Cultivar traits are also problematic. Prospects for improving adaptation to elevated [CO_2], heat stress, drought and water deficits, and nutrient use

efficiency are highly uncertain and many improvements would likely interact with crop management.

- Examine impacts beyond economic yield, especially as related to soil and water resources. Impacts of climate change were predominantly evaluated for economic yield. Crop models can describe a large number of processes and output a large number of rate and state variables besides economic yield. A few studies are available which deal with carbon and tillage. This again may relate to the lack of crop models that were considered suitable for simulating effects of tillage and residue management.
- Assess impacts in terms of risk, preferably using probability distributions rather than simple statistics such as coefficients of variation. A widespread concern with climate change is whether the frequency of extreme adverse events such as droughts, heat waves or hurricanes will increase, further threatening the stability of agricultural production. Among specific components of risk, variability due to drought was considered the most often. Given the potential importance of heat stress, surprisingly few papers partially or fully considered heat stress. For biotic factors affecting variability, no papers provided quantitative analyses of effects of pests, but two papers modeled effects of rice blast, and one paper partially examined effects of red rice as a weed in cultivated rice.
- To provide a balanced assessment of climatic risk in relation to other sources of variation, simulate effects of other sources of variability such as sowing dates and seed rates.
- Most papers only considered variation in weather conditions as the source of stochastic variation, yet values of model input variables or parameters are estimates, often with uncertainty that is stochastic due to variation in farmer behavior, machinery performance, spatial variability of individual fields, effects of diseases, pests or weeds, and numerous other factors. To accurately simulate expected variability, a modelling study should assess the potential impact of the major factors that have a large stochastic component. Thus, rather than assume a fixed or automatically determined planting date and a constant seed rate, a study might sample a range of dates and seed rates, mimicking these intrinsically variable aspects of crop management. Soil profile descriptions are another source of uncertainty in simulations. The profiles typically provide information on initial water, nutrient levels and organic matter concentrations as well as soil properties that are less dynamic, including drainage and runoff characteristics, bulk density, water holding capacity, and maximum depth for root development. However,

the values for these properties have large uncertainties due both to within field variability and to measurement error. Nonetheless, only 21 papers tested more than one soil profile description for a single location, map polygon or grid cell, and of these, only ten papers quantified effects on variability, considering only productivity and drought-related risk. The net result is that impacts of climate change on risk likely were overestimated relative to factors ranging from variability in initial plant populations to sub-optimal weed and water management due to labor constraints.

- Diverse methods, scenarios, and models have been used to characterize the potential impacts of climate change on crop yield and other associated aspects of agricultural production. The single action that might most benefit research on potential impacts of climate change is to establish a coordinated resource for crop management, climate, and soil data, building on resources. The impact studies advocate strong interdisciplinary collaborations.
- Accurate prediction of climate extremes in the simulation models and projection at finer resolutions so that it helps to develop appropriate preparedness policies. Human resource development to perform ground-level data collections and modelling in a regional scale.
- Long-term field experiments to understand fundamental climate processes should be conducted.
- Land use changes are the major driving agent for the water resources compared to climatic threats.
- Basin-scale water modelling should be processed through spatiotemporally downscaling.
- Location-specific indigenous model for simulating country-specific climatic impact is essential.
- Inclusion of biophysical and socioeconomic determinant of the crop productivity in the simulation model.
- Data repository development and meta-data reliability assurance.
- Knowledge gaps should be filled, and information must be reached in intact form by the end-users in time.

Future lines of crop simulation studies on climate risk assessment on crops

There is an urgent need for inter-comparison of simulation studies using different crop simulation models. Internationally, such efforts have been made as the Agricultural Model Inter-comparison and Improvement Project (www.agmip.org) launched in October 2010. It aims to build a transdisciplinary modelling framework to provide more robust estimates of climate impacts on crop yields including estimates of associated uncertainties. This will set new global standards for modelling crop yields. Multi-model ensemble simulations provide a valuable way of estimating the range of possible outcomes, and the importance of different physical and biological processes considered in different crop models. As there is clearly no best model, determining multi-model averages is a promising practice that can help to better replicate observed results (White *et al.*, 2011). Although this result would seem to offer attractive predictive potential, caution is needed. Similarity to observations may simply reflect a fortuitous cancelling of errors among models, a situation that cannot be relied on under future conditions where key responses absent from some models may become dominant. It is important to review the big defici-encies in models that are used in understanding crop growth and development processes. At present the big deficiency is the plant responses in the models to increased carbon dioxide concentrations over a wide range of crops and environments, and the responses of leaf area and grain setting to stress factors; need for a bigger effort to generate and compile high-quality field data for model testing and evaluation and need to develop better methods to estimate larger-scale regional productivity. Models are built to deal with plant processes studied at the scale of single plots and there is a need to find ways of getting those models to account for the variable landscape and environmental conditions across larger areas at regional or national levels. There is a need that the model should perform to the optimum in sensitive environments.

Finally, crop–climate modelers must be careful to report their level of certainty. This is the most immediate and important research requirement, especially when results are being used to inform decisions. There is an urgent need for more quality, transparency and consistency in the modelling methods used for assessing climate change impacts on crop production.

A number of yield-limiting factors are not considered in simulations, including phosphorus limitation, negative effects of weed competition, diseases and insect pests, and incidence of catastrophic weather events in most of the simulation studies. In addition, crop varieties, crop hybrids, and crop management are not adapted to future climatic conditions but are held constant as for the baseline conditions. A full understanding of the possible responses of the crop-environment system to the predicted climate change can only be gained through consideration of crop-climate interactions at the site, and at regional and national levels. The disadvantages of the site-specific modelling scale are caused by a lack of information on the spatial patterns of change. So, spatial pattern of change in simulation study is also important.

❑❑❑

18 Rothamsted Carbon (RothC) Model and its Application in Agriculture

P. Bhattacharya[a], *J. Somasundaram*[b], *N.K. Sinha*[b] *M. Mohanty*[b] *and R.S. Chaudhary*[b]

[a] ICAR- Central Rice Research Institute, Cuttack -753 006, India

[b] ICAR- Indian Institute of Soil Science, Nabibagh, Berasia Road, Bhopal-462 038, India

Introduction

Soil organic carbon plays an important role in soil fertility and can contribute to the mitigation of climate change through the sequestration of carbon (C) in soils. Soil organic matter (SOM) takes part in many environmental functions and, depending on the conditions, it can be a source or a sink of the greenhouse gases. Presently, the changes in soil organic carbon (SOC) stock can arise because of the climatic changes or changes in the land use and land management. A promising method in the estimation of SOC changes is modelling, one of the most used models for the prediction of changes in soil organic carbon stock on agricultural land being the Rothamsted carbon model which is widely known as *RothC*. The simplicity and availability of the input data makes RothC a convenient model for testing the efficiency to predict the development of SOC stock in diverse ecosystems including agriculture systems (Table 1).

In RothC model soil organic carbon is split into four compartments *viz.* decomposable plant material, resistant plant material, microbial biomass and humified organic matter. Apart from this there is also a small amount of inert organic matter which is resistant to decomposition. The relative proportion of decomposable plant material and resistant plant material depends on the type of vegetation.

It uses a monthly time step to calculate total organic carbon (t ha^{-1}), microbial biomass carbon (t ha^{-1}) and ^{14}C (from which the equivalent radiocarbon age of the soil can be calculated) on years to centuries timescale (Jenkinson and Coleman, 1994). It needs few inputs and which are easily obtainable. RothC is designed to run in two modes: 'forward' in which known inputs are used to calculate changes in soil organic matter and 'inverse', when inputs are calculated from known changes in soil organic matter. Mean monthly air temperature and mean annual rainfall were estimated from the nearest standard meteorological station for each site. Using standard meteorological data, weather files for each were created following the method of Coleman *et al.* (1997).

Table 1: Application of RothC model in various land use systems

Country	Climate	Ecosystem	Site comments	Performance	Reference
Rotha-msted, UK	Temperate	Forest	Geescroft wild-erness, naturally regener-ated forest	Good reproduction of increase in soil C and radiocarbon content of soil C	Coleman *et al.* (1997)
Rotha-msted, UK	Temperate	Fallow	Highfield Bare Fallow, grazed until 1959, fallow thereafter	Measured decline in soil C initially faster than RothC, but gap narrowed	Coleman *et al.* (1997)
Rotha-msted, UK	Temperate	Grassland	Park-grass experiment, grassland for 300 years	Reasonable fit to changes in soil C in unmanured and fertilised plots. Poor reproduction in plots treated with FYM and fishmeal	Coleman *et al.* (1997)
Ruzynê, Czech Republic	Cool, temperate	Arable	Ruzynê expe-riment, continuous cultivation for 100s of years	Good agreement with measurements for fertilised plots. Poor agreement in plots receiving N and FYM due to variab-ility in measurements	Coleman *et al.* (1997)
Tam-worth, Australia	Temp-erate, sub-tropical	Arable	Tamworth expe-riment, cultivat-ion for 130 years	Poor fit to measurements	Coleman *et al.* (1997)
Waite, Australia	Temp-erate, sub-tropical	Arable		Decline in soil C due to cultivation reproduced well	Coleman *et al.* (1997)
China	Cool, temperate	Arable	Longterm fertili-sation trial	Accurate simulation of changes in soil C	Yang *et al.* (2003)
China	Cool, temperate	Arable	Longterm fertilisation trial	Accurate simulation of changesin soil C	Yang *et al.* (2003)
China	Cool, temperate	Arable	Longterm expe-riments set up in early 1980s	Accurate simulation of changes in soil C	Guo *et al.* (2007)

Country	Climate	Ecosystem	Site comments	Performance	Reference
Kenya	Tropical	Arable	Two experiments, 26 and 13 years long	Good agreement with measurements, but poor for intercropping	Kamoni *et al.* (2007b)
India	Tropical	Arable	Long Term Fertilizer experiments in black soils	Good Agreement with observed results.	Bhattacharyya *et al.* (2012)

Source: Adapted from STRIVE Report Series No. 20

Input data

The RothC model (Figure 1) requires three types of data sets:

(a) ***Climatic data:*** monthly rainfall (mm), monthly evapotranspiration (mm) and average monthly mean air temperature (°C). The minimal distance between the meteorological station and the corresponding modelling site, the similarity of the environmental conditions compared to the corresponding site, and weather data accessibility and completeness are the decisive criteria in the process of the relevant meteorological station selection.

At meteorological stations, the monthly rainfall (mm) and average monthly mean air temperature (°C) are generally calculated based on daily data. The monthly potential evapotranspiration (mm) is calculated on the basis of Pennmann equation. The missing monthly weather data is complemented in the final data processing with the data observed at the same weather station and identical month in preceding or following year.

(b) ***Soil data:*** Clay content (%), soil organic carbon (SOC), initial soil organic carbon stock (t ha^{-1}) and depth of the soil layer (cm) are considered. The SOC stock (t ha^{-1}) is calculated according to the following equation

SOC Stock = SOC × BD × SD; where: SOC is concentration of soil organic carbon (%); BD is bulk density (g cm^3) and SD is topsoil depth (cm). The initial SOC content is used for running the RothC model to equilibrium under constant environmental conditions. In each region, first of all RothC is supposed to run to equilibrium (10 000 years) followed by fitting carbon inputs to match the initial SOC stock and then the distribution in compartments (decomposable plant material, resistant plant material, microbial biomass and humified organic matter) with different decomposition rates.

The data of carbon in all these compartments received in equilibrium mode (initial soil state) are used to run the model in short term mode (for the modelling of SOC in the time period). The data of topsoil clay content (clay fraction < 0.001 mm) are obtained from the available topsoil data. The depth of the soil layer for modeling is generally set to 20 cm for agricultural crops.

(c) ***Land use and land management data:*** Soil cover, monthly input of plant residues (t ha^{-1}), monthly input of farmyard manure (FYM) (t ha^{-1}) and residue quality factor.

The management data are obtained on the basis of universal questionnaires which had been addressed to all users of the selected sites. The questionnaire contained the data about crops cultivated in specific time period, tillage practices during the period, organic fertilizer application rates and crop yields.

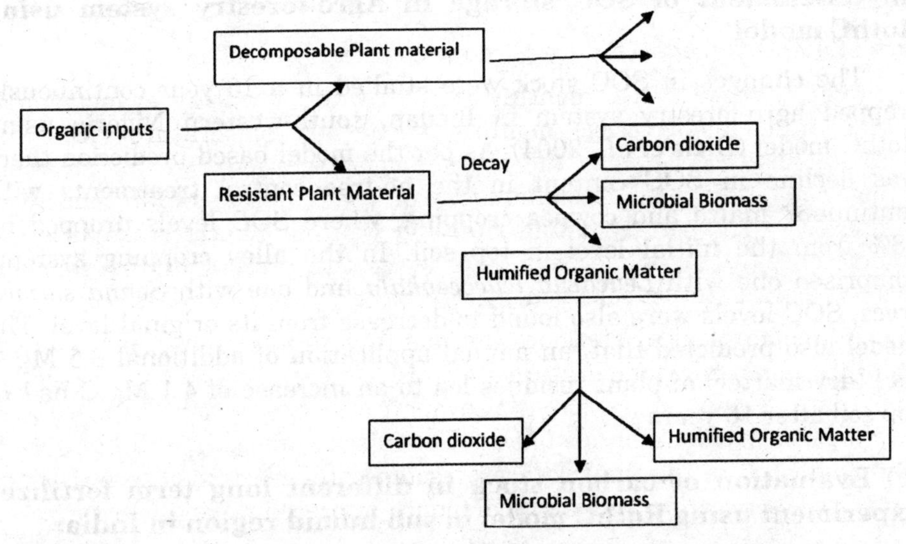

Figure 1: Schematic diagram of Rothamsted carbon (RothC) model

Application of RothC model

(a) Evaluation of SOC storage in cotton based cropping system using RothC model

The performance of RothC model in simulating SOC storage was evaluated in an irrigated Vertisol in semi-arid, subtropics in Australia. The model was assessed using data sets of long-term (1994-2012) cotton cropping systems under different tillage management practices (Senapati *et al.*, 2014). The experimental treatments were continuous cotton/ conventional tillage, continuous cotton/minimum tillage and cotton-wheat (*Triticum aestivum* L.) rotation/minimum tillage. The RothC model satisfactorily simulated the dynamics of SOC in cotton based cropping systems under minimum tillage, whereas the model performance was poor under intensive conventional tillage. The model RothC over estimated SOC storage in cotton cropping under conventional intensive tillage management system. This over estimation could not be attributed to the overestimation of soil C inputs, or errors in initial quantification of SOC pools for model initialization, or the ratio of incoming decomposable plant materials to resistant plant materials. Among other different factors affecting SOC dynamics and its modelling under intensive tillage in tropics and sub-tropics, we conclude that factors for tillage and soil erosion might be needed when modelling SOC dynamics using RothC under intensive tillage management system in the tropics and the sub-tropics.

(b) Assessment of SOC storage in Agro-forestry system using RothC model

The changes in SOC stock were studied in a 16 year continuously cropped agro-forestry system in Ibadan, south-western Nigeria using RothC model (Diels *et al.*, 2004). As per the model based prediction there was decline in SOC content in the no-tree control treatments with continuous maize and cowpea cropping, where SOC levels dropped by 48% from the initial level in top soil. In the alley cropping systems comprised one with *Leucaena leucocephala* and one with *Senna siamea* trees, SOC levels were also found to decrease from its original level. The model also predicted that, an annual application of additional 8.5 Mg C ha^{-1} (dry matter) of plant residues led to an increase of 4.1 Mg C ha^{-1} in top soil after 16 years.

(c) Evaluation of carbon stock in different long term fertilizer experiment using RothC model in sub-humid region in India:

The C stock was estimated in four long term fertilizer managed sites representing sub-humid moist (Sarol and Nabibagh), sub-humid dry (Panjri) and semi-arid (Teligi) climate in India using RothC model (Bhattacharyya *et al.*, 2011). It was predicted that RothC could simulate changes in total C in surface soil layers. The root mean square error (RMSE) considered as modelling error ranged from 11.50-15.01, 4.70-11.60, 2.14-6.52 and 1.45-13.74 in the surface layers at Sarol, Nabibagh, Panjri, and Teligi sites, respectively. It was simulated that total C increased in all the treatments in the sub-humid site of Sarol and Nabibagh; while application of manures alone or in combination increased carbon stock appreciably in Teligi and Panjri. The total carbon content did not show any positive accrual in non-fertilized control treatment and in NPK treatment.

(d) Simulating SOC changes in different land use systems in Brazilian Amazon using RothC

The C stock was estimated to evaluate the accuracy of the RothC model in estimating soil organic C (SOC) changes under forest-to-pasture conditions in the Brazilian Amazon (Cerri *et al.*, 2004). The data were taken from 11 site-specific 'forest to pasture' chronosequences with RothC model. The models predicted that forest clearance and conversion to well managed pasture would cause an initial decline in soil C stocks (0-20 cm depth), followed by a slow rise to levels exceeding those under native forest. The models also provided reasonable estimates (coefficient of correlation = 0.8) of the microbial biomass C in the 0-10 cm soil layer for three chronosequences, when compared with available measured data.

(e) Assessment of soil organic carbon changes in hillside systems by RothC model

A study was conducted to estimate SOC changes in three regions of the Sierra Norte of Oaxaca in plots under agricultural, forest and grassland systems (Gonzalez *et al.*, 2010). The measured and simulated SOC changes were compared under three grouping criteria: individual plot, land use system and without consideration to system. The model predicted satisfactorily SOC stocks in these systems over the four-year period ranging from 1-25 Mg C ha^{-1} accumulated in the top 30cm of soil.

(f) Use of RothC to simulate changes of organic carbon stock in the arable Mollisol

The validation of RothC with SOC stock in two long term experiments were carried out at Balcarce, Argentina (Studdert *et al.*, 2011). There were nineteen different crop sequences with and without N fertilizer and under conventional tillage, were used. Model performance was evaluated based on statistical tools including simple least square linear regression, several indicators based on the difference between observed and simulated values, and the fitting of observed and simulated values to an exponential model as a function of time under cropping. The overall performance of RothC was very good since mean difference between observed and simulated was lower than 2 Mg C ha^{-1} (2% of the mean observed value), and the root mean square error (RMSE) was lower than RMSE at 95% probability. However, RothC tended to underestimate SOC stocks especially when N had been applied and at high levels of SOC stock. Despite there was no statistical difference ($P > 0.05$) between the models resulting from fitting observed and simulated values to the exponential model. This indicated that simulation was poorer in the short term after a pasture (higher SOC stocks) than in the long run.

Limitation of RothC model

The measured and predicted values in process based modelling system including RothC model often suffers from discrepancies in observations. The probable or causes/reasons of these discrepancies are the following:

- Some of the experiments ran for many years before the first soil samples were collected; replication is often inadequate, particularly in the older experiments; replicate samples have sometimes been bulked to reduce the number of analyses needed; stored soil samples have been lost.
- Events such as erosion or a change in depth of cultivation could mask changes brought about by the original experimental treatments, as can changes in depth and technique of sampling.
- Experimental treatments were often altered during the course of the experiment. All these problems make it difficult to assess the statistical significance of any discrepancy between modelled and measured data.
- The microbial biomass carbon is considered as one of the sensitive indicator of carbon cycling and storage in a particular ecosystem. The estimation of this parameter should be cautiously done as erroneous estimation of this parameter could produce wrong prediction.
- The RothC model predicts the amount of C stock in soil in particular period of time. These predictions can change markedly from site-to-site, climatic regions, soil texture and management over many years. so, these factors should be taken into considerations and model should be modified accordingly to get proper prediction.

References

Abd El Maabou, M.S., Medan, M.A., Edriss, M., Abou -Hadid, A.F., 2004. Climate change and productivity of some wheat cultivars under rainfed and supplementary irrigation conditions. In: Cantero-Martínez C. (ed.), Gabiña D. (Ed.). *Mediterranean rainfed agriculture: Strategies for sustainability*. Zaragoza: CIHEAM, p. 139-146 (Option s Méditerranéen n es: Série A. Séminaires Méditerranéens; n. 60).

Agarwal, D. A., Humphrey, M., Beekwilder, N. F., Jackson, K. R., Goode, M. M., van Ingen, C., 2010. A data-centered collaboration portal to support global carbon-flux analysis, Concurrency Comput. Pract. Exper. 22 (17), 2323-2334, doi:10.1002/cpe.1600.

Aggarwal, B. B., *et al.* 2007. *Curcumin: The Indian solid gold*. New York: Springer.

Aggarwal, P.K., 2008. Global Climate Change and Indian Agriculture: Impacts, adaptation and mitigation. *Indian Journal of Agricultural Science* 78(10): 911 -919.

Aggarwal, P.K., Banerjee, B., Daryaei, M.G., Bhatia, A., Bala, A., Rani, S., Chander, S., Pathak, H. Kalra, N. 2006b. InfoCrop: A dynamic simulation model for the assessment of crop yields, losses due to pests, and environmental impact of agro-ecosystems in tropical environments. II. Performance of the model. *Agricultural Systems* 89, 47-67.

Aggarwal, P.K., Kalra, N., Chander, S. Pathak, H. 2004. InfoCrop: A generic simulation model for annual crops in tropical environments. Indian Agricultural Research Institute, New Delhi, p 132.

Aggarwal, P.K., Kalra, N., Chander, S., Pathak, H. 2006. Infocrop: A dynamic simulation model for tl e assessment of crop yields, losses due to pests, and environmental impact of agro-ecosystems in tropical environments. I. Model description. *Agricultural Systems* 89: 1-25.

Aggarwal, P.K., Kalra, N., Singh, A.K., Sinha, S.K., 1994. Analyzing the limitations set by climatic factors, genotype, and water and nitrogen availability on productivity of wheat. I. The model documentation, parameterization and validation. *Field Crops Research* 38, 73-91.

Aggarwal, P.K., Kumar, N. Pathak, H., 2010. Impacts of climate change on growth and yield of rice and wheat in the Upper Ganga Basin. WWF Report 2010 India.

Aggarwal, P.K., Pathak, H., 2009. Climate change and soil health: impact, mitigation and adaptation. Souvenir, Platinum Jubilee Celebrations of the Indian Society of Soil Science. pp. 35-38.

Aggarwal, P.K., Pathak, H., 2009. Conservation cultivation to combat climate change. *Indian Farming* 59 (3):5-10.

Aggarwal, P.K., Rani, S., 2009. Assessment of climate change: Impacts on wheat production in India. In: PK Aggarwal (ed.) *Global climate change and Indian Agriculture.* pp. 5-12.

Ahamed, T.A., Tian, L., Zhang, Y., Ting, K.C., 2011. Review of remote sensing methods for biomass feedstock production. *Biomass and Bioenergy* 35:2455-2469.

Ahmad, M.U.D., Masih, I., Giordano, M., 2014. Constraints and opportunities for water savings and increasing productivity through resource conservation technologies in Pakistan. *Agriculture, Ecosystems & Environment* 187, 106-115.

Ainsworth, E.A., 2008. Rice production in a changing climate: a meta analysis of responses to elevated carbon dioxide and elevated ozone concentration. *Global Change Biology* 14: 1642-1650.

Ainsworth, E.A., Long, S.P., 2005. What have we learned from 15 years of free-air CO_2 enrichment (FACE)? A meta-analytic review of the responses of photosynthesis, canopy properties and plant production to rising CO_2. *New Phytologist* 165, 351-372.

Ainsworth, E.A., Ort, D., 2010. How Do We Improve Crop Production in a Warming World? *Plant Physiology* 154: 526-530.

Ainsworth, E.A., Rogers, A., 2007. The response of photosynthesis and stomatal conductance to rising (CO_2): mechanisms and environmental interactions. *Plant Cell Environment*. 30, 258-270.

Alagarswamy, G., Singh, P., Hoogenboom, G., Wani, S.P., Pathak, P., Virmani, S.M., 2000. Evaluation and application of the CROPGRO-Soybean simulation model in vertic inceptisol. *Agricultural Systems* 63, 19-32.

Al-Bakri, A.G., Othman, G., Afifi, F.U., 2010. Determination of the anti-biofilm, antiadhesive, and anti-MRSA activities of seven Salvia species. *Pharmacognosy Magazine.* 6:264-270.

Alexandrov, V.A., Hoogenboom, G., 2000. The impact of climate variability and change on crop yield in Bulgaria. *Agricultural and Forest Meteorology* 104(4):315-327.

Allen, L.H., 1990. Plant response to raising CO2 and potential interaction with air pollutants. *Journal of Environmental Quality* 19: 15-34.

Allen, M.F., Klironomos, J.N., Treseder, K.K., Oechel, W.C., 2005. Responses of soil biota to elevated CO_2 in a chaparral ecosystem. *Ecological Applications* 15, 1701-1711.

Allen, R.G., Pereira, L.A., Raes, D., Smith, M., 1998. Crop evapotranspiration. FAO Irrigation and Drainage Paper 56. FAO, Rome, Italy, 293 p.

Allen, R.G., Pereira, L.S., Raes, D., Smith, M., 1998. Crop evapotranspiration: guidelines for computing crop water requirements. *Irrigation and Drainage Paper* n.56. FAO, Rome, Italy, 300 pp.

Allison, S.D., Wallenstein, M.D., Bradford, M.A., 2010. Soil-carbon response to warming dependent on microbial physiology. *Nature Geoscience* 3, 336-340.

Anderegg, W.R.L., Kane, J.M., Anderegg, L.D.L., 2013. Consequences of widespread tree mortality triggered by drought and temperature stress. *Nature Climate Change* 3: 30-36.

Anderson, D.W., de Jong, E., Verity, G.E., Gregorich, G.E., 1986. The effects of cultivation on the organic matter of soils of the Canadian prairies. In: *Transactions of the XIII Congress of International Society of Soil Science,* Hamburg, vol. 7, pp. 1344-1345.

Andrews, J. A., Schlesinger, W. H., 2001. Soil CO_2 dynamics, acidification, and chemical weathering in a temperate forest with experimental CO_2 enrichment. *Global Biogeochemical Cycles* 15:149-162.

Angulo, C., R€otter, R. P., Lock, R., Enders, A., Fronzek, S., Ewert, F., 2013. Implication of crop model calibration strategies for assessing regional impacts of climate change in Europe. *Agric. For. Meteorol.* 170: 32-46.

Annamalai, H., Hafner, J., Sooraj, K.P., Pillai, P., 2013. Global warming shifts the monsoon circulation, drying South Asia. *Journal of Climate* 26(9): 2701-2718.

Annamalai, H., Hamilton, K., Sperber, K. R., 2007. The South Asian summer monsoon and its relationship with ENSO in the IPCC AR4 simulations. *Journal of Climate* 20(6).

Anonymous, 2013. Economic Survey of India 2012-13: Sustainable Development and Climate change. Chapter 12. Pp. 254-268.

Anonymous, 2014a http://climate.nasa.gov/evidence/.

Anonymous, 2014b. http://www.climate-leaders.org/climate-change-resources/india-and-climate-change/sectoral-approach.

Anonymous, 2014c. http://mars.jrc.ec.europa.eu

Antonellini, M., Mollema, P., Giambastiani, B., Banzola, E., Bishop, K., Caruso, L., Minchio, A., Pellegrini, L., Sabia, M., Ulazzi, E., Gabbianelli, G., 2008. Salt water intrusion in the coastal aquifer of the southern Po Plain, Italy. *Hydrogeology Journal* 16, 1541-1556.

Arnell, W., 1999. The Effect of Climate Change on Hydrological Regimes in Europe: A Continental Perspective. *Global Environmental Change* 9, 5-23.

Arnoldus, H.M.L., 1977. Methodology used to determine the maximum potential average annual soil loss due to sheet and rill erosion in Morocco. *FAO Soils Bulletin,* 34:39-51.

Arnone, J.A., Hirschel, G., 1997. Does fertilizer application alter the effects of elevated CO_2 on Carex leaf litter quality and *in situ* decomposition in an alpine grassland? *Acta Oecologica* 18, 201-206.

ASCE., 1993. Criteria for evaluation of watershed models. *Journal of Irrigation and Drainage Engineering* 119: 429-442.

Asseng, S., Foster, I., Turner, N.C., 2011. The impact of temperature variability on wheat yields. *Global Change Biology*; 17(2):997-1012.

Attri, S.D., Rathore, L.S., 2003 Simulation of impact of projected climate change on wheat in India', *International Journal of Climatology* 23, 693-705.

Azzali, S., Menenti, M., 2000. Mapping vegetation-soil - climate complexes in southern Africa using temporal Fourier transform of NOAA-AVHRR NDVI data. *International Journal of Remote Sensing* 21: 975-996.

Bach, H., Schneider, K., Verhoef, W., Stolz, R., Mauser, W., Leeuwen, H., Schouten, L., Borgeaud, M., 2001. Retrieval of geo- and biophysical information from remote sensing through advanced combination of a land surface process model with inversion techniques in the optical and microwave spectral range. In 8th International Symposium Physical Measurements & Signatures in Remote Sensing, pp. 639-647 (France: Centre Paul Langevin, Aussois).

Badini, O., Stockle, C. O., Franz, E. H., 1997. Application of crop simulation modelling and GIS to agroclimatic assessment in Burkina Faso. *Agriculture, Ecosystems & Environment* 64 (3): 233-244.

Baggs, E.M., Richter, M., Cadisch, G., Hartwig, U.A., 2003. Denitrification in grass swards is increased under elevated atmospheric CO_2. *Soil Biology and Biochemistry* 35, 729-732.

Bajracharya, R.M., Lal, R., Kimble, J.M., 1998. Soil organic carbon distribution in aggregates and primary particle fractions as influenced by erosion phases and landscape position. In: Lal R, Kimble JM, Follett RF, Stewart BA, editors. Soil processes and the carbon cycle. Boca Raton, FL: CRC Press; pp. 353- 368.

Bajracharya, R.M., Lal, R., Kimble, J.M., 2000. Erosion effects on carbon dioxide concentration and carbon flux from an Ohio alfisol. *Soil Science Society of America Journal* 64(2):694- 700.

Baker, J.T., Allen Jr., L.H., Boote, K.J., Jones, P.H., Jones, J.W., 1989. Response of soybean to air temperature and carbon dioxide concentration. *Crop Science* 29, 98-105.

Balasubramanian, T.N., Palaniappan, S.P., 1995. Agriculture and Climate Change. In: *Agricultural Inputs and Environment.* 208-232. Scientific Publishers, Jodhpur.

Ball, A.S., 1997. Microbial decomposition at elevated CO_2 levels: effect of litter quality. *Global Change Biology* 3, 379-386.

Bandyopadhyay, K.K., Chopra, U.K., Pradhan, S., Singh, R., 2012. Assessing trade-off between wheat production and global warming potential using

InfoCrop- an indigenous generic simulation model. In. Malik, N.K., Singh, G.R., Awanish and Pruthi, N.K. (eds.) *Climate change and Agriculture.* C.C.R. (P.G.) College, Muzaffarnagar-251001 (UP), pp 25-28.

Bandyopadhyay, K.K., Prakash, A.H., Shankaranarayanan, K., Dharajothi, B., Gopalakrishnan, N., 2009. Assessing trade-off between Bt cotton production and global warming potential using InfoCrop-an indigenous generic simulation model. Abstract. National Symposium on "Bt cotton: Opportunities and Prospects". 17-19 November, 2009, pp 52.

Barber, S.A., 1995. Soil nutrient bioavailability: a mechanistic approach, 2nd edn. Wiley, New York.

Baret, F., Guyot, G., Major, D., 1989. Crop biomass evaluation using radiometric measurements. *Photogrammetria,* 43:241-256.

Barker, T., Bashmakov, I., Bernstein, L., Bogner, J.E., Bosch, P.R., Dave, R., Davidson, O.R., Fisher, B.S., Gupta, S., Halsnæs, K., Heij, G.J., Kahn Ribeiro, S., Kobayashi, S., Levine, M. D., Martino, D.L., Masera, O., Metz, B., Meyer, L.A., Nabuurs, G.J., Najam, A., Nakicenovic, N., Rogner, H.H., Roy, J., Sathaye, J., Schock, R., Shukla, P., Sims, R.E.H., Smith, P., Tirpak, D.A., Urge-Vorsatz, D., Zhou, D., 2007. Technical Summary. In: B. Metz, O.R. Davidson, P.R. Bosch, R. Dave and L.A. Meyer (eds.), Climate Change 2007: Mitigation. Contribution of Working Group III to the Fourth Assessment Report of the Intergovernmental Panel on Climate Change, Cambridge University Press, Cambridge, United Kingdom and New York, NY, USA.

Barrett, D.J., Richardson, A.E., Gifford, R.M., 1998. Elevated atmospheric CO_2 concentrations increase wheat root phosphatase activity when growth is limited by phosphorus. *Aust. Journal of Plant Physiology* 25, 87-93.

Barrett, J.R., Nearing, A.M., 1998. Humanization of decision support using information from simulations, In Agricultural Systems Modelling and Simulation Eds, Robert M. Peart; R. Bruce Curry, Marcel Dekker, INC.

Basak, J.K., Ali, M.A., Islam, M.N., Rashid, M.A., 2010. Assessment of the Effect of Climate Change on Boro Rice Production in Bangladesh using DSSAT Model, *Journal of Civil Engineering (IEB),* 38(2), pp 95-108.

Bassirirad, H., 2000. Kinetics of nutrient uptake by roots: responses to global change. *New Phytologist* 147,155-169.

Basso, B., Gargiulo, O., Paustian, K., Robertson, G.P., Porter, C., Grace, P.R., Jones, J.W., 2011. Procedures for initializing soil organic carbon pools in the DSSAT-century model for agricultural systems. *Soil Science Society of America Journal* 75:69-78.

Bauman, B., Kropff, M., Tuong, T., Wopereis, M., Berge, H., Laar, H., 2001. ORYZA 2000: Modelling lowland rice. Manila, Philippines' IRRI.

Bechini, L., Bocchi, S., Maggiore, T., Confalonieri, R., 2006. Parameterization of a crop growth and development simulation model at sub-model components level. An example for winter wheat (*Triticum aestivum* L.). *Environmental Modelling & Software* 21, 1042-1054.

Bell, M.A., Fischer, R.A., 1994. Using yield prediction models to assess yield gains: A case study for wheat. *Field Crops Research* 36: 161-166.

Bell, T.H., Klironomos, J.N., Hugh, A.L.H., 2010. Seasonal responses of extra cellular enzyme activity and microbial biomass to warming and N addition. *Soil Science Society of America Journal* 74, 820-828.

Bellocchi, G., Silvestri, N., Mazzoncini, M., Menini, S., 1996. Using the cropsyst model in continuous rainfed maize under alternative management options. *Italian Journal of Agronomy* 6(1): 43-56.

Ben Nouna, B., Katerji, N., Mastrorilli, M., 2000. Using the CERES-Maize model in a semi-arid Mediterranean environment. Evaluation of model performance. *European Journal of Agronomy* 13 (4), 309-322.

Benbi, D.K., Kaur, R., 2009. Modelling Soil Processes in Relation to Climate Change. *Journal of the Indian Society of Soil Science* 57:433-444.

Benli, B., Pala, M., Stöckle, C., Oweis, T., 2007. Assessment of winter wheat production under early sowing with supplemental irrigation in a cold highland environment using CropSyst simulation model. *Agricultural Water Management* 93, 45-53.

Berc, J, *et al.* 2003. Conservation Implications of Climate Change: Soil Erosion and Runoff from Croplands: A Report from the Soil and Water Conservation Society. Ankeny (IA): Soil and Water Conservation Society.

Berhe, A.A., Harte, J., Harden, J.W., Torn, M.S., 2007. The significance of the erosion induced terrestrial carbon sink. *Journal of Biosciences* 57, 337-346.

Berntson, G.M., Bazzaz, F.A., 1997. Nitrogen cycling in microcosms of yellow birch exposed to elevated CO_2: simultaneous positive and negative below-ground feedbacks. *Global Change Biology* 3, 247-258.

Beyer, L., Köbbemann, C., Finnern, J., Elsner, D., Schluß, W., 1993. Colluvisols under cultivation in Schleswig-Holstein, 1: Genesis, definition and geoecological significance. *Journal of Plant Nutrition and Soil Science* 156: 197-202.

Bezemer, T.M., Jones, T.H., 1998. Plant-insect herbivore interactions in elevated CO_2: quantitative analyses and guild effects. Oikos, 82: 212-222.

Bhandari, A.L., Ladha, J.K., Pathak, H., Padre, A.T., Dawe, D., Gupta, R.K., 2002. Yield and soil nutrient changes in a long-term rice-wheat rotation in India. *Soil Science Society of America Journal* 66, 167-170.

Bhatia, A., Aggarwal, P.K., Jain, N., Pathak, H., 2012. Greenhouse gas emission from rice and wheat growing areas in India: spatial analysis and upscaling. *Greenhouse Gases: Science and Technology* 2(2), 115-125.

Bhatia, A., Pathak, H., Aggarwal, P.K., Jain, N., 2010. Trade-off between productivity enhancement and global warming potential of rice and wheat in India. *Nutrient Cycling in Agro ecosystems* 86, 413-424.

Bhatia, V.S., Singh, P., Kesava Rao, A.V.R., Srinivas, K., Wani, S. P., 2009. Analysis of water non-limiting and water limiting yields and yield gaps of groundnut in India using CROPGRO-Peanut model. *Journal of Agronomy and Crop Science* 195, 455-463.

Bhattacharyya, T., Pal, D.K., Deshmukh, A.S., Deshmukh, R.R., Ray, S.K., Chandran, P., Mandal, C., Telpande, B., Nimje, A.M., Tiwary P., 2011. Evaluation of RothC model using four Long Term Fertilizer Experiments in black soils, India. Agriculture, Ecosystems and Environment, 144, 222-234.

Bhuvaneswari, K., Geethalakshmi, V., Lakshmanan, A., Anbhazhagan, R., Sekhar, D.N.U., 2014. Climate change impact assessment and developing adaptation strategies for rice crop in western zone of Tamil Nadu. [June 2013] Climate change impact on rice in Tamil Nadu. *Journal of Agrometeorology* 16 (1): 38-43.

Bindi, M., Fibbi, L., Gozzini, B., Orlandini, S., Miglietta, F., 1996. Modelling the impact of future climate scenarios on yield and yield variability of grapevine. *Climate Research* 7:213-224.

Blaikie, P.M., Brookfield, H., 1987. Land Degradation and Society. London: Methuen.

Blondlot, A., Gate, P., Poilve, H., 2005. Providing operational nitrogen recommendations to farmers using satellite imagery. In Stafford, J.V. (Ed.), *Precision Agriculture* '05. Wageningen University Publishers, Wageningen, pp. 345-351.

Bloom, A.J., Burger, M., Rubio Asensio, J.S., Cousins, A.B., 2010. Carbon dioxide inhibits nitrate assimilation in wheat and Arabidopsis. *Science* 328, 899-903.

Boer, G., Flato, G., Ramsden, D., 2000. A transient climate change simulation with greenhouse gas and aerosol forcing: Projected climate for the 21st century. *Climate Dynamics* 16:427-450.

Boer, G.J., Flato, G., Reader, M.C., Ramsden, D., 2000. A transient climate change simulation with green house gas and aerosol forcing: experimental design and comparison with the instrumental record for the 20th century. *Climate Dynamics* 16(6):405-425.

Bollasina, M.A., Ming, Y., Ramaswamy, V., 2011. Anthropogenic aerosols and the weakening of the South Asian summer monsoon. *Science,* 334, 502-505.

Booltink, H.W.G., van Alphen, B.J., Batchelor, W.D., Paz, J.O., Stoorvogel, J.J., Vargas, R., 2001. Tools for optimizing management of spatially-variable fields. *Agricultural Systems* 70, 445-476.

Booltink, H.W.G., Verhagen, J., 1997. Using decision support systems to optimize barley management on spatial variable soil. In: Kropff, M., *et al.* (Eds.), Applications of Systems Approaches at the Field Level, vol. 2. Kluwer Academic Publishers, Dordrecht, Netherlands, pp. 219-233.

Boomiraj, K., Chakrabarti, B., Aggarwal, P.K., Choudhary, R., Chander, S., 2010. Assessing the vulnerability of Indian mustard to climate change. *Agriculture, Ecosystems & Environment* 138, 265-273.

Boote, K.J., Jones, J.W., Pickering, N.G., 1996. Potential uses and limitations of crop models. *Agronomy Journal* 88: 704-716.

Botes, J.H.F., Bosch, D.J., Oosthuizen, L.K., 1996. A simulation optimization approach for evaluating irrigation information. *Agricultural Systems* 51: 165-184.

Bouman, B.A.M., 1992. Linking physical remote sensing models with crop growth simulation models, applied for sugar beet. *International Journal of Remote Sensing* 13, 2565-2581.

Bouman, B.A.M., 1996. The "School of de Wit" crop growth simulation models: A pedigree and historical overview. *Agricultural Systems* 52(2-3), 171-198.

Bouman, B.A.M., Van Diepen, C.A., Vossen, P., Van Der Val., T., 1997. Simulation and systems analysis tools for crop yield forecasting. In P.S. Teng *et al.* (Eds.). Applications of Systems Approaches at the Farm and Regional Levels. *Kluwer Academic Publishers.* Pp. 325-340.

Bouniols, A., Cabelguenne, M., Jones, C.A., Chalamet, A., Charpenteau, J.L., Marty, J.R., 1991. Simulation of soybean nitrogen nutrition for a silty clay soil in southern France. *Field Crops Research* 26, 19-34.

Bouraoui, F., Grizzetti, B., Granlund, K., Rekolainen, S., Bidoglio, G., 2004. Impact of climate change on the water cycle and nutrient losses in a Finnish catchment. *Climatic Change.* 66, 109-126.

Bouwman, A.F., 1989. Modelling soil organic matter decomposition and rainfall erosion in two tropical soils after forest clearing for permanent agriculture. *Land Degradation and Rehabilitation*; 1:125-40.

Brassard, J.P., Singh, P., 2008. Impacts of climate change and CO_2 increase on agricultural production and adaptation options for Southern Quebec, Canada. *Mitigation and Adaptation Strategies for Global Change 13:241-265.*

Brinkman, R., 1982. Clay transformations: aspects of equilibrium and kinetics. In: Soil Chemistry. B. Physicochemical Models. Developments in Soil Science 5B. G.H. Bolt (Ed.). 2nd ed. Elsevier, Amsterdam. pp. 433-458.

Brinkman, R., 1990. Resilience against climate change? Soil minerals, transformations and surface properties, Eh, pH. In: Scharpenseel *et al.* (Eds.). pp. 51-60.

Brinkman, R., Sombroek, W. G., 1996. The effects of global change on soil conditions in relation to plant growth and food production. In: F. Bazzaz and W. Sombroek (eds.), Global climate change and agricultural production. Direct and indirect effects of changing hydrological, pedological and plant physiological processes, John Wiley & Sons Ltd., England, Chapter 3.

Brisson, N., Bona, S., Bouniols, A., 1989. A soybean crop simulation model: validation and adaptation to varieties cultivated in southern Europe. *Agronomie* 9, 27-36.

Brisson, N., Launay, M., Mary, B., Beaudoin, N., 2009. Conceptual basis, formalisations and parameterization of the STICS crop model. Éditions Quæ, Versailles, France.

Brouder, S.M., Volenec, J.J., 2008. Impact of climate change on crop nutrient and water use efficiencies. *Plant Physiology* 133:705-724.

Brown, R.A., Rosenberg, N.J., 1997. Sensitivity of crop yield and water use to change in a range of climatic factors and CO_2 concentrations: a simulation study applying EPIC to the central USA. *Agricultural and Forest Meteorology* 83:171-203.

Butler, E.E., Huybers, P., 2013. Adaptation of US maize to temperature variations. *Nature Climate Change* 3: 68-72.

Byjesh, K., Kumar, S.N., Aggarwal, P.K., 2010. Simulating impacts, potential adaptation and vulnerability of maize to climate change in India. *Mitigation and Adaptation Strategies for Global Change* 15, 413-431.

Cai, W., Sullivan, A., Cowan, T. 2011. Interactions of ENSO, the IOD, and the SAM in CMIP3 Models. *Journal of Climate* 24:1688-1704.

Campbell, G.S., 1985. *Soil Physics with Basic.* Elsevier, Amsterdam.

Canadell, J., Jackson, R.B., Ehleringer, J.R., Mooney, H.A., Sala, O.E., Schulze, E.D., 1996. Maximum rooting depth of vegetation types at the global scale. *Oecologia,* 108:583-595.

Carberry, P.S., Muchow, R.C., McCown, R.L., 1992. A simulation model of kenaf for assisting fibre industry planning in Northern Australia. IV. Analysis of climatic risks. *Australian Journal of Agricultural Research* 44: 713-730.

Cardon, Z.G., 1996. Effects of root exudation and rhizode position on ecosystem carbon storage under elevated CO_2. *South African Journal of Plant and Soil* 87:277-288.

Carney, K. M., Hungate, B.A., Drake, B. G., Megonigal, J.P., 2007. Altered soil microbial community at elevated CO_2 leads to loss of soil carbon. Pro. *National Academy of Sciences* USA 104, 4990-4995.

Carter, T. R., Saarikko, R,. 1996. Estimating regional crop potential in Finland under a changing climate. Agric. For. Meteorol. 79:301-313.

Casaa, R., Varella, H., Buis, S., Guérif, M., De Solan, B., Baret, F., 2012. Forcing a wheat crop model with LAI data to access agronomic variables: Evaluation of the impact of model and LAI uncertainties and comparison with an empirical approach. *European Journal of Agronomy* 37: 1-10.

Cassman, K.G., Dobermann, A., Walters, D.T., 2002. Agro-ecosystems, nitrogen-use efficiency, and nitrogen management. Ambio 31: 132-140.

Castrignano, A., Colucci, R., Degiorgio, D., Rizzo, V., Stelluti, M., 1997. Tillage effects on plant extractable soil-water in a silty clay vertisol in southern Italy. *Soil and Tillage Research* 40 (3 -4), 227-237.

Cerri, C.E.P., Paustian, K., Bernoux, M., Victoria, R.L., Mellilo, J.M., Cerri, C.C., 2004. Modelling changes in soil organic matter in Amazon forest to pasture conversion, using the Century model. Global Change Biology, 10, 815-832.

Cervigni, Raffaello, Valentini, R., Santini, M., 2013. Toward Climate-Resilient Development in Nigeria. Directions in Development. Washington, DC: World Bank. doi:10.1596/978-0-8213-9923-1. License: Creative Commons Attribution CC BY 3.0.

Challinor, A. J., 2011. *Agriculture: forecasting food. Nat. Clim. Change* 1:103-104.

Challinor, A. J., Slingo, J. M., Wheeler, T. R., Doblas - Reyes, F. J., 2005c. Probabilistic hindcasts of crop yield over western India. Tellus 57A 498-512.

Challinor, A. J., Wheeler, T. R., 2008a. Use of a crop model ensemble to quantify CO_2 stimulation of water-stressed and well-watered crops. *Agric. For. Meteorol*, 148 1062-1077.

Challinor, A., 2009.Towards development of adaptation options using climate and crop yield forecasting at seasonal to multidecadal time scales. *Environ. Sci. Pol.* 12, 453-465.doi:10.1016/j.envsci.2008.09.008.

Chander, C., Ahuja, L.R., Peairs, F.B., Aggarwal, P.K., Kalra, N., 2006. Modelling the effect of Russian wheat aphid, *Diuraphis noxia* (Mordvilko) and weeds in winter wheat as guide to management. *Agricultural Systems* 88, 494-513.

Chander, G., Huang, C., Yang, L., Homer, C., Larson, C., 2009. Developing consistent landsat data sets for large area applications: the mrlc 2001 protocol. *IEEE Geoscience* and *Remote Sensing Letters* 6: 777-781.

Chaturvedi, I., Dutta, T.K., Singh, P K., 2013. Herbals additives: for goat production (ISBN: 978-3-659-40851-9). (Book Publication (2013) by LAP LAMBERT Academic Publishing, AV Akademikerverlag GmbH & Co. KG, Germany.

Chaturvedi, R. K., Joshi, J., Jayaraman, M., Bala, G., Ravindranath, N.H., 2012. Multi-model climate change projections for India under Representative Concentration Pathways (RCPs): A preliminary analysis *Current Science* 103, 791-802.

Chaturvedi, R.K., Joshi, J., Jayaraman, M., Bala, G., Ravindranath, N.H., 2012. Multi-model climate change projections for India under representative concentration pathways. *Current Science* 103, 791-802.

Cheeroo-Nayamuth, B.F., Robertson, M.J., Wegener, M.K., Nayamuth, A.R.H., 2000. Using a model to assess potential and attainable sugarcane yield in Mauritius. *Field Crops Research* 66: 225-243.

Cheeroo-Nayamuth, F.C., Robertson, M.J., Wegener, M.K., Nayamuth, A.R.H., 2000. Using a simulation model to assess potential and attainable sugar cane yield in Mauritius. *Field Crops Research* 66(3):225-243.

Cheng, W., Chander, K., Inubushi, K., 2000a. Effect of elevated CO_2 and temperature on microbial biomass nitrogen and nitrogen mineralization in submerged soil microcosms. *Soil Microorganisms* 54:51-59.

Cheng, W., Chander, K., Inubushi, K., 2000b. Effect of elevated CO_2 and temperature on methane production and emission from submerged soil microcosms. *Nutrient Cycling in Agroecosystems* 58:339-347.

Cheng, W.X., 1999. Rhizosphere feedbacks in elevated CO_2. Tree Physiol. 19, 313-320.

Cherchi, A., Alessandri, A., Masina, S., Navarra, A., 2010. Effect of increasing CO_2 levels on monsoons. *Climate Dynamics* 37, 83-101.

Ching, P. C., Barber, S.A., 1979. Evaluation of temperature effects on K uptake by corn. *Agronomy Journal* 71, 1040-1044.

Choudhury, B.U., Das, A., Ngachan, S.V., Bordoloi, L.J., Chowdhury, P., 2012a.Trend Analysis of Long Term Weather Variables in Mid Altitude Meghalaya, North-East India. *Journal of Agricultural Physics* 12(1): 12-22.

Chung, C.E., Ramanathan, V., 2006. Weakening of North Indian SST gradients and the monsoon rainfall in India and the Sahel. *Journal of Climate* 19, 2036-2045.

Clevers, J. G. P. W., 1997. A simplified approach for yield prediction of sugar beet based on optical remote sensing data. *Remote Sensing of Environment* 61:221-228.

Coleman K., Jenkinson D.S., 2005. ROTHC-26.3 A model for the turnover of carbon in soil. Model description and windows users guide. Available at http://www.rothamsted.bbsrc.ac.uk/aen/carbon/ mod26_3_win.pdf

Coleman K., Jenkinson D.S., Crocker G.J., Grace P.R., Klir J., Korschens M., Poulton P.R., Richter D.D.,1997. Simulating trends in soil organic carbon in long-term experiments using RothC-26.3. Geoderma, 81, 29-44.

Coleman, K., Jenkinson, D.S., 1999. RothC-26.3 - A Model for the turnover of carbon in soil: Model description and windows users guide: November 1999 issue. Lawes Agricultural Trust Harpenden.

Collins, A. L., Walling, D. E., Leeks, G.J. L., 1997. `Source type ascription for fluvial suspended sediment based on a quantitative composite fingerprinting technique', *Catena* 29, 1-27.

Collins, M.R., Knutti, Arblaster J., Dufresne, J.L., Fichefet, T., Friedlingstein, P., Gao, X., Gutowski, W.J., Johns, T., Krinner, G., Shongwe, M., Tebaldi, C., Weaver, A.J., Wehner, M., 2013. Long-term Climate Change: Projections, Commitments and Irreversibility. In: Climate Change 2013: The Physical Science Basis. Contribution of Working Group I to the Fifth Assessment Report of the Intergovernmental Panel on Climate Change [Stocker, T.F., D. Qin, G.-K. Plattner, M. Tignor, S.K. Allen, J. Boschung, A. Nauels, Y. Xia, V. Bex and P.M. Midgley (eds.)]. Cambridge University Press, Cambridge, United Kingdom and New York, NY, USA.

Colson, J., Bouniols, A., Jones, J.W., 1995. Soybean reproductive development: adapting a model for European cultivars. *Agronomy Journal* 87, 1129-1139.

Cotrufo, M.F., Ineson, P., 1995. Effects of enhanced atmospheric CO_2 and nutrient supply on the quality and subsequent decomposition of the fine roots of *Betula pendula* Roth and *Picea sitchensis* (Bong.) Carr. Plant Soil 170, 267-277.

Cotrufo, M.F., Ineson, P., Rowland, A.P., 1994. Decomposition of tree leaf litters grown under elevated CO_2: Effect of litter quality. Plant Soil 163, 121-130.

Cotrufo, M.F., Ineson, P., Scott, Y., 1998. Elevated CO_2 reduces the nitrogen concentration of plant tissues. *Global Change Biology* 4, 43-54.

Cowling, S.A., Field, C.B., 2003. Environmental control of leaf area production: Implications for vegetation and land-surface modelling. *Global Biogeochemical Cycles* 17: 1007.

Coxwell, C.C., Bock, C.E., 1995. Spatial variation in diurnal surface temperature and the distribution and abundance of an alpine grasshopper. *Oecologia* 104: 433?439.

Cuculeanu, V., Marcia, A., Simota, C., 1999. "Climate change impact on agricultural crops and adaptation options in Romania". *Climate Research.* Vol.12, pp. 153-160.

Curtis, P.S., Zak, D.R., Pregitzer, K.S., Teeri, J.A., 1994. Above and below ground response of *Populus grandidentata* to elevated atmospheric CO_2 and soil N availability. Plant Soil 165, 45-51.

Daily, G.C., 1995. Restoring value to the world's degraded lands. Science 269: 350-354.

Das, A., Patel, D. P., Munda, G.C., Ghosh, P. K., Ngachan, S.V., Choudhury, B.U., Ramkrushna, G.I., Saha, R., Rajkhowa, D.J., Panwar, A.S., Kumar, R., Kumar, M., Buragohain, J., 2011. Conservation agriculture in rice and maize based cropping systems: package and practices for NEH Region. Research Bulletin no. 75. ICAR Research Complex for NEH Region, Umiam, Meghalaya, pp. 36.

Das, D.K., Hati, K.M., 2013. Impact of climate change and global warming on soil physical processes. In Souvenir of the 78th Annual Convention of Indian Society of Soil Science, October 23-26, 2013, Jodhpur. Pp. 37-43.

Dash, S.K., Hunt, J.C.R., 2007. Variability of climate change in India, *Current Science* 93(6), 782-788.

Dash, S.K., Jenamani, R.K., Kalsi, S.R., Panda, S.K., 2007. Some evidence of climate change in twentieth-century India, *Climatic Change* 85, 299-321.

Datta, S.C., 2008. Theory and principal of simulation modelling in soil-plant system. Capital Publishing Company. ISBN No. 8185589550.

Davidson, E.A., Janssens, I.A., 2006. Temperature sensitivity of soil carbon decomposition and feedbacks to climate change. *Nature* 440, 165-173.

Day, F. P., Weber, E.P., Hinkle, C.R., Drake, B.G., 2000. Effects of elevated CO_2 on fine root length and distribution in an oakpalmetto scrub ecosystem in central Florida. *Global Change Biology* 2, 143-148.

de Graaff, M.A., Van Groenigen, K.J., Six, J., Hungatez , B., Kessal , C.V., 2006. Interactions between plant growth and soil nutrient cycling under elevated CO_2: a meta-analysis. *Global Change Biology* 12, 2077-2091.

De Wit, C.T., 1965. Photosynthesis of leaf canopies. Agricultural Research Report. 663. Wagenigen.

De Wit, C.T., 1982a. Simulation of living systems. In: F.W.T. Penning de Vries & H.H. van Laar (Eds): Simulation of plant growth and crop production. Simulation Monographs, PUDOC, Wageningen, pp. 3-8.

De wit, C.T., Penning de Vries, F.W.T., 1982. La Productivite des Paturages Saheliens. Agricultural Research Report 918. PUDOC, Wageningen, The Netherlands.

DeFries, R.S., Field, C.B., Fung, I., Collatz, G.J., Bounoua, L., 1999. Combining satellite data and biogeochemical models to estimate global effects of human-induced land cover change on carbon emissions and primary productivity. *Global Biogeochemical Cycles* 13: 803-815.

Delecolle, R., Maas, S. J., Guerif, M., Baret, F., 1992. Remote sensing and crop production models: present trends. *ISPRS Journal of Photogrammetry and Remote Sensing* 47: 145- 161.

DeLucia, E. H., Hamilton, J. G., Naidu, S. L., Thomas, R. B., Andrews, J. A., Finzi, A., Lavine, M., Matamala, R., Mohan, J. E., Hendrey, G. R., Schlesinger, W. H., 1999. Net primary production of a forest ecosystem with experimental CO_2 enrichment. Sci. 284, 1177-1179.

Deng, Z.W., 2010. Impact of climate warming and drying on food crops in northern China and the countermeasures. *Acta Ecologica Sinica,* 30(22): 6278-6288.

Dhillon, S. S., Roy, J., Abrams, M., 1996. Assessing impact of elevated CO_2 on soil microbial activity in a Mediterranean model ecosystem. Plant Soil 187,333-342.

Diaz, S., Grime, J.P., Harris, J., McPherson, E., 1993. Evidence of a feedback mechanism limiting plant response to elevated carbon dioxide. *Nature* 364, 616-617.

Diels, J., Vanlauwe, B., Van der Meersch, M.K., Sanginga, N., Merckx, R., 2004. Long-term soil organic carbon dynamics in a subhumid tropical climate: ^{13}C data in mixed C_3/C_4 cropping and modeling with ROTHC. Soil Biology and Biochemistry, 36, 1739-1750.

Djumaniyazova, Y., Sommer, S., Ibragimov, I., Ruzimov, J., Lamers, J.P.A., Vlek, P.L.G., 2010. Simulating water use and N response of winter wheat in the irrigated floodplains of Northwest Uzbekistan. *Field Crops Research* 116, 239-251.

Don, N.C., Hang, N.T.M., Araki, H., Yamanishi, H., Koga, K., 2006. Salinization processes in an alluvial coastal lowland plain and effect of sea water level rise. *Environmental Geology* 49,743-751.

Donatelli, M., Stockle, C., Ceotto, E., Rinaldi, M., 1997. Evaluation of CropSyst for cropping systems at two locations of northern and southern Italy. *European Journal of Agronomy* 6: 35-45.

Doorenbos, J., Kassam, A.H., 1979. Yield response to water. Irrigation and DrainagePaper n. 33. FAO, Rome, Italy, 193 pp.

Doos, B.R., 2002. The problem of predicting global food production. *Ambio* 31: 417-424.

Doraiswamy, P. C., Cook, P. W., 1995. Spring wheat yield assessment using NOAA AVHRR data. *Canadian Journal of Remote Sensing* 21: 43-51.

Doraiswamy, P.C., Sinclair, T.R., Hollinger, S., Akhmedov, B., Stern, A., Prueger, J., 2005. Application of MODIS derived parameters for regional crop yield assessment. *Remote Sensing of Environment* 97: 192-202.

Dotaniya, M.L., 2012. Crop Residue Management in Rice-Wheat Cropping System. Lap Lambert Academic Publisher, Germany. pp .116. ISBN no. 978-3-659-29388-7.

Dotaniya, M.L., Datta, S.C., Biswas, D.R., Meena, B.P., 2013a. Effect of Solution Phosphorus Concentration on the Exudation of Oxalate Ions by Wheat (*Triticum aestivum* L.). *Proceedings of the National Academy of Sciences, India Section B: Biological Sciences DOI* 10.1007/s40011-012-0153-7.

Dotaniya, M.L., Meena, H. M., Lata, M., Kumar, K., 2013b. Role of phytosiderophores in iron uptake by plants. *Agricultural Science Digest* 33(1), 73-76.

Douville, H., Ribes, A., Decharme, B., Alkama, R., Sheffield, J., 2013. Anthropogenic influence on multidecadal changes in reconstructed global evapotranspiration. *Nature Climate Change* 3: 59-62.

Drigo, B., Kowalchuk, G.A., van Veen, J.A., 2008. Climate change goes underground: effects of elevated atmospheric CO_2 on microbial community structure and activities in the rhizosphere. *Biology and Fertility* of *Soils* 44, 667-679.

Duxbury, J.M., 2005. Reducing green house warming potential by carbon sequestration in soils: Opportunities, limits and tradeoffs. In: R. Lal, N. Uphoff, B. A. Stewart and D. O. Hansen (editors), Climate change and global food security, *Taylor & Francis* pp. 435-450.

Dzotsi, K.A., Jones, J.W., Adiku, S.G.K., Naab, J.B., Singh, U., Porter, C.H., Gijsman, A.J., 2010. Modelling soil and plant phosphorus within DSSAT. *Ecological Modelling,* 221: 2839-2849.

Easterling, W., Apps, M., 2005. Assessing the consequences of climate change for food and forest resources: a view from the IPCC. *Climate change* 70:165-189.

Easterling, W.E., Aggarwal, P.K., Batima, P., Brander, K.M., Erda, L., Howden, S.M., Kirilenko, A., Morton, J., Soussana, J.F., Schmidhuber, J., 2007. Food, fibre and forest products. In S Solomon, D Qin, M Manning, Z Chen, M Marquis, KB Averyt, M Tignor, HL Miller, eds, Climate Change

2007: The Physical Science Basis. Contribution of Working Group I to the Fourth Assessment Report of the Intergovernmental Panel on Climate Change. Cambridge University Press, Cambridge, UK, and New York, pp 273-313.

Ebrayi, K.N., Pathak, H., Kalra, N., Bhatia, A., Jain, N., 2007. Simulation of nitrogen dynamics in soil using InfoCrop model. *Environmental Monitoring and Assessment* 131, 1-3.

Edwards, N.T., Norby, J.R., 1999. Below-ground respiratory responses of sugar maple and red maple seedlings to atmospheric CO_2 enrichment and elevated temperatures. *Plant Soil* 206, 85-97.

Edwards, W.M., Owens, L.B., 1991. Large storm effects on total soil-erosion. *Journal of Soil and Water Conservation* 46(1):75- 8.

Elanchezhian, R., Haris, A.A., Biswas, S. Chhabra, V., 2012. Simulation of yield and its component traits of rice (*Oryza sativa* L.) varieties grown in Indo-Gangetic plains of Bihar under projected climate change. *Indian Journal of Plant Physiology* 17: 195-202.

Enzler, S.M. (2013) http://www.lenntech.com/greenhouse-effect/global-warming-history.htm

Erickson, J. D., 1984. The LACIE experiment in satellite aided monitoring of global crop production. In The Role of Terrestrial Vegetation in the Global Carbon Cycle: Measurements by Remote Sensing, edited by G.M. Woodwell (Chichester : Wiley), pp. 191-217.

Evensen, G., 2003. The Ensemble Kalman filter: theoretical formulation and practical implementation. *Ocean Dynamics* 53: 343-367.

Ewert, F., van Ittersum, M. K., Heckelei, T., Therond, O., Bezlepkina, I., Andersen, E., *et al.* 2011. Scale changes and model linking methods for integrated assessment of agri-environmental systems. Agric. Ecosyst. Environ. 142:6-17.

F.A.O., 2001. Conservation Agriculture Case Studies in Latin America and Africa. Introduction. FAO Soils Bulletin No. 78. Rome: FAO.

F.A.O., 2009. FAO Statistical Databases. Food and Agricultural Organization of the United Nations (FAO). http://faostat.fao.org.

F.A.O., 2011. The State of food and agriculture. Women in agriculture. Closing the gender gap for development. Rome: Food and Agriculture Organization of the United Nations. 2.

F.A.O.S.T.A.T., 2009. http://faostat.fao.org.

Falloon, P., Smith. P., 2003. Accounting for changes in soil carbon under the Kyoto Protocol: Need for improved long-term data sets to reduce uncertainty in model projections. *Soil Use and Management* 19:265-269.

Fang, H., Liang, S., Hoogenboom, G., 2011. Integration of MODIS products and a crop simulation model for crop yield estimation. *International Journal of Remote Sensing* 32: 1039-1065.

FAO, 2010. "Climate-Smart" Agriculture: Policies, Practices and Financing for Food Security, Adaptation and Mitigation, 41 pp. Rome.

Felkner, J., Tazhibayeva, K., Townsend, R., 2008. "Crop Production: Combining Socio- Economic and Biophysical Data and Models," working paper, University of Chicago, 2008.

Feng, L., Bouman, B.A.M., Tuong, T.P., Cabangon, R.J., Li, Y., Lu, G., Feng, Y., 2007. Exploring options to grow rice using less water in northern China using modelling approach. I. Field experiments and model evaluation. *Agricultural Water Management* 88:1-13.

Finzi, A.C., Norby, R.J., Calfapietra, C., Gallet-Budynek, A., Gielen, B., Holmes, W.E., Hoosbeek, M.R., Iversen, C.M., Jackson, R.B., Kubiske, M.E., Ledford, J., Liberloo, M., Oren, R., Polle, A., Pritchard, S., Zak, D.R., Schlesinger, W.H., Ceulemans, R., 2007. Increases in nitrogen uptake rather than nitrogen-use efficiency support higher rates of temperate forest productivity under elevated CO_2. *Proceedings of the National Academy of Sciences of the USA* 104, 14014-14019.

Fleming, R. J., 2010. Gilbert N. Plass: Climate science in perspective. American Scientist, 98, 60-67.

Fourcaud, T., Ji, J.N., Zhang, Z.Q., Stokes, A., 2008. Understanding the impact of root morphology on overturning mechanisms: a modelling approach. *Annals of Botany* 101: 1267-1280.

Fournier, F., 1960. Climat et Erosion. Universitaries France, Paris.

France, J., Thornley, J.H.M., 1984. Mathematical Models in Agriculture. Butterworth & Co. Ltd., UK.

G.L.A.S.O.D., 1990. Global assessment of soil degradation. World map. Wageningen, Netherlands: ISRIC.

Gabrielle, B., Denoroy, P., Gosse, G., Justes, E., Andersen, M.N., 1998. Development and evaluation of a CERES-type model for winter oilseed rape. *Field Crops Research* 57 (1), 95-111.

Gadgil, S., Rao, P.R.S., Rao, K.N., 2002. Use of climate information for farm-level decision making: rainfed groundnut in southern India. Agric. Syst. 74, 431-457.

Ganguly, S., Samanta, A., Schull, M. A., Shabanov, N.V., Milesi, C., Nemani, R.R., Knyazikhin, Y., 2008. Generating vegetation leaf area index Earth system data record from multiple sensors. Part 2: Implementation, analysis and validation. *Remote Sensing of Environment* 112:4318-4332.

Ganguly, S., Verma, J., Hu,Z., Xing, H., Jena, D., 2014. Performance enhancement of InAlN/GaN HEMTs by KOH surface treatment Appl. Phys. Express 7 034102. doi:10.7567/APEX.7.034102.

Gardi, C., 2001. Lnause, agronomic management and water quality in a small northern Italian watershed. *Agriculture, Ecosystems & Environment* 87 (1): 1-12.

Gaydon, D.S., Probert, M.E., Buresh, R.J., Meinke, H., Suriadi, A., Dobermann, A., Bouman, B.A.M., Timsina, J., 2012. Rice in cropping systems - Modelling transitions between flooded and non-flooded soil environments. *European Journal of Agronomy* 39: 9-24.

Geethalakshmi, V., Lakshmanan, A., Rajalakshmi, D., Jagannathan, R., Sridhar, G., Ramaraj, A.P., Bhuvaneswari, K.L., Gurusamy, Anbhazhagan, R., 2011. 'Climate change impact assessment and adaptation strategies to sustain rice production in Cauvery basin of Tamil Nadu'. *Current Science,* 101(3): 342-347.

Gerwitz, A., Page, E.R., 1974. An empirical mathematical model to describe plant root systems. *Journal of Applied Ecology* 11:773-781.

Ghadiri, H., Rose, C.W., 1991b. Sorbed chemical-transport in overland-flow: 2. Enrichment ratio variation with erosion processes. *Journal of Environmental Quality* 20(3):634- 41.

Ghadiri, H., Rose, CW., 1991a. Sorbed chemical-transport in overland-flow: 1. A nutrient and pesticide enrichment mechanism. *Journal of Environmental Quality* 20(3):628- 33.

Ghahramani, A., Ishikawa, Y., Gomi, T., Shiraki, K., Miyata, S., 2011. Effect of ground cover on splash and sheetwash erosion over a steep forested hill slope: a plotscale study. *Catena* 85, 34-47.

Giannakopoulos, C., Le Sager, P., Bindi, M., Moriondo, M., Kostopoulou, E., Goodess, C.M., 2009. Climatic changes and associated impacts in the Mediterranean resulting from global warming. *Global and Planetary Change* 68:209-224.

Gielen, B., Calfapietra, C., Lukac, M., Wittig, V.E., De Angelis, P., Janssens, I.A., Moscatelli, M.C., Grego, S., Cotrufo, M.F., Godbold, D.L., *et al.*, 2005. Net carbon storage in a poplar plantation (POPFACE) after three years of free-air CO2 enrichment. *Tree Physiology* 25: 1399-1408.

Gifford, R.M., Evans, L.T., 1981. Photosynthesis, carbon partitioning and yield. Annu. Rev. *Plant Physiology* 34:485-509.

Giménez, A., 2006. Climate Change and Variability in the Mixed Crop/ Livestock Production Systems of the Argentinean, Brazilian and Uruguayan Pampas. Published by The International START Secretariat 2000 Florida Avenue, NW Washington, DC 20009 USA.

Giorgi, F., Bi, X., 2005. Regional changes in surface climate inter annual variability for the 21st century from ensembles of global model simulations. *Geophysical Research Letters* 32: L13701.

Gleick, P.H., 1999. Studies from the Water Sector of the National Assessment. *Journal of the American Water Resources Association* 35, 1429-1442.

Godwin, D.C., Jones, C.A., 1991. Nitrogen dynamics in soil plant systems. In: Hanks, J., Ritchie, J.T. (Eds.), Modelling plant and soil systems, *American Society of Agronomy* No. 31, pp. 297-302.

Godwin, D.C., Meyer, W.S., Singh, U., 1994. Simulation of the effect of chilling injury and nitrogen supply on floret fertility and yield in rice. *Australian Journal Experimental Agriculture* 34, 921-926.

Gommes, R.A., 1983. Pocket computers in agrometeorology. Plant production and Protection Paper n. 45.FAO, Rome, Italy.

Gonzáleza Lucila, Etchevers Jorge D., González Juan M., Fernando Paz, 2010. Soil organic carbon changes at the plot level in hillside systems. Agriculture, Ecosystems and Environment, 139, 508-515.

Gordon, C., Cooper, C., Senior, C.A., Banks, H., Gregory, J.M., Johns, T.C., Mitchell, J.F.B., Wood, R.A., 2000. The simulation of SST, sea ice extents and ocean heat transports in a version of the Hadley Centre coupled model without flux adjustments. *Climate Dynamics* 16:147-168.

Gorissen, A., van Ginkel, J.H., Keurentjes, J.J.B., van Veen, J.A., 1995. Grass root decomposition in retarded when grass has been grown under elevated CO2. *Soil Biology and Biochemistry* 27, 117-120.

Goss, D.W., Sanabria, J., Kellogg, R.L., Berc, J.L., 2001. A national assessment of soil carbon sequestration on cropland: description of an analytical approach. In: Lal R, editor. Soil carbon sequestration and the greenhouse effect. Special Publication, vol. 57. Madison, WI: *Soil Science Society of America;* 2001.

Goswami, B.N., Venugopal, V., Sengupta, D., Madhusoodanan, M.S., Prince, K., Xavier, 2006. Increasing Trend of Extreme Rain Events over India in a Warming Environment. *Science,* 314. 5804, 1442 - 1445.

Grace, J.A., Amin, N., Singh, N.C., Theunissen, F.E., 2003. Selectivity for conspecific song in the zebra finch auditory forebrain. *Journal of Neurophysiology* 89: 472-487.

Gregorich, E.G., Greer, K.J., Anderson, D.W., Liang, B.C., 1998. Carbon distribution and losses: erosion and deposition effects. *Soil & Tillage Research* 47 (3- 4) 291-302.

Gregory, P.J., Ingram, J.S.I., 2000. Global environmental change and future crop production. *Agriculture, Ecosystems & Environment* 82: 3-14.

Gregory, P.J., Ingram, J.S.I., Brklacich, M., 2005. Climate change and food security. Philosophical Transactions of the Royal Society; B 360:2139-2148.

Grenz, J.H., Manschadi, A.M., Voil, P., Meinke, H., Sauerborn, J., 2006. Simulating crop-parasitic weed interactions using APSIM: model evaluation and application. *European Journal of Agronomy* 24, 257-267.

Griffiths, B.S., Ritz, K., Bardgett, R.D., Cook, R., Christensen, S., Ekelund, F., Sørensen, S.J., Bååth, E., Bloem, J., de Ruiter, P.C., Dolfing, J., Nicolardot, B., 2000. Ecosystem response of pasture communities to fumigation-induced microbial diversity reductions: an examination of the biodiversity-ecosystem function relationship. *Oikos* 90, 279-294.

Guerena, A., Ruiz-Ramos, M., Díaz-Ambrona, CH., Conde, J., Mínguez, M.I., 2001. Assessment of climate change and agriculture in Spain using climate models. *Agronomy Journal* 93:237-249.

Guhathkurta, P., Rajeevan, M., 2006. Trends in the rainfall pattern over India. National Climate Centre, Report No.2, India Met Department, Pune.

Guo, R., Lin, Z., Mo, X., Yang, C., 2010. Responses of crop yield and water use efficiency to climate change in the North China Plain. *Agricultural Water Management* 97 (2010) 1185-1194.

Haase, S., Neumann, G., Kania, A., Kuzyakov, Y., Romheld, V., Kandeler, E., 2007. Elevation of atmospheric CO_2 and N-nutritional status modify nodulation, nodule-carbon supply, and root exudation of *Phaseolus vulgaris* L. *Soil Biology and Biochemistry* 39, 2208-2221.

Hadley-Centre, 2006. Effects of climate change in the developing countries. UK Meteorological Office.

Haefner, J.W., 2005. Modelling Biological Systems: Principles and Applications, Springer.

Haile, M., 2005. Weather patterns, food security and humanitarian response in sub-Saharan Africa. Philosophical Transactions of the Royal Society B: Biological Sciences; 360:2169-2182.

Hall-Beyer, M., 2003. Comparison of single-year and multiyear NDVI time series principal components in cold temperate biomes. *IEEE Transactions on Geoscience and Remote Sensing* 41, 2568-2574.

Hamar, D., Ferencz, C., Lichtenberg, J., Tarcsai, G., Frencz-A' rkos, I., 1996. Yield estimation for corn and wheat in the Hungarian Great Plain using Landsat MSS data. *International Journal of Remote Sensing* 17, 1689-1699.

Hara, M., Okada, S., Yagi, H., Moriyama, T., Shigehara, K., Sugimori, Y., 2003. Developing and evaluation of the noise reduction filter for the time-series satellite image. *Journal of The Japan Society of Photogrammetry and Remote Sensing* 5, 48- 59.

Harden, J.W., Fries, T.L., Pavich, M.J., 2002. Cycling of beryllium and carbon through hillslope soils in Iowa. *Biogeochemistry* 60: 317-335.

Harden, J.W., Sharpe, J.M., Parton, W.J., Ojima, D.S., Fries, T.L., Huntington, T.G., Dabney, S.M., 1999. Dynamic replacement and loss of soil carbon on eroding cropland. *Global Biogeochemical Cycles* 13: 885-901.

Haris, A.A., Biswas, S., Chhabra, V., Elanchezhian, R., Bhatt, B.P., 2013. Impact of climate change on wheat and winter maize over a sub-humid climatic environment. *Current Science* 104 (2), 206-214.

Harrison, P.A., Butterfield, R.E., 1996. Effect of climate change on Europe-wide winter wheat and sunflower productivity. *Climate Research* 7, 225-241.

Hatfield, J.L., Boote, K.J., Kimball, B.A., Izaurralde, R.C., Ort, D., Thomson, A., Wolfe, D.W., 2011. Climate impacts on agriculture: implications for crop production. *Agronomy Journal* 103:351-370.

Hawkes, C.V., Hartley, I.P., Ineson, P., Fitter, A.H., 2008. Soil temperature affects carbon allocation within arbuscular mycorrhizal networks and carbon transport from plant to fungus. *Global Change Biology* 14, 1181-1190.

Hayhoe, K., Wake, C., Huntington, T.G., Luo, L., Schwartz, M.D., Sheffield, J., Wood, E.F., Anderson, B., Bradbury, J., DeGaetano, T.T., Wolfe, D., 2007. Past and future changes in climate and hydrological indicators in the U.S. Northeast. *Climate Dynamics* 28: 381-407.

Hebbar, K.B., Venugopalan, M.V., Seshasai, M.V.R., Rao, K.V., Patil, B.C., Prakash, A.H., Kumar, V., Hebbar, K.R., Jeyakumar, P., Bandhopadhyay, K.K., Rao, M.R.K., Khadi, B.M., Aggarwal, P.K., 2008. Predicting cotton production using InfoCrop-cotton simulation model, remote sensing and spatial agro-climatic data. *Current Science* 95(11), 1570-1579.

Heinemeyer, A., Ineson, P., Ostle, N., Fitter, A.H., 2006. Respiration of the external mycelium in the arbuscular mycorrhizal symbiosis shows strong dependence on recent photosynthates and acclimation to temperature. *New Phytologist* 171, 159-170.

Helsel, D.R., Hirsch, R.M., 2002. Statistical Methods in Water Resources. Chapter A3. Book 4, Hydrologic Analysis and Interpretation. Techniques of Water-Resources Investigations of the United States Geological Survey.

Hobbs, P. R., Govaerts, B., 2010. How conservation agriculture can contribute to buffering climate change. In Climate Change and Crop Production, edited by M. P. Reynolds, 151-176. CABI Series in Climate Change. New York, NY: CABI International.

Hobbs, P.R., Gupta, R.K., 2004. Problems and challenges of no-till farming for the rice-wheat systems of the Indo- Hobbs P.R. and Gupta R.K. (2003) Gangetic Plains in South Asia. In Sustainable Agriculture and the Rice-Wheat System Eds. Lal R, Hobbs P, Uphoff N, Hansen D.O Paper 6. pp. 101-119. Columbus, OH; New York: NY: Ohio State University; Marcel Dekker, Inc.

Hodge, A., 1996. Impact of elevated CO_2 on mycorrhizal associations and implications for plant growth' *Biology and Fertility of Soils* 23(4), 388-398.

Hodges, T., Botner, D., Sakamoto, C., Haug, J.H., 1987. Using the CERES-MAIZE model to estimate production for the U.S. Cornbelt. Agricultural and Forest Meteorology, 40: 293-303.

Holland, J.M., 2004. The environmental consequences of adopting conservation tillage in Europe: reviewing the evidence. *Agriculture, Ecosystems & Environment* 103. 1-25. 2004.

Hoosbeek, M.R., Vos, J.M., Meinders, M.B.J., Velthorst, E.J., Scarascia-Mugnozza, G.E. 2007. Free atmospheric CO_2 enrichment (FACE) increased respiration and humification in the mineral soil of a poplar plantation. *Geoderma* 138, 204-212.

Houlès, V., Guérif, M., Mary, B., 2007. Elaboration of a nitrogen nutrition indicator for winter wheat based on leaf area index and chlorophyll content for making nitrogen recommendations. *European Journal of Agronomy* 27, 1-11.

Howden, S.M., Soussana, J.F., Tubiello, F.N., Chhetri, N., Dunlop, M., Meinke, H., 2007. Adapting agriculture to climate change. *Proceedings of the National Academy of Sciences USA* 104: 19691-19696.

http://cliveg.bu.edu/modismisr/index.html. The GEOV2 LAI/.

http://earthexplorer.usgs.gov.

http://unctad.org/en/

http://www.geoland2.eu/portal/.

http://www.lenntech.com/greenhouse-effect/global-warming-history.htm

Huete, A., Didan, K., Miura, T., Rodriguez, E.P., Gao, X., Ferreira, L.G., 2002. Overview of the radiometric and biophysical performance of the MODIS vegetation indices. *Remote Sensing of Environment* 83, 195- 213.

Huhne, C., Slingo, J., 2012. Climate: Observations, projections and impacts - India. Met Office - Hadley Centre and Leeds University, UK. Pp. 1-163.

Hundal, S.S., Kaur, P., 1997. Application of the CERES-Wheat model to yield predictions in the irrigated plains of the Indian Punjab. *Journal of Agricultural Science* 129, 13-18.

Hungate, B.A., Dijkstra, P., Johnson, D.W., Hinkle, C.R., Drake, and B.G. 1999. Elevated CO_2 increases nitrogen fixation and decreases soil nitrogen mineralization in Florida scrub oak. *Global Change Biology* 5, 781-789.

Hungate, B.A., Dukes, J.S., Shaw, M.R., Luo, Y., Field, C.B., 1999. Atmospheric science: nitrogen and climate change. *Science* 302, 1512-1513.

Hussaini, M. S., Halilu, A., 2013. Evaluation of DSSAT Crop Model for the Prediction of Irrigated Pearl Millet Yield. *International Journal of Engineering Sciences,* 2(9) 450-453.

Hyvonen, R., Hyvönen, R., Ågren, G.I., Linder, S., Persson, T., Cotrufo, M.F., Ekblad, A., Freeman, M., Grelle, A., Janssens, I.A., Jarvis, P.G., Kellomäki, S., Lindroth, A., Loustau, D., Lundmark, T., Norby, R.J., Oren, R., Pilegaard, K., Ryan, M.G., Sigurdsson, B.D., Strömgren, M., Oijen, M.V., Wallin, G., 2007. The likely impact of elevated CO_2, nitrogen deposition, increased temperature and management on carbon sequestration in temperate and boreal forest ecosystems: a literature review. *New Phytologist* 173, 463-480.

I.B.S.N.A.T., 1988. Experimental design and data collection procedures for IBSNAT. Tech. Rep. 1. 3rd ed. Dept of Agronomy and Soil Science, College of Tropical Agriculture and Human Resources. University of Hawaii, Honolulu.

I.B.S.N.A.T., 1990. Documentation for IBSNAT crop model input and output files. Version 1.1 for the Decision Support System for Agrotechnology Transfer (DSSAT V2.1). IBSNAT Tech. Rep. 5. Dept of Agronomy and Soil Science, College of Tropical Agriculture and Human Resources. University of Hawaii, Honolulu.

I.R.R.I., 2002. Rice Almanac: Source Book for the Most Important Economic Activity on Earth, Ed 3. CABI Publishing, Oxon, UK.

Ines, A.V.M., Honda, K., Gupta, A.D., Droogers, P., Clemente, R.S., 2006. Combining remote sensing-simulation modelling and genetic algorithm optimization to explore water management options in irrigated agriculture. *Agricultural Water Management* 83(3): 221-232. http://dx.doi.org/10.1016/j.agwat.2005.12.006.

Ines, A.V.M., Narendra, N., Das, James, W., Hansen, Eni, G., Njoku, 2013. Assimilation of remotely sensed soil moisture and vegetation with a crop simulation model for maize yield prediction. *Remote Sensing of Environment* 138 () 149-164.

Inman-Bamber, N.G., Culverwell, T.L., Mcglinchey, M.G., 1993. Predicting yield responses to irrigation of sugarcane from a growth model and field records. *Proceedings of the South African Sugar Technologists Association* 67: 66-72.

IPCC, 2000. Special Report on Emission Scenarios, A Special Report of Working Group III of the Intergovernmental Panel on Climate Change, Cambridge University Press, Cambridge.

IPCC, 2001. Africa. In: J.J. McCarthy, O. Canziani, N.A. Leary, D.J. Dokken and K.S. White (editors), Climate Change 2001: Impacts, Adaptation and Vulnerability. Contribution of Working Group II to the Third Assessment Report of the Intergovernmental Panel on Climate Change. Cambridge University Press, Cambridge, Chapter 10.

IPCC, 2001. Climate Change 2001: synthesis report. R. T. Watson, and the Core Writing Team, editors. Cambridge University Press, Cambridge, UK.

IPCC, 2007. Climate Change 2007: The Physical Science Basis. Contribution of Working Group II to the Fourth Assessment Report of the Intergovernmental Panel on Climate Change, [Parry, L. *et al.* (eds)], Cambridge University Press, Cambridge, United Kingdom and New York, NY, USA, p. 976.

IPCC, 2007. Climate Change 2007: The Physical Science Basis. Contribution of Working Group I to the Fourth Assessment Report of the Intergovernmental Panel on Climate Change. Cambridge, UK: Cambridge University Press.

IPCC, 2007a. Climate Change 2007: Climate Change Impacts, Adaptation and Vulnerability. Contribution of Working Group II to the Fourth Assessment Report of the Intergovernmental Panel on Climate Change, Cambridge University Press, Cambridge.

IPCC, 2007a. Summary for policy makers. In: Climate Change 2007: The physical science basis. Contribution of Working Group I to the Fourth Assessment Report of the Intergovernmental Panel on Climate Change, Eds., Solomon, S., Qin, D., Manning, M., Chen, Z.J., Marquis, M., Averyt, K.B., Tignor M. and Miller, H.L., pp. 1-18. Cambridge University Press, Cambridge, UK and New York, NY, USA.

IPCC, 2007b. Climate Change 2007: Mitigation of Climate Change Contribution of Working Group III to the Fourth Assessment Report of the Intergovernmental Panel on Climate Change, Cambridge University Press, Cambridge.

IPCC, 2007b. Climate Change 2007: The Physical Science Basis. Contribution of Working Group I to the Fourth Assessment Report of the Intergovernmental Panel on Climate Change [Solomon, S., Qin D., Manning M., Chen Z., Marquis M., Averyt K.B., Tignor M. and Miller H.L. (eds.)]. Cambridge University Press, Cambridge, United Kingdom and New York, NY, USA, pp. 996.

IPCC, 2007b. Summary for Policymakers. In: Climate Change 2007: Impacts, Adaptation and Vulnerability. Contribution of Working Group II to the Fourth Assessment Report of the Intergovernmental Panel on Climate Change, Eds., M.L. Parry, O.F. Canziani, J.P. Palutikof, P.J. van der Linden and C.E. Hanson, Cambridge University Press, Cambridge, UK, 7-22.

IPCC, 2007b. Working Group II Fourth Assessment Report. Cambridge University Press, Cambridge.

IPCC, 201.1 IPCC Special Report on Renewable Energy Sources and Climate Change Mitigation. Prepared by Working Group III of the Intergovernmental Panel on Climate Change [O. Edenhofer, R. Pichs-Madruga, Y. Sokona, K. Seyboth, P. Matschoss, S. Kadner, T. Zwickel, P. Eickemeier, G. Hansen, S. Schlömer, C. von Stechow (eds)]. Cambridge University Press, Cambridge, United Kingdom and New York, NY, USA, 1075 pp.

IPCC, 2013. Climate Change 2013: The Physical Science Basis, Summary for Policymakers, Working Group I Contribution to the IPCC Fifth Assessment Report pp. 1-36. http://www.climatechange2013.org/ images/ uploads/WGIAR5-PM.

IPCC, 2013. Climate Change 2013: The Physical Science Basis. Contribution of Working Group I to the Fifth Assessment Report of the Intergovernmental Panel on Climate Change [Stocker, T.F., Qin, D., Plattner, G.K., Tignor, M., Allen, S.K., Boschung, J., Nauels, A., Xia, Y., Bex, V. and Midgley, P.M. (eds.)]. Cambridge University Press, Cambridge, United Kingdom and New York, NY, USA.

Izaurralde, R.C., McGill, W.B., Bryden, A., Graham, S., Ward, M., Dickey, P., 1998. Scientific challenges in developing a plan to predict and verify carbon storage in Canadian Prairie soils. In: Lal R, Kimble J, Follett RF, Stewart BA, editors. Management of carbon sequestration in soil. Boca Raton, FL: CRC Press; p. 433- 46.

Jacinthe, P.A., Lal, R., 2001. A mass balance approach to assess carbon dioxide evolution during erosional events. *Land Degradation and Development* 12: 329-339.

Jackson, R.D., 1982. Canopy temperature and crop water stress. Adv. Irrig. 1, 43-85.

Jagtap, S.S., Abamu, F.J., Kling, J.G., 1999. Long-term assessment of nitrogen and variety technologies on attainable maize yields in Nigeria using CERES-maize. *Agricultural Systems* 60, 77-86.

Jakubauskas, M., Legates, D., Kastens, J., 2002. Crop identification using harmonic analysis of time-series AVHRR NDVI data. *Computers and Electronics in Agriculture* 37, 127- 139.

Jalota, S.K., Kaur, H., Ray, S.S., Tripathi, R., Vashisht, B.B., Bal, S.K., 2012. Mitigating future climate change effects by shifting planting dates of crops in rice-wheat cropping system. *Regional Environmental Change* 12:913-922, DOI 10.1007/s10113-012-0300-y.

Jalota, S.K., Singh, K B., Chahal, G.B.S., Gupta, R.K., Chakraborty,S., Sood, A., Ray, S.S., Panigrahy, S., 2009. Integrated effect of transplanting date, cultivar and irrigation on yield, water saving and water productivity of rice (*Oryza sativa* L.) in Indian Punjab: Field and simulation study. *Agricultural Water Management* 96: 1096-1104.

Jalota, S.K., Sood, A., Chahal, G.B.S., Choudhury, B.U., 2006. Crop water productivity of cotton (*Gossypium hirsutum* L.) - wheat (*Triticum aestivum* L.) system as influenced by deficit irrigation, soil texture and precipitation. *Agricultural Water Management* 84, 137-146.

Jalota, S.K., Sukhvinder Singh, Chalal, G.B.S., Ray, S.S., Panigraghy, S., 2010. Soil texture, climate and management effects on plant growth, grain yield and water use by rainfed maize-wheat cropping system: Field and simulation study. *Agricultural Water Management* 97: 83-90.

Jansen, E., Overpeck, K.R., Briffa, J.C., Duplessy, F., Joos, V., Masson-Delmotte, D., Olago, B., Otto-Bliesner, W.R., Peltier, S., Rahmstorf, R., Ramesh, D., Raynaud, D., Rind, O., Villalba, R., Zhang, D., 2007. Paleoclimate. In: Climate Change 2007: The Physical Science Basis. Contribution of Working Group I to the Fourth Assessment Report of the Intergovernmental Panel on Climate Change (eds., Solomon, S., D. Qin, M. Manning, Z. Chen, M. Marquis, K.B. Averyt, M. Tignor and H.L. Miller). Cambridge University Press, Cambridge, United Kingdom and New York, NY, USA.

Jenkinson, D.S., Coleman, K., 1994. Calculating the annual input of organic matter to soil from measurements of total organic carbon and radiocarbon. European Journal of Soil Science, 45, 167-174.

Jennifer, Y., King, S.R., Tscherko, D., 2006. Response of soil microbial biomass and enzyme activities to the transient elevation of carbon dioxide in a semi-arid grassland. *Soil Biology & Biochemistry* 38: 2448-2460.

Jensen, L.S., Mueller, T., Nielsen, N.E., Hansen, S., Crocker, G.J., Grace, P.R., Klir, J., Korschens, M., Poulton, P.R., 1997. Simulating trends in soil organic carbon in long-term experiments using the soil-plant-atmosphere model daisy. *Geoderma* 81, 5-28.

Jeong, K.W., Jang, T.I., Jeong, H.S., Park, S.W., 2014. Assessment of growth and yield components of rice irrigated with reclaimed wastewater. *Agricultural Water Management* 138, 17-25.

Jiang, Q., 2009. An agro-economic model to analyse climate change impacts on farmers' income. In: Australian Agricultural & Resource Economics Society's Annual Conference" pp.1-15. Cairns, Australia.

Johns, T.C., Gregory, J.M., Ingram, W.J., Johnson, C.E., Jones, A., Lowe, J.A., Mitchell, J.F.B., Roberts, D.L., Sexton, D.M.H., Stevenson, D.S., Tett, S.F.B., Woodage, M.J., 2003. Anthropogenic climate change for 1860-2100 simulated with the HadCM3 model under up-dated emissions scenarios. *Climate Dynamics* 20: 583-612.

Johnson, D.W., Hungate, B.A., Dijkstra, P., Hymus, G., Drake, B.G., 2001. Effects of elevated carbon dioxide on soils in a Florida scrub oak ecosystem. *Journal of Environmental Quality* 30, 501-507.

Jones, C.A., Kiniry, J.R., 1986. CERES-Maize: A simulation model of maize growth and development.Texas A&M University Press, College Station, TX.

Jones, H.G., Serraj, R., Loveys, B.R., Xiong, L., Wheaton, A., Price, A.H., 2009. Thermal infrared imaging of crop canopies for the remote diagnosis and quantification of plant responses to water stress in the field. *Functional Plant Biology* 36, 978-989.

Jones, J.W., 1993. Decision support systems for agricultural development. In: Penning de Vries, F., Teng, P., Metselaar, K. (Eds.), Systems Approaches for Agricultural Development. Kluwer Academic Press, Boston, pp. 459-471.

Jones, J.W., Hoogenboom, G., Porter, C.H., Boote, K.J., Batchelor, W.D., Hunt, L.A., Wilkens, P.W., Singh, U., Gijsman, A.J., Ritchie, J.T., 2003. The DSSAT cropping system model. *European Journal of Agronomy* 18:235-265.

Jones, P.G., Thornton, P.K., 2003. The potential impacts of climate change on maize production in Africa and Latin America in 2055. *Global Environmental Change* 13, 51-59.

Jongen, M., Jones, M.B., Hebeisen, T., Blum, H., Hendrey, G.R., 1995. The effects of elevated CO_2 concentrations on the root growth of *Lolium perenne* and *Trifolium repens* grown in a FACE system. *Global Change Biology* 1, 361-371.

Joutei, A.B., Roy, J., Van Impe, G., Lebrun, P., 2000. Effect of elevated CO_2 on the demography of a leaf-sucking mite feeding on bean. *Oecologia* 123: 75-81.

Kamel, A., Schroeder, K., Sticklen, J., Rafea, A., Salah, A., Schulthess, U., Ward, R., Ritchie, J.T., 1995. Integrated wheat crop management based on generic task knowledge based systems and CERES numerical-simulation. AI Applications 9 (1), 17-28.

Kamoni, P.T., Gicheru, P.T., Wokabi, S.M., Easter, M., Milne, E., Coleman, K., Falloon, P., Paustian, K., Killian, K., Kihanda, F.M., 2007. Evaluation of two soil carbon models using two Kenyan long term experimental datasets. *Agriculture, Ecosystems and Environment*, 122, 95-104.

Kandeler, E., Mosier, A., Morgan, J., Milchunas, D., King, J., Rudolph, S., Tscherko, D., 2006. Response of soil microbial biomass and enzyme activities to transient elevation of carbon dioxide in a semi-arid grassland. *Soil Biology & Biochemistry* 38, 2448- 2460.

Kang, S., Running, S. W., Lim, J., Zhao, M., Park, C., & Loehman, R. (2003). A regional phenology model for detecting onset of greenness in temperate mixed forests, Korea: An application of MODIS leaf area index. *Remote Sensing of Environment* 86, 232- 242.

Karberg, N.J., Pregitzer, K.S., King, J.S., Friend, A.L., Wood. J.R. 2005. Soil carbon dioxide partial pressure and dissolved inorganic carbonate chemistry under elevated carbon dioxide and ozone. *Oecologia* 142, 296-306.

Kattenburg, A., F. Giorgi, H. Grassl, G.A. Meehl, J.B.F. Mitchell, R.J. Stouffer, T. Tokioka, A.J. Weaver, and T.M.L. Wigley. 1995. Climate models-projections of future climate, p. 290-349. In J.T. Houghton, L.G. Meira Fiho, B.A. Callander, N. Harris, A. Kattenburg, and K, Maskell (eds.), Intergovernmental Panel on Climate Change. Cambridge University Press, New York.

Katz, R.W., Brown, B.G., 1992. Extreme events in a changing climate: variability is more important than averages. *Climate Change* 21:289-302.

Kaur, P., Hundal, S.S., 1999. Forecasting growth and yield of groundnut (*Arachis hypogaea*) with a dynamic simulation model 'PNUTGRO' under Punjab conditions. *Journal of Agricultural Science* 133, 167-173.

Kay, B.D., 1998. Soil structure and organic carbon: a review, pp. 169-198 in R. Lal, J. M. Kimble, R. F. Follett and B. A. Stewart (eds.), Soil Processes and the Carbon Cycle, CRC Press, Boca Raton, FL.

Keating, B.A., Carberry, P.S., Hammer, G.L., Probert, M.E., Robertson, M.J., Holzworth, D., Huth, N.I., Hargreaves, J.N.G., Meinke, H., Hochman, Z., Mclean, G., Verburg, K., Snow, V., Dimes, J.P., Silburn, M., Wang, E., Brown, S., Bristow, K.L., Asseng, S., Chapman, S., Mccown, R.L., Freebairn, D.M., Smith, C.J., 2003. An overview of APSIM, a model designed for farming systems simulation. *European Journal of Agronomy* 18, 267-288.

Keeling, C.D., Whorf, T. P., 2004. Atmospheric carbon dioxide record from SIO air sampling network. In Trends: A compendium of data on global change. Carbon dioxide Information Analysis Center, Oak Ridge National Laboratory, U. S. Department of energy, Oak Ridge, Tenn., U. S. A.

Keeling, C.D., Whorf, T.P., 2005. Atmospheric CO2 records from sites in the SIO air sampling network. In: Trends: A Compendium of Data on Global Change. Carbon Dioxide Information Analysis Center, Oak Ridge National Laboratory, U.S. Department of Energy, Oak Ridge, TN, http://cdiac.esd.ornl.gov/trends/co2/sio-keel-fl ask/sio-keel-fl ask.html.

Keeling, R.F., Piper, S.C., Bollenbacher, A.F., Walker, J.S., 2009. Atmospheric CO_2 records from sites in the SIO air sampling network. In Trends: A Compendium of Data on Global Change (Oak Ridge, TN:

Carbon Dioxide Information Analysis Center, Oak Ridge National Laboratory, U.S. Department of Energy).

Kendall, M.G., 1995. "Rank Correlation Methods" Charles Griffin, London.

Kimball, B.A., Kobayashi, K., Bindi, M., 2002. Responses of agricultural crops to free-air CO_2 enrichment. *Advances in Agronomy* 77: 293-368.

Kimble, J.M., Lal, R., Mausbach, M., 2001. Erosion effect on soil organic carbon pool in soils of Iowa. In: Stott DE, Mohtar RH, Steinhardt GC, editors. Sustaining the global farm. The 10th International Soil Conservation Organization Meeting, May 24- 29, 1999, West Lafayette, IN, USA. USDA-ARS; p. 472-5.

Kinnell, P.I.A., 2004. Raindrop-impact-induced erosion processes and prediction: a review. *Hydrological Processes* 19, 2815-2844.

Knight, J., Harrison, S., 2013. The impacts of climate change on terrestrial Earth surface systems. *Nature Climate Change* 3: 24-29.

Knutson, T.R., Delworth, T.L., Dixon, K.W., Stouffer, R.J., 1999. Model assessment of regional surface temperature trends (1949-97). *Journal of Geophysical Research* 104, 30981-30996.

Kovacs, G.T., Ne´meth, T., 1995. Testing simulation models for the assessment of crop production and nitrate leaching in Hungary. *Agricultural Systems* 49, 385-397.

Kripalani, R., Oh, J., Kulkarni, A., Sabade, S., Chaudhari, H., 2007. South Asian summer monsoon precipitation variability: Coupled climate model simulations and projections under IPCC AR4. Theory and Application. *Climatolalogy* 90, 133-159.

Krishnakumar, S.K., Patwardhan, A., Kulkarni, K., Kamala, K., Rao, K., Jones, R., 2011. Simulated projections for summer monsoon climate over India by a high-resolution regional climate model (PRECIS). *Current Science* 101, 312-325.

Krishnamurthy.V. Ajayamohan, R.S., 2010. Composite Structure of Monsoon Low Pressure Systems and Its Relation to Indian Rainfall, *Journal of Climate* 23, DOI: 10.1175/2010JCLI2953.1.

Krishnan, P., Swain, D.K., Chandra Bhaskar, B., Nayak, S.K., Dash, R.N., 2007. Impact of elevated CO_2 and temperature on rice yield and methods of adaptation as evaluated by crop simulation studies. *Agriculture, Ecosystems & Environment* 122: 233-242.

Kropff, M.J., Teng, P.S., Aggarwal, P.K., Bouman, B., Bouma, J., van Laar, H.H., (Eds.), 1996. Applications of Systems Approaches at the Field Level, vol. 2. Kluwer Academic Publishers, The Netherlands, p. 465.

Kucharik, C.J., Serbin, S.P., 2008. Impacts of recent climate change on Wisconsin corn and soybean yield trends. *Environmental Research Letters* 3: 034003.

Kulkarni, A., 2012. Weakening of Indian summer monsoon rainfall in warming environment. *Theoretical and Applied Climatology.*

Kumar M., 2011. Evidences, Projections and Potential Impacts of Climate Change on Food Production in Northeast India. *Indian Journal of Hill Farming* 24(1&2): 1-10.

Kumar, P., Singhal, V.K., 2011b. Chromosome number, male meiosis and pollen fertility in selected angiosperms of the cold deserts of Lahaul-Spiti and adjoining areas (Himachal Pradesh, India). *Plant Systematics and Evolution* 297: 271-297.

Kumar, R., Singh, R.D. and Sharma, K.D. 2005. Water Resources of India. *Current Science* 89(5): 794-811.

Kumar, S.N., Kasturibai, K.V., Rajagopal, V., Aggarwal, P.K., 2008. Simulating coconut growth, development and yield with the InfoCrop-coconut model. *Tree Physiology* 28, 1049-1058.

Kuusk, A., 1998. Monitoring of vegetation parameters on large areas by the inversion of a canopy reflectance model. *International Journal of Remote Sensing* 19: 2893-2905.

Lacointe, A., 2000. Carbon allocation among tree organs: A review of basic processes and representations in functional-structural tree models. *Annals of Forest Science* 57, 521-533.

Lactin, D. J., Johnson, D.L., 1998. Environmental, physical and behavioural determinants of body temperature in grasshopper nymphs (Orthoptera: Acrididae). *The Canadian Entomologist* 130: 551-577.

Lal, M., 2003. Global climate change: India's monsoon and its variability. *Journal of Environmental Study and Policy* 6: 1-34.

Lal, R., 1987. Managing the soils of sub-Saharan Africa. *Science* 236, 1069-1076.

Lal, R., 1995. Global soil erosion by water and carbon dynamics, in pp. 131-142, R. Lal, J. Kimble, E. Levine and B.A. Stewart (eds.), Soil Management and Greenhouse Effect, CRC/Lewis, Boca Raton, FL.

Lal, R., 1995. Global soil erosion by water and carbon dynamics. In: Lal R, Kimble JM, Levine E, Stewart BA, editors. Soils and global change. Boca Raton, FL: CRC Press, p. 131- 42.

Lal, R., 1997. `Residue management, conservation tillage and soil restoration for mitigating greenhouse effect by CO_2-enrichment', *Soil and Tillage Research* 43, 81-107.

Lal, R., 1998. Soil erosion impact on agronomic productivity and environment quality. *Critical Reviews in Plant Sciences* 17, 319-464.

Lal, R., 2001. Soil conservation for C-sequestration. Pages 459-465 in Steinhardt GC, ed. sustaining the Global Farm: Selected Papers from the 10th International Soil Conservation Organization Meeting held May 24-29, 1999 at Purdue University and the USDA-ARS National Soil Erosion Research Laboratory. (6 March 2007; http://topsoil.nserl.purdue. edu/ fpa dmin/isco99/pdf/ISCOdisc/SustainingTheGlobalFarm/K010-R%20Lal.pdf)

Lal, R., 2001. Soil erosion and carbon dynamics on grazing lands. In The potential of U.S. grazing lands to sequester carbon and mitigate the greenhouse effect, eds. Follett, R.F., Kimble, J.M., Lal, R., Boca Raton, Florida, CRC/Lewis Publishers, pp 231-247.

Lal, R., 2003. Soil erosion and the global carbon budget. *Environment International* 29: 437- 450.

Lal, R., 2004. Soil carbon sequestration in India. *Climatic Change* 65:277-296.

Lal, R., 2004. Soil carbon sequestration to mitigate climate change. *Geoderma* 123, 1 -22.

Lal, R., 2005. Enhancing crop yields in the developing countries through restoration of the soil organic carbon pool in agricultural lands. *Land Degradation and Development* 17, 197-209.

Lau, K. M., Tsay, S. C., Hsu, C., Chin, M., Ramanathan, V., Wu, G. X., Li, Z., Sikka, R., Holben, B., Lu, D., Chen, H., Tartari, G., Koudelova, P., Ma, Y., Huang, J., Taniguchi, K., Zhang, R., 2008. The Joint Aerosol-Monsoon Experiment: A new Challenge for monsoon climate research, *Bulletin of the American Meteorological Society* 89,369-383, doi: 10.1175/BAMS-89-3-369.

Lau, K., Kim, M., Kim, K., 2006. Asian summer monsoon anomalies induced by aerosol direct forcing: The role of the Tibetan Plateau. *Climate Dynamics* 26, 855-864.

Leakey, A.D.B., 2009. Rising atmospheric carbon dioxide concentration and the future of C4 crops for food and fuel. *Proceedings of the Royal Society B: Biological Sciences* 276: 2333-2343.

Leakey, A.D.B., Ainsworth, E.A., Bernacchi, C.J., Rogers, A., Long, S.P., Ort, D.R. 2009. Elevated CO_2 effects on plant carbon, nitrogen, and water relations; six important lessons from FACE. *Journal of Experimental Botany* 60, 2859-2876.

Leakey, A.D.B., Bernacchi, C.J., Dohleman, F.G., Ort, D.R., Long, S.P., 2004. Will photosynthesis of maize (*Zea mays*) in the US Corn Belt increase in future [CO_2] rich atmosphere? An analysis of diurnal courses of CO_2 uptake under free-air concentration enrichment. *Global Change Biology* 10: 951-962.

Leakey, A.D.B., Uribelarrea, M., Ainsworth, E.A., Naidu, S.L., Rogers, A., Ort, D.R., Long, S.P., 2006. Photosynthesis, productivity, and yield of maize are not affected by open-air elevation of CO_2 concentration in the absence of drought. *Plant Physiology* 140: 779-790.

Legates, D.R., McCabe, G.J., 1999. Evaluating the use of "goodness-of-fit" measures in hydrologic and hydroclimatic model validation. *Water Resources Research* 35: 233-241.

Lekkerkerk, L.J.A., van de Geijn, S.C., van Veen, J.A., 1990. Effects of elevated atmospheric CO_2-levels on the carbon economy of a soil planted with wheat. In A.F. Bouwman (ed.), Soils and the Greenhouse Effect, John Wiley and Sons, New York, pp. 423-429.

Lenka, S., Singh, A.K., 2011. Simulating interactive effect of irrigation and nitrogen on crop yield and water productivity in maize-wheat cropping system. *Current Science* 101(11):1451-1461.

Lewis, J.D., Thomas, R.B., Strain, B.R., 1994. Effect of elevated CO_2 on mycorrhizal colonizination of loblolly pine (*Pinus taeda* L.) seedlings. *Plant Soil* 165, 81-88.

Li, C. 2000. Modelling trace gas emissions from agricultural ecosystems. *Nutrient Cycling in Agroecosystems* 58:259-76.

Li, C. 2001 Conception and method of Biogeochemistry - Development of the DNDC model, *Quaternary Science (in Chinese)* 21(2), 89-99, 2001.

Li, C., 2000. Loss of soil carbon threatens Chinese agriculture: a comparison on agro-ecosystem carbon pool in China and the U.S. Quaternary *Sciences*, 20(4): 345-350.

Li, C., Frolking S and Harriss R. 1994. Modelling carbon biogeochemistry in agricultural soils. *Global Biogeochemical Cycles* 8(3):237-54.

Li, C., Zhuang, Y., Frolking, S., Galloway, J., Harriss, R., Moore III, B., Schimel, D., Wang, X., 2003. Modelling soil organic carbon change in croplands of China, Ecological Applications, 13(2), 327-336, 2003.

Li, Q. Y., Newbury-Ecob, R. A., Terrett, J. A., Wilson, D. I., Curtis, A. R., Yi, C. H., Gebuhr, T., Bullen, P. J., Robson, S. C., Strachan, T., Bonnet, D., Lyonnet, S., Young, I. D., Raeburn, J. A., Buckler, A. J., Law, D. J., Brook, J. D., 1997. Holt-Oram syndrome is caused by mutations in TBX5, a member of the Brachyury (T) gene family. *Nature Genetics* 15, 21-29.

Liu, H. L., Yang, J. Y., Tan, C. S., Drury, C. F., Reynolds, W. D., Zhang, T. Q., Bai, Y. L., Jin, J., He, P., Hoogenboom, G., 2011. Simulating water content, crop yield, and nitrate-N loss under free and controlled tile drainage with subsurface irrigation using the DSSAT model. *Agricultural Water Management* 98(6): 1105-1111.

Liwenga, E.T., 2008. Adaptive livelihood strategies for coping with water scarcity in the drylands of central Tanzania. *Physics and Chemistry of the Earth* 33, 775-779.

Lobell, D. B., Burke, M. B., 2010. On the use of statistical models to predict crop yield responses to climate change. Agric. For. Meteorol. 150:1443-1452.

Lobell, D.B., Asner, G.P., 2003. Climate and management contributions to recent trends in U.S. agricultural yields. *Science* 299: 1032.

Lobell, D.B., Asner, G.P., Ortiz-Monasterio, J.I., Benning, T.L., 2003. Remote sensing of regional crop production in the Yaqui Valley.

Lobell, D.B., Burke, M.B., 2010. On the use of statistical models to predict crop yield responses to climate change. Agricultural and Forest Meteorology, 150 (11): 1443-1452.

Loladze, I., 2002. Rising atmospheric CO_2 and human nutrition: toward globally imbalanced plant stoichiometry. *Trends in Ecology & Evolution* 17, 457-461.

Long, S. P., Ainsworth, E. A., Leakey, A.D.B., Nosberger, J ., Ort, D. R., 2006. Food for thought: Lower-than-expected crop yield stimulation with rising CO_2 concentrations. *Science* 312, 1918-1921.

Long, S.P., Ort, D.R., 2010. More than taking the heat: crops and global change. *Current Opinion in Plant Biology* 13: 1-8.

Loomis, R.S., NG, E., Hunt, W.F., 1976. Productivity of root crops. In Burris, R.H. and Black, C.C. eds. CO_2 metabolism and the productivity of plants. Univ. Park Press, Baltimore, MD. 269-286.

Ludwig, F., Milroy, S.P. Asseng, S., 2009. Impacts of recent climate change on wheat production systems in Western Australia. *Climatic Change* 92:495-517.

Luo, Q., Beellotti, W., Williams, M., Bryan, B., 2005. Potential impact of climate change on wheat yield in South Australia. *Agricultural and Forest Meteorology* 132: 273-285.

Luo, Y., Jackson, R.B., Field, C.B., Mooney, H.A. 1996. Elevated CO_2 increases belowground respiration in California grasslands. *Oecologia* 108:130-137.

Luo, Y., Ogle, K., Tucker, C., Fei, S., Gao, C., LaDeau, S., *et al.* 2011. Ecological forecasting and data assimilation in a data-richera. *Ecological Applications.* 21, 1429-1442. doi: 10.1890/09-1275.1 .

Luo, Y., Teng, P.S., Fabellar, N.G., Tebeest, D.O., 1997. A rice-leaf blast combined model for simulation of epidemics and yield loss. *Agricultural Systems* 53 (1), 27-39.

Luo, Z., Wang, E., Sun, O. J., 2010. Soil carbon change and its responses to agricultural practices in Australian agroecosystems: A review and synthesis. *Geoderma* 155, 211-223.

Lyons, W.B., Nezat, C.A., Carey, A.E., Hicks, D.M., 2002. Organic carbon fluxes to the ocean from high-standing islands. *Geology* 30: 443-446.

Ma, L., Ascough, J.C., Ahuja, L.R., Shaffer, M.J., Hanson, J.D., Rojas, K.W. 2000. Root zone water quality model sensitivity analysis using Monte Carlo simulation. *Transactions of the ASAE* 43 (4):883-895.

Ma, S., Zhang, S., Wang, F., Liu, Y., Liu, Y., *et al.* 2013. Highly Efficient and Specific Genome Editing in Silkworm Using Custom TALENs. *PLoS One* 7(9): e45035.

Maas, S.J., 1988. Use of remotely-sensed information in agricultural crop growth models. *Ecological Modelling* 41, 247-268.

MacCarthy, D.S., Diancoumba1, M., Freduah, B.S., Adiku, S.G.K., Agali, A., Hathie, I, Lizaso, J., Fatondji, D., Adams, M., Tigana, L., Koomson, E., Traore, P.C.S., Traore, S., Amikuzuno, J., Naab, J.B., Diarra, D.Z., Sarr, B., N'diaye, O., Sanon, M., 2013. An Integrated Assessment of Climate Change Impact on Crop Production in the Nioro du Rip Basin of Senegal I: Crop Modelling. *The Agricultural Model Intercomparison and Improvement Project,* UK aid.

Macdonald, R. Hall, F., 1980. Global crop forecasting. *Science* 208, pp. 670-679.

Mackay, A.D., Barber, S.A., 1985. Soil moisture effect on potassium uptake by corn. *Agronomy Journal* 77, 524-527.

MacRobert, J.F., Savage, M.J., 1998. The use of a crop simulation model for planning wheat irrigation in Zimbabwe. In: Tsuji, G.Y., Hoogenboom, G., Thornton, P.K. (Eds.), Understanding Options for Agricultural Production. Kluwer Academic Publishers, Dordrecht, the Netherlands, pp. 205-220.

Mall, M., Kreda, S.M., Mengos, A., Jensen, T.J., Hirtz, S., Seydewitz, H.H., Yankaskas, J., Kunzelmann, K., Riordan, J.R., Boucher, R.C. 2004. The Delta F508 mutation results in loss of CFTR. Function and mature protein in native human colon. *Gastroenterology* 126, 32-41.

Mall, R.K., Lal, M., Bhatia, V.S., Rathore, L.S., Singh, R., 2004. Centre for Systems Simulation, Indian Agricultural Research Institute, New Delhi 110012, India *Agricultural and Forest Meteorology* 121 (2004) 113-125.

Mandal, D., Patra, S., Sharma, N.K., Dogra, P., 2010. The threat of soil erosion to agricultural sustainability in north-western Himalayan region of India. In: National Seminar on Development in Soil Science. Indian Institute of Soil Science, Bhopal, November 14-17, 2010. (Abstract No. 181).

Mandal, D., Patra, S., Sharma, N.K., Dogra, P., 2012. Erosion productivity relationship for evaluating vulnerability and resilience of soils under different agro-climatic regions of India. Annual Report. CSWCRTI, Dehradun.

Manjunath, K.R., Potdar, M.B., 2002. Large area operational wheat yield model development and validation based on spectral and meteorological data. *International Journal of Remote Sensing* 23, 3023-3038.

Marin, F.R., Thorburn, P.J., Nassif, D.S.P., Costa, G., Rezend, R.S., Andrade Junior, S.S., 2013. Evaluation of APSIM-Sugar and DSSAT-Canegro for a widely grown Brazilian sugarcane cultivar. *Proceedings of International Society of Sugarcane Technology*, 28: (In Press).

Marschner, H., 1995. Mineral Nutrition of Higher Plants, 2nd ed. London, UK: Academic Press.

Maselli, F., Rembold, F., 2001. Analysis of GAC NDVI data for cropland identification and yield forecasting in Mediterranean African countries. Photogrammatic. *Engineering* and *Remote Sensing* 67, 593-602.

Maslin, M., 2004. Global Warming, a very short introduction. Oxford University Press, Oxford 2004.

Massey, H.F., Jackson, M.L., 1952. Selective erosion of soil fertility constituents. *Soil Science Society of America Journal* 16:353-6.

Mathison, C., Wiltshire, A., Dimri, A.P., Falloon, P., Jacob, D., Kumar, P., *et al.* 2013. Regional projections of North Indian climate for adaptation studies. *Science of the Total Environment* 468-469(Suppl.):S4-S17.

May, W., 2004. Simulation of the variability and extremes of daily rainfall during the Indian summer monsoon for present and future times in a global time slice experiment. *Climate Dynamics* 22: 183-204.

Mbabaliye, T., Wojtkowski, P.A., 1994. Problems and perspectives on the use of a crop simulation model in an African research station. *Experimental Agriculture* 30, 441-446.

McCarty, G.W., Ritchie, J.C., 2002. Impact of soil movement on carbon sequestration in agricultural ecosystems. *Environmental Pollution* 116(3):423- 30.

McCown, R.L., Hammer, G.L., Hargreaves, J.N.G., Holzworth, D.P., Freebairn, D.M., 1996. APSIM: A novel software system for model development, model testing, and simulation in agricultural systems research. *Agricultural Systems* 50:255-271.

McFarlane, N.A., Boer, G.J., Blanchet, J.P., Lazare, M., 1992. The Canadian Climate Centre second-generation general circulation model and its equilibrium climate. *Journal of Climate* 5:1013-1044.

McPhee, M.J., 2009. Mathematical modelling in agricultural systems: A case study of modelling fat deposition in beef cattle for research and industry. 18th World IMACS / MODSIM Congress, Cairns, Australia.

Meade, R.H., Yuzyk, T.R., Day, T.J., 1990. Movement and storage of sediment in rivers of the United States and Canada. Pages 255-280 in Wolman MG, Riggs HC. eds Surface Water Hydrology, vol. 1: The Geology of North America. Boulder (CO): *Geological Society of America*.

Mearns, L.O., Rosenzweig, C., Goldberg, R., 1996. 'The Effect of Changes in Daily and Interannual Climatic Variability on CERES-Wheat Yields. A Sensitivity Study', *Climatic Change* 32, 257-292.

Meehl, G., Arblaster, J., Collins, W., 2008. Effects of black carbon aerosols on the Indian monsoon. *Journal of Climatology* 21, 2869-2882.

Mendelsohn, R., Dinar, A., 1999. Climate change, agriculture, and developing countries: does adaptation matter? *World Bank Research Observer* 14, 277- 293.

Meyer-Roux, J., 1990. PreÂ sentation de Projet Pilote de TeÂ leÂ deÂ tection AppliqueÂ e aux Satistiques Agricoles. Proceedings of the conference on the application of remote sensing to agricultural statistics, Varese, Italy, 10- 11 October 1989, (EUR 12581 EN), pp. 25- 38.

Mikan, C.J., Zak, D.R., Kubiske, M.E., Pregitzer, K.S., 2000. Combined effects of atmospheric CO_2 and N availability on the belowground carbon and nitrogen dynamics of aspen mesocosms. *Oecologia* 124, 432-445.

Mirza, M.Q., 2002. Global warming and changes in the probability of occurrence of floods in Bangladesh and implications. *Global Environmental Change* 12: 127-138.

Mishra, U., Jastrow, J. D., Matamala, R., Hugelius, G., Koven, C. D., Harden, J. W., Ping, C.L., Michaelson, G. J., Fan, Z., Miller, R. M., McGuire, A. D., Tarnocai, C., Kuhry, P., Riley, W.J., Schaefer, K., Schuur, E.A.G.,

Jorgenson, M.T., Hinzman, L.D., 2013. Empirical estimates to reduce modelling uncertainties of soil organic carbon in permafrost regions: a review of recent progress and remaining challenges. *Environmental Research Letters* 8 (2013) 035020 (9 pp).

Mitchell, J.F.B., Johns, T.C., Senior, C.A., 1998. Transient response to increasing greenhouse gases using models with and without flux adjustment. Hadley Centre Technical Note 2. Available from Met. Office, London Road, Bracknell, RG12 2SZ, UK.

Mitchell, P.D., Lakshminarayan, P.G., Otake, T., Babcock, B.A., 1998. The impact of soil conservation policies on carbon sequestration in agricultural soils of the central United States, pp. 125-142 in R. Lal, J. M. Kimble, R. F. Follett and B. A. Stewart (eds.), Management of Carbon Sequestration in Soil, CRC Press, Boca Raton, FL.

Mo, X.G., Lin, Z.H., Liu, S.X., 2007. Climate change impacts on the ecohydrological processes in the Wuding River basin. *Acta Ecologica Sinica* 27 (12), 4999-5007.

Mohanty, M., Probert, M.E., Sammi Reddy, K., Dalal, R.C., Singh, M., Subba Rao, A., Menzies, N.W., 2011. Simulating soybean-wheat cropping system of Madhya Pradesh, central India: APSIM model testing a devaluation. *Agriculture, Ecosystems & Environment* 152, 68-78.

Molina, J.A.E., Smith, P., 1998. Modelling carbon and nitrogen processes in soils, advances in agronomy. In: Advances in Agronomy, vol. 62. San Diego: Academic Press; p. 253- 98.

Monastersky, R., 2013. Global carbon dioxide levels near worrisome milestone. *Nature* 497, 13-14 (02 May 2013) *Digital Object Identifier* 10.1038/497013a.

Monreal, C.M., Zentner, R.P., Robertson, J.A., 1997. An analysis of soil organic matter dynamics in relation to management, erosion and yield of wheat in long-term crop rotation plots. *Canadian Journal of Soil Science* 77(4):553- 63.

Monteith, J.L., 1965. Evaporation and environment. 19th Symposia of the Society for Experimental Biology, vol. 19. University Press, Cambridge, pp. 205-234.

Monteith, J.L., 1977. Climate and crop efficiency of crop production in Britain. *Philosophical Transactions of the Royal Society B: Biological Sciences* 281, 277-329.

Moors, E.J., Groot, A., Biemans, H., Terwisscha van Scheltinga, C., Siderius, C., Stoffel, M., *et al.* 2011. Adaptation to changing water resources in the Ganges basin, northern India. *Environmental Science & Policy*; 14:758-69.

Moran, M. S., Clarke, T. R., Inoue, Y., Vidal, A., 1994. Estimating crop water de® cit using the relation between surface-air temperature and spectral vegetation Index. *Remote Sensing of Environment* 49, 246- 263.

Moriasi, D.N., Arnold, J.G., Van Liew, M.W., Bingner, R.L., Harmel, R.D., Veith, T.L., 2007. Model evaluation guidelines for systematic quantification of accuracy in watershed simulations. *Transactions of the ASAE* 50(3): 885-900.

Moriondo, M., Giannakopoulos, C., Bindi, M., 2011. Climate change impact assessment: the role of climate extremes in crop yield simulation. *Climatic Change* 104:679-701.

Moriondo, M., Maselli, F., Bindi, M., 2007. A simple model of regional wheat yield based on NDVI data. *European Journal of Agronomy* 26, 266-274.

Moss, R.H., Edmonds, J.A., Hibbard, K.A., Manning, M.R., Rose, S.K., van Vuuren, D.P., *et.al.,* 2010. The next generation of scenarios for climate change research and assessment. *Nature* 463: 747-756.

Moulin, S., Bondeau, A., Dele'colle, R., 1998. Combining agricultural crop models and satellite observations from field to regional scales. *International Journal of Remote Sensing* 19, 1021-1036.

Muchow, R.C. Carberry, P.S., 1993b. Designing improved plant types for the semi-arid tropics: Agronomists' viewpoint. In Penning De Vries FWT. et al. (eds.). Systems Approaches for Agricultural Development. Vol 2. Proc. Int. Symp., Bangkok, Thailand. Kluwer Academic Press. 37-61.

Muchow, R.C., Carberry, P.S., 1993a. A simulation model of kenaf for assisting fibre industry planning in Northern Australia. V. Impact of different crop traits. *Australian Journal of Agricultural Research* 44: 713-730.

Munda, G.C., Das, A., Patel, D.P., 2009. Evaluation of transplanted and ratoon crop for double cropping of rice (*Oryza sativa* L.) under organic input management in mid altitude sub-tropical Meghalaya. *Current Science* 96(12):1626-1627.

Murdoch, P.S., Baron, J.S., Miller, T.L., 2000. Potential Effects of Climate Change on Surface- Water Quality in North America. *American Water Resources Association* 36, 347-366.

N.R.C., 2010. Advancing the Science of Climate Change. National Research Council. The National Academies Press, Washington, DC, USA.

Nakicenovic, N., Alcamo, J., Davis, G., de Vries, H.J.M., Fenhann, J., Gaffin, S., Gregory, K., Grübler, A., Jung, T.Y., Kram, T., La Rovere, E.L., Michaelis, L., Mori, S., Morita, T., Pepper, W., Pitcher, H., Price, L., Riahi, K., Roehrl, A., Rogner, H.H., Sankovski, A., Schlesinger, M.E., Shukla, P., Smith, S., Swart, R., van Rooijen, S., Victor, N., Dadi, Z., 2000. Special report on emissions scenarios, international panel on climate change. Cambridge University Press, Cambridge, UK

Narolia, G.P., Jajoria, D.K., Dotaniya, M.L., 2013. Role of Phosphorus, PSB and Zinc in Isabgol (*Plantago ovata* F.). LAP Lambert Academic Publishing, Germany, pp. 124. ISBN 978-3-659-32731-5.

Natarajan, A., Rajendra, H., Naidu, L.G.K., Raizada, A., Adhikari, R.N., Patil, S.L., Rajan, K., Dipak, S., 2010. Soil and plant nutrient loss during the recent floods in North Karnataka: Implications and ameliorative measures. *Current Science* 99, 1333-1340.

NATCOM, 2004. India's Initial National Communication to the United Nations Framework Convention on Climate Change. Ministry of Environment and Forests, Government of India, pp 392.

National Action Plan on Climate Change, 2008. Prime Minister's Council on Climate Change, Government of India, p 49.

Naylor, R.L., Battisti, D.S., Vimont, D.J., Falcon, W.P., Burke, M.B., 2007. Assessing risks of climate variability and climate change for Indonesian rice agriculture. *Proceedings of the National Academy of Sciences* 104(19): 7752.

Nearing, M.A., 2001. Potential changes in rainfall erosivity in the U.S. with climate change during the 21st century. *Journal of Soil and Water Conservation* 56(3):229-232.

Nelson, G. C., Rosegrant, M. W., Koo, J., Robertson, R., Sulser, T., Zhu, T., *et al.* 2009. Climate change: impact on agriculture and costs of adaptation. Food Policy Report. International Food Policy Research Institute, Washington DC.

Nemecek, T., Huguenin-Elie, O., Dubois, D., Gaillard, G., 2005. Ökobilanzierung von Anbausystemen im Schweizerischen Acker- und Futterbau. Schriftenreihe der FAL 58. FAL Reckenholz, Zürich.

Ngwira, A.R., Aune, J.B., Thierfelder, C., 2014. DSSAT modelling of conservation agriculture maize response to climate change in Malawi. *Soil & Tillage Research* 143 (2014) 85-94.

Niyogi, D., Chang, H., Chen, F., Gu, L., Kumar, A., Menon, S., Pielke Sr, R.A., 2007. Potential impacts of aerosol-land-atmosphere interaction on the Indian monsoonal rainfall characteristics, *Natural Hazards* 42, 345-359.

Norby, R.J., 1994. Issues and perspectives for investigating root responses to elevated atmospheric carbon dioxide. *Plant Soil* 165, 9-20.

Norby, R.J., Wullschleger, S.D., Gunderson, C.A., Johnson, D.W., Ceulemans, R., 1999. Tree responses to rising CO_2: implications for the future forest. *Plant Cell and Environment* 22:683-714.

Nord, E.A., Lynch, J.P., 2009. Plant phenology: a critical controller of soil resource acquisition. *Journal of Experimental Botany* 60, 1927-1937.

Novotny, V., Olem, H., 1994. Water quality, prevention, identification, and management of diffuse pollution. Van Nostrand Reinhold, New York, 1954 pp.

Ogle, S.M., Breidt, F.J., Easter, M., Williams, S., Paustian, K., 2007. An empirically-based approach for estimating uncertainty associated with modelling carbon sequestration in soils. *Ecological Modelling* 205:453-463.

Oleson, J., Porter, V., 2009. Deliverable 10: Adaptation and mitigation. PICCMAT project report. http://www.climatechangeintelligence. baastel. be/piccmat/index.php

Olioso, A., Inoue, Y., Ortega-Farias, S., Demarty, J., Wigneron, J.P., Braud, I., *et al.*, 2005. Future directions for advanced evapotranspiration modelling: assimilation of remote sensing data into crop simulation models and SVAT models. *Irrigation and Drainage Systems* 19, 377-412.

Ortiz, R., Sayre, K.D., Govaerts, B., Gupta, R., Subbarao, G.V., Ban, T., Hodson, D., Dixon, J.M., Ortiz-Monasterio, J.I., Reynolds, M., 2008. Climate change: can wheat beat the heat? *Agriculture, Ecosystems & Environment* 126: 46-58.

Ortiz-Monasterio, I., Wassmann, R., Govaerts, B., Hosen, Y., Katayanagi, N., Verhulst, N., 2010. Greenhouse gas mitigation in the main cereal systems: rice, wheat, and maize. In: Reynolds M, ed. Climate change and crop production. Wallingford, UK: CABI Publishing, 151-176.

Osborne, T., Rose, G., Wheeler, T., 2013. Variation in the global-scale impacts of climate change on crop productivity due to climate model uncertainty and adaptation. Agric. For. Meteorol. 170:183-194.

Owens, L.B., Malone, R.W., Hothem, D.L., Starr, G.C., Lal, R., 2002. Sediment carbon concentration and transport from small watersheds under various conservation tillage practices. *Soil and Tillage Research 67*(1):65 - 73.

Ozalkan, C., Sepetoglu, H.T., Daur, I., Sen, O.F., 2010. Relationship between some plant growth parameters and grain yield of chickpea (*Cicer arietinum* L.) during different growth stages. *Turkish Journal of Field Crops* 15, 79-83.

Pala, M., Matar, A., Mazid, A., 1996. Assessment of the effects of environmental factors on the response of wheat to fertilizer in on-farm trials in a Mediterranean type environment. *Experimental Agriculture* 32: 339-349.

Pala, M., Stöckle, C.O., Harris, H.C., 1996. Simulation of durum wheat (*Triticum turgidum* ssp. *durum*) growth under different water and nitrogen regimes in a Mediterranean environment using CropSyst. *Agricultural Systems* 51, 147-163.

Palis, R.G., Ghandiri, H., Rose, C.W., Saffigna, P.G., 1997. Soil erosion and nutrient loss: 3. Changes in the enrichment ratio of total nitrogen and organic carbon under rainfall detachment and entrainment. *Australian Journal of Soil Research* 35(4):891- 905.

Palmroth, S., Maier, C., McCarthy, H., Oishi, C., Kim, H.S., Johnsen, K., Katul, G., Oren, R., 2005. Contrasting responses to drought of forest floor CO_2 efflux in a loblolly pine plantation and a nearby oak-hickory forest. *Global Change Biology* 11, 421-434.

Pannkuk, C.D., Stockle, C.O., Papendick, R.I., 1997. Evaluating Cropsyst simulations of wheat management in a wheat-fallow region of the US Pacific Northwest. Washington State University, Pullman, WA, USA. USDA-ARS.

Pannkuk, C.D., Stöckle, C.O., Papendiek, R.I., 1998. Evaluating CropSyst simulations of wheat management in a wheat-fallow region of the US Pacific Northwest. *Agricultural Systems* 57, 121-134.

Park, J., Tateishi, R., Matsuoka, M., 1999. A proposal of the Temporal Window Operation (TWO) method to remove high-frequency noises in AVHRR NDVI time series data. *Journal of the Japan Society of Photogrammetry and Remote Sensing* 38, 36- 47.

Parry, M., Rosenzweig, C., Inglesias, A., Livermore, M., Gischer, G., 2004. Effects of climate change on global food production under SRES emissions and socio-economic scenarios. *Global Environmental Change* 14: 53-67.

Parry, M.L., Canziani, O.F., Palutikof, J.P., van der Linden, P.J., Hanson, C.E., Eds., 2007. Climate Change 2007: Impacts, Adaptation and Vulnerability. Contribution of Working Group II to the Fourth Assessment Report of the Intergovernmental Panel on Climate Change, Cambridge University Press, Cambridge, UK, 982pp.

Parton, W.J., Schimel, D.S., Cole, C.V., Ojima, D.S., 1987. Analysis of factors controlling soil organic matter levels in Great Plains grasslands. *Soil Science Society of America Journal* 51:1173-9.

Parton, W.J., Scurlock, J.M.O., Ojima, D.S., Gilmanov, T.G., Scholes, R.J., Schimel, D.S., Kirchner, T., Menaut, J.C., Seastedt, T., Moya, E.G., Kamnalrut, A., Kinyamario, J.I., 1993. Observations and modelling of biomass and soil organic matter dynamics for the grassland biome worldwide. *Global Biogeochemical Cycles* 7(4):785 - 809.

Parton, W.J., Scurlock, J.M.O., Ojima, D.S., Schimel, D.S., Hall, D.O., 1995. Impact of climate change on grassland production and soil carbon worldwide. *Global Change Biology* (1)13-22.

Passioura, J.B., 1996. Simulation models: science, snake oil, education, or engineering. *Agronomy Journal* 88: 690-694.

Patel, S.K., 2014. Impact of Climate Change under A2 Scenario on Potato Yield for Anand District, Gujarat Thematic Area: Climate Change and Agriculture. *Journal of Energy Research and Environmental Technology,* Volume 1, Number 1; November, 2014 pp. 6-10.

Paterson, E., Hall, J.M., Rattray, E.A.S., Griffiths, B.S., Ritz, K., Killham. K., 1997. Effect of elevated CO_2 on rhizosphere carbon flow and soil microbial processes. *Global Change Biology* 3, 363-377.

Pathak, H., Prasad, S., Bhatia, A., Singh, S., Kumar, S., Jain, M.C., Kumar, U., 2003. Emission of methane from rice wheat systems of Indo-Gangetic plains of India. *Agriculture, Ecosystems & Environment* 97, 309-316.

Patt, A., Suarez, P., Gwata, C., 2005. Effects of seasonal climate forecasts and participatory workshops among subsistence farmers in Zimbabwe. Proceedings of the National Academy of Sciences of the United States of America. 102: 12623-12628.

Paul, E.A., Horwath, W.R., Harris, D., Follett, R., Leavitt, S.W., Kimball, B.A., Pregitzer, K., 1995. Establishing the pool sizes and fuxes in CO2 emissions from soil organic matter turnover, pp. 297-306 in R. Lal, J. M. Kimble, E. Levine and B. A. Stewart (eds.), Soils and Global Change, CRC Press, Boca Raton, FL.

Paustian, K., 1994. Modelling soil biology and biochemical processes for sustainable agriculture research. In: Pankhurst ZE, Doube BM, Gupta VVSR, Grace PR, editors. Soil biota management in sustainable farming systems. Melbourne: CSIRO Information Services; p. 182-96.

Pearson, P.N., Palmer, M.R., 2000. Atmospheric carbon dioxide concentrations over the past 60 million years. *Nature* 406: 695-699.

Pendall, E., Leavitt, S.W., Brooks, T., Kimball, B.A., Pinter, P.J., Wall, G.W., LaMorte, R.L., Wechsung, G., Wechsung, F., Adamsen, F., Matthias, A.D., Thompson, T.L., 2001. Elevated CO_2 stimulates soil respiration in a FACE wheat field. *Basic and Applied Ecology* 2, 193-201.

Peng, S., Huang, J., Sheehy, J.E., Laza, R.C., Visperas, R.M., Zhong, X., Centeno, G.S., Khush, G.S., Cassman, K.G., 2004. Rice yields decline with higher night temperature from global warming. *Proceedings of the National Academy of Sciences USA* 101: 9971-9975.

Penning de Vries, F.W.T., 1980. Simulation models of growth of crops particularly under nutrient stress Dans physiological aspects of crop productivity Pages 213-216 In Proceedings 15th Colloqum of the International Potash Institute. Wageningen The Netherlands Berne, Switzerland International Potash Institute.

Penning de Vries, F.W.T., Jansen, D.M., ten Berge, H.F.M. Bakema, A., 1989. Simulation of ecophysiological processes of growth in several annual crops. In: Simulation Monographs, Pudoc, Wageningen; and International Rice Research Institute, Los Baños, Philippines, p. 271.

Peralta, J. M. Stockle, C.O., 2002. Dynamics of nitrate leaching under irrigated potato rotation in Washington state: a long-term simulation study. *Agriculture*, *Ecosystems* and *Environment* 88: 23-24.

Pert, M.R., Walling, D.E., 1982. Particle size characteristics of fluvial suspended sediment, pp. 397-407 Recent Developments in the Explanation and Prediction of Erosion and Sediment Yield, Proceedings of the Exeter Symposium, July 1982, International Association of Hydrological Sciences(IAHS) Publication No. 137.

Peterson, C., Woods, H.A., Kingsolver, J.G., 2000. Stage-specific effects of temperature and dietary protein on growth and survival of *Manduca sexta* caterpillars. *Physiological Entomology* 25: 35-40.

Pielke, R.A., Avissar, R., Raupach, M., Dolman, A.J., Zeng, X.B., Denning, A S., 1998. Interactions between the atmosphere and terrestrial ecosystems: Influence on weather and climate. *Global Change Biology* 4, 461-475.

Pimentel, D., Harvey, C., Resosudarmo, P., Sinclair, K., Kurz, D., McNair, M., Crist, S., Sphritz, L., Fitton, L., Saffouri, R., Blair, R., 1995. Environmental and economic costs of soil erosion and conservation benefits. *Science* 267: 1117-1123.

Pinnschmidt, H.O., Batchelor, W.D., Teng, P.S., 1995. Simulation of multiple species pest damage in rice using CERES rice. *Agricultural Systems* 48, 193/222.

Pisani, du A., 1987. The CERES-MAIZE model as a potential tool for drought assessment in South Africa. *Water SA* 13 (3), 159-164.

Plummer, S.E., 2000. Perspectives on combining ecological process models and remotely sensed data. Ecological Modelling, 129, pp. 169-186.

Poorter, H., Navas, M.L., 2003. Plant growth and competition at elevated CO_2: on winners, losers and functional groups. *New Phytologist* 157, 175-198.

Pope, V.D., Gallani, M.L., Rowntree, P.R., Stratton, R.A., 2000. The impact of new physical parameterizations in the Hadley Centre Climate Model - HadAM3. *Climate Dynamics* 16:123-146.

Porter, J.R., Semenov, M.A., 2005. Crop responses to climatic variation. *Philosophical Transactions of the Royal Society of London. Series B: Biological Sciences* 360:2021-2035.

Pradhan, Y., Dotaniya, M.L., 2013. Water management in European Dill (*Anethum graveolens* L.). LAP Lambert Academic Publishing, Germany, pp. 117. ISBN 978-3-659-36942-1.

Prasad, R.N., Sharma, U.C., 1993. Approaches towards wasteland development with special reference to NE India. *Indian Journal of Hill Farming* 6, 215-218.

Pre'vot, L., Dechambre, M., Taconet, O., Vidal-Madjar, D., Normand, M., Galle, S., 1993. Estimating the characteristics of vegetation canopies with the airborne radar measurements. *International Journal of Remote Sensing* 14, 2803-2818.

Pregitzer, K.S., Zak, D.R., Curtis, P.S., Kubiske, M.E., Teeri, J.A., Vogel, C.S., 1995. Atmospheric CO_2, soil nitrogen and turnover of fine roots. *New Phytologist* 129: 579-585.

Pregitzer, K.S., Zak, D.R., Maziasz, J., DeForest, J., Curtis, P.S., Lussenhop, J., 2000. Interactive effects of atmospheric CO_2 and soil-N availability on fine roots of *Populus tremuloides*. *Ecological Applications* 10, 18-13.

Prestwidge, D.B., Robertson, M.J., Muchow, R.C., Hammer, G.L., Hughes, R.M., Keating, B., Kingston, G., Liu, D.L., Wood, A.W., 1994. SUGARBAG: A database system for sugarcane crop growth, climate, soils, and management data. Tropical Agronomy Technical Memorandum Number 84a. CSIRO, St Lucia, Brisbane, Australia.

Prévot, L., Chauki, H., Troufleau, D., Weiss, M., Baret, F., Brisson, N., 2003. Assimilating optical and radar data into the STICS crop model for wheat. *Agronomie* 23: 297-303.

Priestley, C.H.B., Taylor, R.J., 1972. On the assessment of surface heat flux and evaporation using large scale parameters. *Monthly Weather Review* 100, 81-92.

Pritsker, A.A.B., 1995. Introduction to Simulation and SLAM II, 4th ed., Halsted Press, John Wiley & Sons, New York.

Probert, E.J., Dawson, G.F., Cockrill, A., 2005. "Evaluating preferences within the composting industry in Wales using a conjoint analysis approach". Resources, Conservation and Recycling, 45 (2), pp. 128-141.

Probert, M.E., Dimes, J.P., Keating, B.A., Dalal, R.C., Strong, W.M., 1998. APSIM's water and nitrogen modules and simulation of the dynamics of water and nitrogen in fallow systems. *Agricultural Systems* 56, 1-26.

Pruski, F.F., Nearing, M.A., 2002. Runoff and soil loss responses to changes in precipitation: a computer simulation study. *Journal of Soil and Water Conservation* 57(1):7-16.

Rabbinge, R., 1986. The bridge function of crop ecology. *Netherlands Journal of Agricultural Science* 34:239-251.

Rabbinge, R., 1993. The ecological background of food production. In: Rabbinge, R. (Ed)., Crop Production and Sustainable Agriculture, CIBA Foundation Symposium, John Wiley and Sons, N.Y., U.S.A.

Rabbinge, R., De Wit, C.T., 1989. Systems, models and simulation. Pages 3-15 in: Simulation and Systems Management in Crop Protection. Rabbinge, R., Ward, S.A. and Van Laar, H.H., Eds. Pudoc, Wageningen, 420 p.

Rabbinge, R., Rossing, W.A.H., van der, Werf W., 1994. Systems approaches in pest management: the role of production ecology. In: Proceedings of the Fourth International Conference on Plant Protection in the Tropics. Ed. by Rajan A, Ibrahim Y, Malaysian Plant Protection Society, Kuala Lumpur, 25-46.

Rajeevan, M., Bhate, J., Jaswal, A. K., 2008. Analysis of variability and trends of extreme rainfall events over India using 104 years of gridded daily rainfall data. *Geophysical Research Letters* 35(18).

Rajendiran, S., Coumar, M.V., Kundu, S., Ajay, Dotaniya, M.L. Rao, A., Subba, 2012. Role of phytolith occluded carbon of crop plants for enhancing soil carbon sequestration in agro-ecosystems. *Current Science* 103 (8), 911-920.

Ramaraj, A.P., Jagannathan, R., Dheebakaran, G.A., 2009. Assessing Predictability of PRECIS Regional Climate Model for Downscaling of Climate Change Scenarios, ISPRS Archives XXXVIII-8/W3 Workshop Proceedings: Impact of Climate Change on Agriculture held during 17-18 December 2009, Space Application Centre, Ahmedabad, India. pp. 80-85.

Randles, C., Ramaswamy, V., 2008. Absorbing aerosols over Asia: A Geophysical Fluid Dynamics Laboratory general circulation model sensitivity study of model response to aerosol optical depth and aerosol absorption. *Journal of Geophysical Research* 113, D21203.

Rao, G.G.S.N., Kesava Rao, A.V.R., Ramakrishna, Y.S., Victor, U.S., 1999. Resource Characterization of Drylands: Climate. In a book, Fifty Years of

Dryland Agricultural Research in India (Eds. HP Singh *et al.*), CRIDA, Hyderabad.

Rao, G.G.S.N., Rao, A.V.M.S., Rao, V.U.M., 2009. Trends in Rainfall and Temperature in Rainfed India in Previous Century In Global Climate Change and Indian Agriculture-case studies from ICAR Network Project (PK Aggarwal ed.), ICAR, New Delhi Pub., pp.71-73.

Rao, L.S., Usha, R.P., Deshmukh, P.S., Kumar, P.A., Panguluri, S.K., 2007. RAPD and ISSR fingerprinting in cultivated chickpea (*Cicer arietinum* L.) and its wild progenitor Cicer reticulatum Ladizinsky. *Genetic Resources and Crop Evolution* 54: 1235-1244.

Rasmussen, M.S., 1992. Assessment of millet yields and production in northern Burkina Faso using integrated NDVI from the AVHRR. *International Journal of Remote Sensing* 13, 3431-3442.

Reader, M.C., Boer, G.J., 1998. The modification of greenhouse gas warming by the direct effect of sulphate aerosols. *Climate Dynamics* 14:593-607.

Refsgaard, J.C., 1997. Parameterisation, calibration, and validation of distributed hydrological models. *Journal of Hydrology* 198 (1): 69-97.

Reilly, J.M., Schimmelpfenning, D., 1999. Agricultural impact assessment, vulnerability, and the scope for adaptation. *Climatic Change* 43, 745-788.

Renald, K.G., Freimund, J.R., 1994. Using monthly precipitation data to estimate the R-factor in the revised USLE. *Journal of Hydrology* 157:287-306.

Reyenga, P.J., Howden, S.M., Meinke, H., Hall, W.B., 2001. Global change impacts on wheat production along an environmental gradient in South Australia. *Environment International* 27: 195-200.

Reynolds, M.P., Ortiz, R., 2010. Adapting crops to climate change: a summary. In: Reynolds, M.P. (ed.) *Climate Change and Crop Production*. CAB International, Wallingford, UK. In press.

Rice, C.W., 2002. Organic matter and nutrient dynamics. In: Lal R, editor. The encyclopedia of soil science. New York: Dekker; pp. 925- 8.

Rillig, M.C., Wright, S.F., Kimball, B.A., Pinter, P.J., Wall, G.W., Ottman, M.J., Leavitt, S.W., 2001. Elevated carbon dioxide and irrigation effects on water stable aggregates in a Sorghum field: a possible role for arbuscular mycorrhizal fungi. *Global Change Biology* 7, 333-337.

Ritchie, J.C., McHenry, J.R., 1990. Application of radioactive fallout cesium-137 for measuring soil-erosion and sediment accumulation rates and patterns-a review. *Journal of Environmental Quality* 19(2):215- 33.

Ritchie, J.T., 1983. Efficient water use in crop production: Discussion on the generality of relations between biomass production and evapotranspiration. In: HM Taylor *et al.* (ed.) Limitations to efficient water use in production. ASA, CSSA, and SSSA, Madison, WI.

Robertson, M.J., Sakala, W., Benson, T., Shamudzarira, Z., 2005. Simulating response of maize to previous velvet bean (*Mucuna pruriens*) crop and nitrogen fertilizer in Malawi. *Field Crops Research* 91, 91-105.

Robinson, D., Conroy, J.P., 1999. A possible plant-mediated feedback between elevated CO_2, denitrification and the enhanced greenhouse effect. *Soil Biology & Biochemistry* 31, 43-53.

Rogers, A., Ainsworth, E., Leakey, A.B.D., 2009. Will elevated carbon dioxide concentration amplify the benefits of nitrogen fixation in legumes? *Plant Physiology* 151, 1009-1016.

Rogers, H.H., Runion, G.B., Krupa, S.V., 1994. Plant responses to atmospheric CO_2 enrichment with emphasis on roots and the rhizosphere. *Environmental Pollution* 83, 155-189.

Ropoport, H.F., Loomis, R.S., 1986. Structural aspects of root thickening in Beta vulgaris L.: Comparative thickening in sugar beet and chard. Bot. *Gazette* 147: 270-277.

Rosenzweig, C., Hillel, D., 1998. Climate change and the global harvest. Potential impacts of the green house effect on agriculture. Oxford University Press, Inc. New York.

Rosenzweig, C., Parry, M. L., 1994. Potential impact of climate change on world food supply. Nature 367:133-138.

Rosenzweig, C., Phillips, J., Goldberg, R., Carroll, J., Hodges, T., 1996. Potential Impacts of Climate Change on Citrus and Potato Production In The US. Agr. Syst., 52(4): 455-479.

Rosenzweig, Cynthia; Jones, J. W.; Hatfield, J. L.; Ruane, A. C.; Boote, K. J.; Thornburn, P.; Antle, J. M.; Nelson, G. C.; Porter, C.; Janssen, S.; Asseng, S.; Basso, B.; Ewert, F.; Wallach, D.; Baigorria, G.; and Winter, J. M., "The Agricultural Model Intercomparison and Improvement Project (AgMIP): Protocols and pilot studies" (2013). Papers in Natural Resources. Paper 445.

Ross, P.J., Bristow, K.L., 1990. Simulating water movement in layered and gradational soils using the Kirchhoff transform. *Soil Science Society of America Journal* 54, 1519-1524.

Rossini, P., Benedetti, R., 1993. On the use of NDVI profiles as a tool for agricultural statistics: the case study of wheat yield estimate and forecast in Emilia Romagna. *Remote Sensing of Environment* 45, 311-326.

Rotter, R. P., Carter, T. R., Olesen, J. E., Porter, J. R., 2011b. Crop-climate models need an overhaul. Nat. Clim. Change 1:175-177.

Rotter, R. P., Palosuo, T., Pirttioja, N. K., Dubrovsky, M., Salo, T., Fronzek, S., *et al.* 2011a. What would happen to barley production in Finland if global warming exceeded 4°C? A model-based assessment. Eur. J. Agron. 35:205-214.

Rouhier, H., Billes, G., Kohen, A.E., Mousseau, M., Bottner, P., 1994. Effect of elevated CO_2 on carbon and nitrogen distribution within a tree (*Castanea sativa* Mill.)-soil system. *Plant Soil* 162, 281-292.

Ruiz-Nogueria, B., Boote, K.J., Sau, F., 2001. Calibration and use of CROPGRO-soybean model for improving soybean management under rainfed conditions. *Agricultural Systems* 68 (2), 151-173.

Ruiz-Vera, U.M., Siebers, M., Gray, S.B., Drag, D.W., Rosenthal, D.M., Kimball, B.A., Ort, D.R., Bernacchi, C.J., 2013. Global Warming Can Negate the Expected CO_2 Stimulation in Photosynthesis and Productivity for Soybean Grown in the Midwestern United States. *Plant Physiology* 162: 410-423.

Runlon, G., Butnor, B., Prior, J.R., Mitchell, S.A., Rogers, H.H., 2012. Effect of atmospheric CO_2 enrichment on soil CO_2 efflux in a young longleaf pine system. *International Journal of Agronomy* DOI: 10.1155/2012/549745.

Rupakumar, K., Pant, G.B., Parthasarathy, B., Sontakke, N.A., 1992. Spatial and subseasonal patterns of the long-term trends of Indian summer monsoon rainfall. *International Journal of Climatology* 12, 257-268.

Rupakumar, K., Sahai, A.K., Krishna, K.K., Patwardhan, S.K., Mishra, P.K., Revadekar, J.V., Kamala, K., Pant, G.B., 2006. High-resolution climate change scenarios for India for the 21st century. *Current Science* 90, 334-345.

Russell, G.L., Rind, D., 1999. Response to CO_2 transient increase in the GISS coupled model-regional cooling in a warming climate. *Journal of Climate* 12, 531-539.

Sabade, S.S., Kulkarni, A., Kripalani, R.H., 2011. Projected changes in South Asian summer monsoon by multi-model global warming experiments, *Theoretical and Applied Climatology* 103, 543-565.

Sadowsky, M.J., Schortemeyer, M., 1997. Soil microbial responses to increased concentrations of atmospheric CO_2. *Global Change Biology* 3, 217-224.

Sadras, V., 2002. Interaction between rainfall and nitrogen fertilization of wheat in environments prone to terminal drought: economic and environmental risk analysis. *Field Crops Research* 77(2-3), 201-215.

Sage, R.F., Kubien, D.S., 2007. The temperature response of C-3 and C-4 photosynthesis. *Plant, Cell & Environment* 30, 1086-1106.

Sakamoto, T., Yokozawa, M., Toritani, H., Shibayama, M., Ishitsuka, N., Ohno, H., 2005. A crop phenology detection method using time-series MODIS data. *Remote Sensing of Environment* 96, 366-374.

Salam, M.U., Jones, J.W., Kobayashi, K., 2001. Predicting nursery growth and transplanting shock in rice. *Experimental Agriculture* 37, 65-81.

Salmi, T., Maatta, A., Anttila, P., Ruoho-Airola, T., Amnell, T., 2002. Detecting Trends of Annual Values of Atmospheric Pollutants by the Mann-Kendall Test and Sen's Slope Estimates -The Excel Template Application Makesens. Finnish Meteorological Institute Publications on Air Quality No. 31, Helsinki, Finland.

Samuel, B.S., Lynch, J.P., 2010. The opening of Pandora's Box: climate change impacts on soil fertility and crop nutrition in developing countries. *Plant Soil.* 335, 101- 115.

Sawada, Y., Mitsuzuka, N., Sawada, H., 1999. Classification of vegetation types by time series data analysis of vegetation index. Proceedings of the

27th Conference of the Remote Sensing Society of Japan in November 1999 (pp. 73-74). In Japanese, with English abstract.

Saxton, K.E., Rawls, W.J., Romberger, J.S., Papendick, R.I., 1986. Estimating generalized soil-water characteristics from texture. *Soil Science Society of America Journal* 50, 1031-1036.

Scherr, S.J., Sthapit, S., 2009. Mitigating climate changes through food and land use. Worldwatch Report 179. Worldwatch Institute, Washington, DC. 48 p.

Schimel, D., Stillwell, M.A., Woodmansee, R.G., 1985a. Biogeochemistry of C, N, and P in a soil catena of the shortgrass steppe. *Ecology* 66(1): 276- 82.

Schimel, D.S., Coleman, D.C., Horton, K.A., 1985b. Soil organic-matter dynamics in paired rangeland and cropland topo sequences in North-Dakota. *Geoderma* 36(3-4): 201-14.

Schlenker, W., Roberts, M. J., 2009. Nonlinear temperature effects indicate severe damage to U.S. crop yields under climate change. *Proceedings of the National Academy of Sciences USA* 106: 15594-15598.

Schneider, S.H., Semenov, S., Patwardhan, A., Burton, I., Magadza, C.H.D., Oppenheimer, M., Pittock, A.B., Rahman, A., Smith, J.B., Suarez, A., Yamin, F., 2007. Assessing key vulnerabilities and the risk from climate change. In: Parry ML, Canziani OF, Palutikof JP, van der Linden PJ, Hanson CE (Eds) Climate change 2007: Impacts, adaptation and vulnerability. Contribution of working group II to the Fourth Assessment Report of the Intergovernmental Panel on Climate Change. Cambridge University Press, Cambridge, pp 779-810.

Schortemeyer, M., Dijkstra, P., Johnson, D.W., Drake, B.G., 2000. Effects of elevated atmospheric CO_2 concentration on C and N pools and rhizosphere processes in a Florida scrub oak community. *Global Change Biology* 6, 383-391.

Seddigh, M., Joliff, G.D., 1984a. Effects of night temperature on dry matter partitioning and seed growth of indeterminate field grown soybean. *Crop Science* 24, 704-710.

Seddigh, M., Joliff, G.D., 1984b. Night temp. Effects on morphology, phenology, yield, and yield components of indeterminate field grown soybean. *Agronomy Journal* 76, 824-828.

Seguin, B., Courault, D., Gue'rif, M., 1994. Surface temperature and evapotranspiration: application of local scale methods to regional scales using satellite data. *Remote Sensing of Environment* 49, 287- 295.

Sellers, P.J., Mintz, Y., Sud, Y.C., Dalcher, A., 1986. A simple biosphere model (sib) for use within general-circulation models. *Journal of the Atmospheric Sciences* 43(6):505-531.

Semenov, M.A., Barrow, E.M., 1997. Use of a stochastic weather generator in the development of climate change scenarios. *Climatic Change* 35:397-414.

Semenov, M.A., Halford, N.G., 2009. Identifying target traitsand molecular mechanisms for wheat breeding under a changing climate. *J. Exper. Bot.* 60:2791-2804.

Senapati Nimai, Hulugalle Nilantha R., Smith Pete, Wilson Brian R., Yeluripati Jagadeesh B., Daniel Heiko, Ghosh Subhadip, Lockwood Peter, 2014. Modelling soil organic carbon storage with RothC in irrigated Vertisols under cotton cropping systems in the sub-tropics. Soil and Tillage Research, 143, 38-49.

Seneviratne, S.I., Nicholls, N., Easterling, D., Goodess, C.M., Kanae, S., Kossin, J., Luo, Y., Marengo, J., McInnes, K., Rahimi, M., Reichstein, M., Sorteberg, A., Vera, C., Zhang, X., 2012: Changes in climate extremes and their impacts on the natural physical environment. In: Managing the Risks of Extreme Events and Disasters to Advance Climate Change Adaptation [Field, C.B., V. Barros, T.F. Stocker, D. Qin, D.J. Dokken, K.L. Ebi, M.D. Mastrandrea, K.J. Mach, G.K. Plattner, S.K. Allen, M. Tignor, and P.M. Midgley (eds.)]. A Special Report of Working Groups I and II of the Intergovernmental Panel on Climate Change, pp. 109-230.

Shah, T., 2009. Taming the anarchy: groundwater governance in South Asia. Resources for Future Press (Washington DC, USA).

Shamudzarira, Z., Robertson, M.J., 2002. Simulating response of maize to nitrogen fertilizer in semi-arid Zimbabwe. *Experimental Agriculture* 38, 79-96.

Shang, Y., Hu, X., Direnzo, J., Lazar, M.A., Brown, M., 2000. Cofactor dynamics and sufficiency in estrogen receptor-regulated transcription. *Cell* 103:843-52.

Sharman, M.J., de Boissezon, H., Gonzales, G., Pouss, K., 1992. Rapid estimates of acreages and potential yield at European scale. Proceedings of the 43rd congress of the International Astronautical Federation. Washington, DC, U.S.A., 28 August- 5 September 1992. IAF, Paper 92-0110 (Washington DC: IAF).

Shaver, T.M., Peterson, G.A., Ahuja, L.R., Westfall, D.G., Sherrod, L.A., Dunn, G., 2002. Surface soil physical properties after twelve years of dryland no till management. *Soil Science Society of America Journal* 66, 1296- 1303.

Silburn, D.M., Loch, R.J., 1989. Evaluation of the creams model: 1. Sensitivity analysis of the soil-erosion sedimentation component for aggregated clay soils. *Australian Journal of Soil Research* 27(3):545- 561.

Singh, A.K., 2005. Crop growth simulation models. http://iasri.res.in/.

Singh, B., El Maayar, M., André, P., Bryant, C.R., Thouez, J.P., 1998. Impacts of a GHG-induced climate change on crop yields: effects of acceleration in maturation, moisture stress and optimal temperature. *Climatic Change* 38:51-86.

Singh, J.P., Govindakrishnan, P.M., Lal, S.S., Aggarwal, P.K., 2005. Increasing the efficiency of agronomy experiments in potato using InfoCrop-potato model. *Potato Research* 48, 131-152.

Singh, K.K., Kumar, R., Mall, R.K., Rathore, L.S., Sanker, U., Gupta, B.R.D., 1999. Soybean (*Glycine max*) yield prediction from current and historical weather data using CROPGRO model. *Indian Journal of Agricultural Sciences* 69 (9), 639-643.

Sirotenko, O.D., 2000. Food production and agricultural models: basic principles of development. Mathematical models of life support systems - Vol. II.

Sivakumar, M.V.K., Stefanski, R., 2011. Climate Change in South Asia. In: Lal, R. *et al.* (eds.), Climate Change and Food Security in South Asia, 13, Springer Science Business Media pp. 13-30.

Smith, D.M., Cusack, S., Colman, A.W., Folland, C.K., Harris, G.R., Murphy, J.M., 2007. Improved surface temperature prediction for the coming decade from a global climate model. Science 317:796-799. doi: 10.1126/science.1139540.

Smith, J.A., Jarman, M., Osborn, M., 1999. Doing interpretative phenomenological analysis. In Qualitative Health Psychology: Theories and Methods, pp. 218-240 [M Murray and K Chamberlain, eds.]. London: Sage Publications.

Smith, P., 2005. Agriculture and Climate Change: An agenda for negotiation in Copenhagen, IFPRI Focus 16. 2005.

Smith, P., Martino, D., Cai, Z., Gwary, D., Janzen, H., Kumar, P., McCarl, B., Ogle, S., O'Mara, F., Rice, C., Scholes, B., Sirotenko, O., 2007: Agriculture. In: B. Metz, O.R. Davidson, P.R. Bosch, R. Dave and L.A. Meyer (editors), Climate Change 2007: Mitigation. Contribution of Working Group III to the Fourth Assessment Report of the Intergovernmental Panel on Climate Change, Cambridge University Press, Cambridge, United Kingdom and New York, NY, USA.

Smith, P., Martino, D., Cai, Z., Gwary, D., Janzen, H.H., Kumar, P., McCarl, B., Ogle, S., O'Mara, F., Rice, C., Scholes, R.J., Sirotenko, O., Howden, M., McAllister, T., Pan, G., Romanenkov, V., Schneider, U., Towprayoon, S., Wattenbach, M., Smith, J.U., 2007a. Greenhouse gas mitigation in agriculture. Philosophical Transactions of the Royal Society, B., 363. doi:10.1098/rstb.2007.2184.

Smith, P., Powlson, D.S., Glendining, M.J., Smith, J.U., 1998. Preliminary estimates of the potential for carbon mitigation in European soils through no-till farming. *Global Change Biology* 4, 679-685.

Smith, S.V., Renwick, W.H., Buddemeier, R.W., Crossland, C.J., 2001. Budgets of soil erosion and deposition for sediments and sedimentary organic carbon across the conterminous United States. *Global Biogeochemical Cycles* 15: 697-707.

Soil Conservation Service, 1993. 'Irrigation water requirements' National Engineering Handbook, part 623. Soil Conservation Service, US Dept. of Agriculture. Washington, USA.

Solomon, S., Qin, D., Manning, M., Alley, R.B., Berntsen, T., Bindoff, N.L., Chen, Z., Chidthaisong, A., Gregory, J.M., Hegerl, G.C., Heimann, M.,

Hewitson, B., Hoskins, B.J., Joos, F., Jouzel, J., Kattsov, V., Lohmann, U., Matsuno, T., Molina, M., Nicholls, N., Overpeck, J., Raga, G., Ramaswamy, V., Ren, J., Rusticucci, M., Somerville, R., Stocker, T.F., Whetton, P., Wood, R.A., Wratt, D., 2007. Technical Summary. In: Climate Change 2007: The Physical Science Basis. Contribution of Working Group I to the Fourth Assessment Report of the Intergovernmental Panel on Climate Change [Solomon, S., D. Qin, M. Manning, Z. Chen, M. Marquis, K.B. Averyt, M. Tignor and H.L. Miller (eds.)]. Cambridge University Press, Cambridge, United Kingdom and New York, NY, USA.

Somerville, R.C.J., Jouzel, J., 2008. The global census and the intergovernmental panel on climate change. In: C. Gautier and J. Louis, Fellous (eds.), Facing climate change together, Cambridge University Press, PP: 12-29.

Sorensen, L.H., 1981. Carbon - nitrogen relationship during the humification of cellulose in soils containing different amounts of clay. *Soil Biology & Biochemistry* 13: 313- 21.

Southworth, J., Randolph, J.C., Habeck, M., Doering, O.C., Pfeifer, R.A., Rao, D.G., Johnston, J.J., 2000. Consequences of future climate change and changing climate variability on maize yields in the Midwestern United States. *Agriculture, Ecosystems & Environment* 82:139-158.

Sposito, G., 1989. The chemistry of soils. Oxford University Press, New York, New York, USA.

Srivastava, A., Naresh, K.S., Aggarwal, P.K., 2010. Assessment on vulnerability of sorghum to climate change in India. *Agriculture, Ecosystems & Environment* 138 (3-4):160-169.

Staddon, P.L., Fitter, A.H., Graves, J.D., 1999. Effect of elevated atmospheric CO_2 on mycorrhizal colonization, external mycorrhizal hyphal production and phosphorus inflow in *Plantago lanceolata* and *Trifolium repens* in association with the arbuscular mycorrhizal fungus Glomus mosseae. *Global Change Biology* 5, 347-358.

Stallard, R.F., 1998. Terrestrial sedimentation and the carbon cycle: coupling, weathering and erosion to carbon burial. *Global Biogeochemical Cycles* 12: 231-257.

Starr, G.C., Lal, R., Kimble, J.M., Owens, L., 2001. Assessing the impact of erosion on soil organic carbon pools and fluxes. Pages 417-426 in Kimble JM, Follett RF, Stewart BA. eds Assessment Methods for Soil Carbon. Boca Raton (FL): CRC.

Starr, G.C., Lal, R., Malone, R., Hothem, D., Owens, L., Kimble, J., 2000. Modelling soil carbon transported by water erosion processes. *Land Degradation & Development* 11: 83-91.

State of Indian Agriculture, 2012-13. Ministry of Agriculture Department of Agriculture and Cooperation, Directorate of Economics and Statistics, New Delhi.

Steduto, P., 2003. Biomass Water-Productivity. Comparing the Growth-Engines of Crop Models. FAO Expert Consultation on Crop Water

Productivity under Deficient Water Supply, 26-28 February 2003, Rome, Italy.

Steduto, P., Hsiao, T.C., Fereres, E., 2007. On the conservative behavior of biomass water productivity. *Irrigation Science* 25: 189-207.

Steduto, P., Hsiao, T.C., Raes, D., Fereres, E., 2009. AquaCrop-The FAO crop model to simulate yield response to water: I. Concepts and underlying principles. *Agronomy Journal* 101(3): 426-437.

Stern, N., 2006. Review on the economics of climate change. HM Treasury, London.

Stevenson, F.J., 1982. Humus chemistry: genesis, composition, reactions. New York: Wiley; 443 pp.

Stewart, L.K., Charlesworth, P.B., Bristow, K.L., Thorburn, P.J., 2006. Estimating deep drainage and nitrate leaching from the root zone under sugarcane using APSIM-SWIM. *Agricultural Water Management* 81, 35-334.

Stocker, T.F., 2013. The closing door of climate targets. *Science* 339: 280-282.

Stöckle, C.O., 1996. GIS and simulation technologies for assessing cropping systems management in dry environments. *American Journal of Alternative Agriculture* 11:115-120.

Stöckle, C.O., Campbell, G.S., 1989. Simulation of crop response to water and nitrogen: an application example using spring wheat. *Transactions of the ASAE* 32, 66-74.

Stockle, C.O., Debaeke, P., 1997. Modelling crop nitrogen requirements: a critical analysis. *European Journal of Agronomy* 7, 161-169.

Stockle, C.O., Nelson, R. Campbell, G.S., 1993. CropSyst, cropping systems simulation model. User's Manual. Dept. of Biological Systems Engineering, Washington State University, Pullman, Washington, USA.

Stockle, C.O., Nelson, R.L., 2000. Cropsyst User's manual (Version 3.0). Biological Systems Engineering Dept., Washington State University, Pullman, WA.

Stockle, C.O., Williams, J.R., Rosenberg, N.J., Jones, C.A., 1992. A method for estimating the direct and climatic effects of rising atmospheric carbon dioxide on growth and yield of crops: Part I: Modification of the EPIC model for climate change analysis. *Agricultural Systems* 38, 225-238.

Stockle, O.C., Donatelli, M., Nelson, R., 2003. Cropsyst a cropping systems simulation model. *European Journal of Agronomy* 18(3-4): 289-307.

Stone, R.C., Meinke, H., 2005. Operational seasonal forecasting of crop performance. Phil. Trans. R. Soc. B, 360:2109-2124.

Stowasser, M., Annamalai, H., Hafner, J., 2009. Response of the South Asian summer monsoon to global warming: Mean and synoptic systems. *Journal of Climatology* 22, 1014-1036.

Studdert Guillermo A., Gloria Monterubbianesi, M., Dominguez German F., 2011. Use of RothC to simulate changes of organic carbon stock in the arable layer of a Mollisol of the southeastern Pampas under continuous cropping. Soil and Tillage Research, 117, 191-200.

Subash, Ram Mohan, 2012. Evaluation of the impact of climatic trends and variability in rice-wheat system productivity using Cropping System Model DSSAT over the Indo Gangetic Plains of India. *Agricultural and Forest Meteorology,* 164, 71-81.

Swain, D.K., Thomas, D., 2010. Climate Change Impact Assessment and Evaluation of Agro-Adaptation Measures for Rice Production in Eastern India. *Journal of Environmental Informatics* 16:94-101.

Tabatabai, M.A., AI Alkhafaji, A., 1980. Comparison of nitrogen and sulfur mineralization in soils. *Soil Science Society of America Journal* 4, 1000-1006.

Tadross, M.A., Hewitson, B.C., *et al.* 2005. The inter annual variability of the onset of the maize growing season over South Africa and Zimbabwe. *J Clim* 18(16):3356-3372.

Tanner, C.B., Sinclair, T.R., 1983. Efficient water use in crop production: research or re-search? In: Taylor, H.M., Jordan, W.R., Sinclair, T.R. (Eds.), Limitations to efficient water use in crop production. Amer. Soc. Agron, Madison, WI.

Tarnawski, S., Hamelin, J., Jossi, M., Aragno, M., Fromin, N., 2006. Phenotypic structure of Pseudomonas populations is altered under elevated pCO_2 in the rhizosphere of perennial grasses - *Soil Biology & Biochemistry* 38: 1193-1201.

Taub, D., Miller, B., Allen, H., 2008. Effects of elevated CO_2 on the protein concentration of food crops: a meta-analysis. *Global Change Biology* 14, 565-575.

Taub, D.R., Wang, X.Z., 2008. Why are nitrogen concentrations in plant tissues lower under elevated CO_2? A critical examination of the hypotheses. *Journal of Integrative Plant Biology* 50, 1365-1374.

Thomas, A., 2008. Agriculture irrigation demand under present and future climate scenarios in China. *Global and Planetary Change* 60 (3-4), 306-326.

Thorburn, P.J., Probert, M.E., Robertson, F.A., 2001. Modelling decomposition of sugarcane surface residues with APSIM-Residue. *Field Crops Research* 70, 223-232.

Thornton, P.K., Bowen, W.T., Ravelo, A.C., Wilkens, P.W., Farmer, G., Brock, J., Brink, J.E., 1997. Estimating millet production for famine early warning: an application of crop simulation modelling using satellite and ground-based data in Burkina Faso. *Agricultural and Forest Meteorology* 83, 95-112.

Thornton, P.K., Saka, A.R., Singh, U., Kumwenda, J.D.T., Brink, J.E., Dent, J.B., 1995. Application of a maize crop simulation model in the central region of Malawi. *Experimental Agriculture* 31, 213-226.

Thorp, K.R., DeJonge, K.C., Kaleita, A.L., Batchelor, W.D., Paz, J.O., 2008. Methodology for the use of DSSAT models for precision agriculture decision support. *Computers and Electronics in Agriculture* 64, 276-285.

Thorp, K.R., Hunsaker, D.J., French, A.N., 2010. Assimilating Leaf Area Index estimates from remote sensing into the simulations of a cropping systems model. *Transaction of the ASABE* 53, 251-262.

Thorp, K.R., Wang, G., West, A.L., Moran, M.S., Bronson, K.F., White, J.W., *et al.,* 2012. Estimating crop biophysical properties from remote sensing data by inverting linked radiative transfer and ecophysiological models. *Remote Sensing of Environment* 124, 224-233.

Timsina, J., Godwin, D., Humphreys, E., Singh, Y., Singh, B., Kukal, S.S., Smith, D., 2008. Evaluation of options for increasing yield and water productivity of wheat in Punjab, India using the DSSAT-CSM-CERES Wheat model. *Agricultural Water Management* 95:1099-1110.

Trnka, M., *et al.* 2011. Agroclimatic conditions in Europe under climate change. Global Change Biology, 17(7), pp. 2298-2318.

Tsuji, G.Y., Hoogenboom, G., Thornton, P.K., 1998. Understanding Options for Agricultural Production. *Kluwer Academic Publishers*, Dordrecht, The Netherlands.

Tsuji, G.Y., Uehara, G., Balas, S., 1994. DSSAT version 3. A Decision Support System for Agrotechnology Transfer, Vols 1-3. Univ. of Hawaii, Honolulu.

Tsuji, L.J.S., Young, J., Kozlovic, D.R., 1998. Lead shot ingestion in several bird species of the western James Bay region of northern Ontario, Canada. Can Field-Nat 112:86-89.

Turner, A.G., Annamalai, H., 2012. Climate change and the South Asian summer monsoon, *Nature Climate Change*, 2, 587-595.

Turner, R.N., Crisp, R.J., Lambert, E., 2007a. Imagining intergroup contact can improve intergroup attitudes. *Group Processes & Intergroup Relations* 10, 427-441.

Tyson, P., Odada, E., Schulze, R., Vogel, C.I., 2002. Regional-global change linkages: Southern Africa. In: P. Tyson, R. Fuchs, C. Fu, L. Lebel, A.P. Mitra, E. Odada, J. Perry, W. Steffen and H. Virji (eds.), Global-regional Linkages in the Earth System. START/ IHDP/IGBP/WCRP. Springer, Berlin, London.

U.N., 2009. United Nations population division of Economics and Social Affairs - World Population prospects (http://esa.un.org/unpp).

U.N.-D.E.S.A., 2004. World Population Prospects: The 2004 Revision Report, Chapter 1. United Nations Department of Economic and Social Affairs/ Population Division.

U.N.F.C.C.C., 2002. Climate Change Information Kit. http://unfccc.int/ essential_background/background_publications_htmlpdf/climate_ changeinformation_kit/items/305.php.

U.N.P.F.A., 2011. State of the world population 2011, United Nations Population Fund.

U.S.-E.P.A., 2002. http://www.epa.gov/.

Ueda, H., Iwai, A., Kuwako, K., Hori, M., 2006. Impact of anthropogenic forcing on the Asian summer monsoon as simulated by eight GCMs. *Geophysical Research Letter* 33, L06703.

Van Duivenbooden, N., Gosseye, P., 1990. Compétition pour des Ressources Limitées: Le Cas de la Cinquième Région du Mali. Raport 2. Productions Végétales, Animales et Halieutigues. CAP-DLO, Wageningen.

Van Groenigen, K.J., Osenberg, C.W., Hungate, B.A., 2011. Increased soil emissions of potent greenhouse gases under increased atmospheric CO_2. *Nature* 475, 214-216.

Van Ittersum, M.K., Leffelaar, P.A., van Keulen, H., Kropff, M.J., Bastiaans, L., Goudriaan, J., 2003. On approachesand applications of the Wageningen crop models. *European Journal of Agronomy* 18:201 -234.

Van Keulen, H., 1976. Evaluation of models. In Arnold, G.W. and De Wit, C.T. (eds.) Critical Evaluation of Systems Analysis. PUDOC, Netherlands.

Van Kraalingen, D.W.G., 1995. The FSE system for crop simulation: version 2.1. Quantitative Approaches in Systems Analysis Report 1: C.T. de Wit Graduate School for Production Ecology and AB-DLO. Wageningen, The Netherlands, p. 58.

Van Laar, H.H., Goudriaan, J., van Keulen, H., 1997. SUCROS 97. Simulation of Crop Growth for Potential and Water-limited Production Situations. Quantitative Approaches in Systems Analysis 14. Wageningen, The Netherlands. CT de Wit Graduate School of Production ecology and AB-DLO. p. 52.

Van Veen, J.A., Liljeroth, E., Lekkerkerk, L.J.A., van de Geijn, S.C., 1991. Carbon fluxes in plantsoil systems at elevated atmospheric CO_2 levels. *Ecological Applications* 1, 175-181.

VanDijk, A., Callis, S.L., Sakamoto, C.M., Decker, W.L., 1987. Smoothing vegetation index profiles: An alternative method for reducing radiometric disturbance in NOAA/AVHRR data. *Photogrammetric Engineering & Remote Sensing* 53, 1059-1067.

Vanveen, J.A., Kuikman, P.J., 1990. Soil structural aspects of decomposition of organic-matter by microorganisms. *Biogeochemistry* 11(3): 213-33.

Vanveen, J.A., Ladd, J.N., Frissel, M.J., 1984. Modelling C and N turnover through the microbial biomass in soil. *Plant Soil* 76(1-3):257-74.

Varshney, R.K., Bansal, K.C., Aggarwal, P.K., Datta, S.K., Craufurd, P.Q., 2011. Agricultural biotechnology for crop improvement in a variable climate: hope or hype? *Trends in Plant Science,* 16: 363-371.

Vazifedoust, M., Van Dam, J.C., Bastiaanssen, W.G.M., Feddes, R.A., 2009. Assimilation of satellite data into agrohydrological models to improve crop yield forecasts. *International Journal of Remote Sensing* 30, 2523-2545.

Ventrella, D., Charfeddine, M., Moriondo, M., Rinaldi, M., Bindi, M., 2012. Agronomic adaptation strategies under climate change for winter durum wheat and tomato in southern Italy: irrigation and nitrogen fertilization. *Regional Environmental Change* 12, 407-419.

Verdin, J., Funk, C., Senay, G., Choularton, R., 2005. Climate science and famine early warning. Philosophical Transactions of the Royal Society B: Biological Sciences; 360:2155-2168.

Viña, A., Anatoly, A., Gitelson, b., Anthony, L., Nguy-Robertson, b., Peng, Y., 2011. Comparison of different vegetation indices for the remote assessment of green leaf area index of crops. *Remote Sensing of Environment* 115 (2011) 3468-3478.

Viovy, N., Arino, O., Belward, A.S., 1992. The best index slope extraction (BISE): A method for reducing noise in NDVI time-series. *International Journal of Remote Sensing* 13, 1585-1590.

Voroney, R.P., Vanveen, J.A., Paul, E.A., 1981. Organic-C dynamics in grassland soils: 2. Model validation and simulation of the long-term effects of cultivation and rainfall erosion. *Canadian Journal of Soil Science* 61(2):211 -24.

Wafula, B.M., 1995. Applications of crop simulation in agricultural extension and research in Kenya. *Agricultural Systems* 49, 399-412.

Walker, T.S., Ryan, J.G., 1990. Village and household economics in India's semi-arid tropics. Johns Hopkins University Press, Baltimore, Maryland, USA.

Wan, S., Norby, R.J., Ledford, J., Weltzin, J.F., 2007. Responses of soil respiration to elevated CO_2, air warming, and changing soil water availability in a model old-field grassland. *Global Change Biology* 13, 2411-2424.

Wan, Y., El-Swaify, S.A., 1997. Flow-induced transport and enrichment of erosional sediment from a well-aggregated and uniformly-textured oxisol. *Geoderma* 75(3-4): 251 -65.

Wang, K., Tang, D., Wang, M., Lu, J., Yu, H., Liu, J., Qian, B., Gong, Z., Wang, X., Chen, J., GU, M., Cheng, Z., 2009. MER3 is required for normal meiotic crossover formation, but not for presynaptic alignment in rice. *Journal of Cell Science* 122(Pt 12):2055-63.

Wang, X., Wiens, M., Schröder, H.C., Schloßmacher, U., Pisignano, D., Jochum, K.P., Müller, W.E.G., 2011. Evagination of cells controls bio-silica formation and maturation during spicule formation in sponges. *PLoS one* 6(6), e20523.

Waring, R.H., Landsberg, J.J., Williams, M., 1998. Net primary production of forests: a constant fraction of gross primary production? *Tree Physiology* 18,129-134.

Washington, R., Todd, M. C., Engelstaedter, S., Mbainayel, S., Mitchell, F., 2006. Dust and the low-level circulation over the Bodélé Depression,

Chad: Observations from BoDEx 2005, J. Geophys. Res., 111, D03201, doi: 10.1029/2005JD006502.

Wassmann, R., Jagadish, S.V.K., Sumfleth, K., Pathak, H., Howell, G., Ismail, A., Serraj, R., Redona, E., Singh, R.K., Heuer, S., 2009. Regional vulnerability of climate change impacts on Asian rice production and scope for adaptation. *Advances in Agronomy* 102: 91-103.

Wegener, M., 1994. Operational urban models: state of the art *Journal of the American Planning Association* 60, 17-29.

Weir, A.H., Bragg, P.L., Porter, J.R., Rayner, J.H., 1984. A winter wheat crop simulation model without water or nutrient limitations. *Journal* of *Agricultural Science Cambridge* 102:371-382.

Weiss, M., Troufleau, D., Baret, F., Chauki, H., Prevot, L., Olioso, A., Bruguier, N., Brisson, N., 2001. Coupling canopy functioning and radiative transfer models for remote sensing data assimilation. *Agricultural and Forest Meteorology* 108, 113-128.

Wen, D., Pimentel, D., 1998. Agriculture in China: water and energy resources, in: Tso T.C., Tuan F., FaustM. (Eds.), Agriculture in China 1949-2030, Beltsville, IDEALS Inc., pp. 479-497.

West, D.W., Taylor, J.A., 1980. The effects of temperature on salt uptake by tomato plants with diurnal and nocturnal water logging of salinized root zones. *Plant Soil.* 56:113-121.

West, T.O.,Wali, M.K., 2002. Modelling regional carbon dynamics and soil erosion in disturbed and rehabilitated ecosystems as affected by land use and climate. *Water, Air & Soil Pollution* 138(1-4):141- 63.

Wetherald, R.T., Manabe, S., 2002. Simulation of hydrologic changes associated with global warming. *Journal of Geophysical Research* 107(D19): 4379.

Whisler, F.D., Acock, B., Baker, D.N., Fye, R.E., Hodges, H.F., Lambert, J.R., Lemmon, H.E., McKinion, J.M., Reddy, V.R. 1986. Crop simulation models in agronomic systems. *Advances in Agronomy* 40, 141-208.

White, J. W., Hoogenboom, G., Kimball, B. A., Wall, G. W., 2011. Methodologies for simulating impacts of climate change on crop production. Field Crops Res. 124:357-368.

Wiemken, V., Laczko, E., Ineichen, K., Boller, T., 2001. Effects of elevated carbon dioxide and nitrogen fertilization on mycorrhizal fine roots and the soil microbial community in Beech-Spruce ecosystems on siliceous and calcareous soil. *Microbial Ecology* 42, 126-135.

Williams, J.R., 1975. Sediment yield prediction with universal equation using runoff energy factor. US Department of Agriculture, (ARS-S-40).

Williams, J.R., 1995. The EPIC. In: V.P. Singh (ed.), Computer models of watershed hydrology. Water Resources Publications, Littleton, CO, pp. 909-1000 (Chapter 25).

Williams, J.R., Jones, C.A., Dyke, P.T., 1984. A modelling approach to determining the relationship between erosion and soil productivity. *Transactions of the ASAE* 27:129-144.

Williams, J.R., Jones, C.A., Kiniry, J.R., Spanel, D.A., 1989. The EPIC crop growth model. *Transactions of the ASAE* 32: 497-511.

Willmott, C.J., 1981. On the validation of models. *Physical Geography* 2: 184-194.

Willmott, C.J., 1984. On the evaluation of model performance in physical geography. In Spatial Statistics and Models, 443-460. G.L. Gaile and C. J. Willmott, (eds.), Norwell, Mass.: D. Reidel.

Wischmeier, W.H., D.D., Smith., 1978. Predicting rainfall erosion losses: A guide to conservation planning. U.S. Department of Agriculture, Agricultural Handbook No. 537, Washington, D.C. 85 pp.

Wolfe, D.W., Comstock, J., Lakso, A., Chase, L., Fry, W., Petzoldt, C., Leichenko, R., Vancura, P., 2011. Agriculture. In: Rosenzweig C., W. Solecki, A. DeGaetano *et al.* (eds.) *Responding to Climate Change in New York State*. pp. 217-254. New York Academy of Sciences. Blackwell Pub., Boston, MA. Chapter 7.

Wolfe, D.W., Gifford, R.M., Hilbert, D., Luo, Y., 1998. Integration of photosynthetic acclimation to CO_2 at the whole-plant level. Glob Change Biol; 4: 879-93.

Wolfe, D.W., Ziska, L., Petzoldt, C., Seaman, A., Chase, L., Hayhoe, K., 2008. Projected change in climate thresholds in the Northeastern U.S.: Implications for crops, pests, livestock, and farmers. *Mitigation and Adaptation Strategies for Global Change* 13: 555-575.

Wood, R.A., Keen, A.B., Mitchell, J.F.B., Gregory, J.M., 1999. Changing spatial structure of the thermohaline circulation in response to atmospheric CO_2 forcing in a climate model. *Nature* 399: 572-575.

Woods, H.A., Kingsolver, J.G., 1999. Feeding rate and the structure of protein digestion and absorption in Lepidopteran midguts. *Archives of Insect Biochemistry and Physiology* 42:74-87.

World Bank, 2009. "Development and Climate Change." World Bank.

Wu, C., Anlauf, R., Ma, Y., 2013. Application of the DSSAT Model to Simulate Wheat Growth in Eastern China. *Journal of Agricultural Science* Vol. 5, No. 5; 2013.

Wylie, B.K., Johnson, D.A., Laca, E., Saliendra, N.Z., Gilmanov, T.G., Reed, B.C., 2003. Calibration of remotely sensed, coarse resolution NDVI to CO_2 fluxes in a sagebrushsteppe ecosystem. *Remote Sensing of Environment* 85, 243-255.

Xiao, X., Boles, S., Frolking, S., *et al.*, 2002. Observation of flooding and rice transplanting of paddy rice field at the site to landscape scales in China using VEGETATION sensor data. *International Journal of Remote Sensing* 23, 3009-3022.

Yang, L., Wang, Y., Dong, G., Gu, H., Huang, J., Zhu, J., Yang, H., Liu, G., Han, Y., 2007. The impact of free-air CO_2 enrichment (FACE) and nitrogen supply on grain quality of rice. *Field Crops Research* 102: 128-140.

Yang, P., Bi, L., Baum, B.A., Liou, K.N., Kattawar, G.W., Mishchenko, M.I., Cole, B., 2013. Spectrally consistent scattering, absorption, and polarization properties of atmospheric ice crystals at wavelengths from 0.2 to 100 μm. *J. Atmos. Sci.*, **70**, 330-347, doi:10.1175/JAS-D-12-039.1.

Yeo, A., 1999. Predicting the interaction between the effects of salinity and climate change on crop plants. *Scientia Horticulturae* 78, 159-174.

Yin, X., Struik, P.C., 2008. Applying modelling experiences from the past to shape crop systems biology: the need to converge crop physiology and functional genomics. *New Phytologist* 179: 629-642.

Yuping, M., Shili, W., Li, Z., Yingyu, H., Liwei, Z., Yanbo, H., Futang, W., 2008. Monitoring winter wheat growth in North China by combining a crop model and remote sensing data. *International Journal of Applied Earth Observation and Geoinformation* 10: 426-437.

Zak, D.R., Pregitzer, K.S., King, J.S., Holmes, W.E., 2000. Elevated atmospheric CO_2, fine roots and the response of soil microorganisms: A review and hypothesis. *New Phytologist* 147, 201-222.

Zalud, Z., Stralkova, R., Pokorny, E., Podesvova, J., 2001. Estimation of winter wheat nitrogen stress using the CERES crop model. *Rostlinna Vyroba* 47 (6), 253-259.

Zalud, Z., Trnka, M., Dubrovsky, M., 2000. Change of spring barley production potential using crop model ceres Barley. *Rostlinna Vyroba* 46(9), 423-428.

Zhang, F., Li, C., Wang, Z., Wu, H., 2006. Modelling impacts of management alternatives on soil carbon storage of farmland in Northwest China. *Biogeosciences* 3, 451-466.

Zhang, J.H., Quine, T.A., Ni, S.J., Ge, F.L., 2006. Stocks and dynamics of SOC in relation to soil redistribution by water and tillage erosion. Global Change Biol. 12, 1834-1841.

Zhang, X., Friedl, M.A., Strahler, A.H., Hodges, J.C.F., Gao, F., Reed, B.C., 2003. Monitoring vegetation phenology using MODIS. *Remote Sensing of Environment* 84: 471-475.

Zhe, L., Ted, H., Zhang, A., Zhou, F., McConkey, B., 2012. Spatially locating soil classes within complex soil polygons - Mapping soil capability for agriculture in Saskatchewan Canada. *Agriculture, Ecosystems & Environment* 152: 59-67.

Zhu, X.G., Long, S.P., Ort, D.R., 2010. Improving photosynthetic efficiency for greater yield. *Annual Review of Plant* Biology 61: 235-261.

Ziska, L.H., 2008. Rising atmospheric carbon dioxide and plant biology: the overlooked paradigm. In Controversies in Science and Technology, from Climate to Chromosomes. Eds. Kleinman, D.L., Cloud-Hansen, K.A. (New Rochele: Liebert, Inc.) , pp. 379-400.

Ziska, L.H., Faulkner, S.S., Lydon, J., 2004. Changes in biomass and root: shoot ratio of field-grown Canada thistle (*Cirsium arvense*) with elevated CO_2: Implications for control with glyphosate. Weed Sci. 52: 584-588.

Zobeck, T.M., Fryrear, D.W., 1986. Chemical and physical characteristics of windblown sediment: 2. Chemical characteristics and total soil and nutrient discharge. *Transactions of the ASAE* 29(4):1037-41.

Zong-Ming, W., Bai, Z., Xiao-Yan, L., Kai-Shan, S., Dian-Wei, L.I.U., Shu-Qing, Z., 2006. Using Crop Syst to Simulate Spring Wheat Growth in Black Soil Zone of Northeast China. *Pedosphere* 16(3): 354-361.

❑❑❑

Subject Index

❑❑❑